Potato
Science and Technology for
Sub-Tropics

About the Editors

Dr. Anand Kumar Singh, Deputy Director General (Horticultural Science), Indian Council of Agricultural Research, and Former Managing Director, National Horticulture Board, Ministry of Agriculture and Farmers Welfare, Govt. of India is an internationally reputed scientist. He did pioneering research on tissue culture of horticultural crops and mango breeding at TERI, New Delhi; IIVR, Varanasi and IARI, New Delhi. He received several national and international awards such as Mombusho Award, Japan; Visiting Scientist, Saga University Japan; DBT Associateship; Gold Medal, HSI & Delhi Agri-Horti. Society; and International Registrar-Mangifera, Commission for Nomenclature & Cultivar Registration, ISHS, UK. He is the Fellow of National Academy of Agricultural Sciences, Horticulture Society of India, Hill Horticulture Development Society, National Academy of Biological Sciences, and International Society of Noni Sciences

Dr. Swarup Kumar Chakrabarti, Director, ICAR-Central Potato Research Institute, Shimla and Former Director, ICAR-CTCRI, Thiruvananthapuram has specialization in the area of genomics, molecular biology, and disease diagnostics. He made significant contribution in the area of potato structural and functional genomics, linkage mapping, marker assisted selection; development of transgenics; and plant disease diagnosis. He is the recipient of Shri L.C. Sikka Endowment Award of NAAS, Dr. S. Ramanujam Award of ICAR-CPRI, IPA-Kaushalaya Sikka Memorial Award, Biotechnology Overseas Associateship DBT, Dr. J.P. Verma Memorial Award, Indian Phytopathological Society and many others. He is the Fellow of National Academy of Agricultural Sciences, New Delhi; Indian Phytopathological Society, New Delhi; Indian Potato Association, Shimla; and Confederation of Horticultural Associations of India, New Delhi

Dr. Brajesh Singh, Principal Scientist & Head, Division of CPB&PHT, ICAR-Central Potato Research Institute has specialized in the area of Plant Physiology. He has made significant contribution in the area of development and refinement of potato storage technologies at farm and cold store level; processing technologies for the entrepreneurs; establishing the quality and nutritional laboratory; and development of protocols for nutritional profiling of potatoes. He is recipient of Associateship of National Academy of Agricultural Sciences and HSI-Dr. JC Anand Gold Medal for his work on Post-Harvest Technology. He is the Fellow of Indian Society for Plant Physiology, New Delhi; The Indian Academy of Horticultural Sciences, New Delhi; and Indian Potato Association, Shimla.

**Dr. Jagdev Sharma**, Principal Scientist, ICAR- Central Potato Research Institute, Shimla has specialisation in soil fertility and plant nutrition. He has more than 22 years of research experience and has made significant contributions in grapevine and potato crop nutrition, pressurised irrigation and diagnosis of nutritional disorders. He is a recipient of 'University Gold Medal' and 'Himotkarsh Gold Medal' for his academic excellence. He has been awarded 'Gaurav Chinh' of Maharashtra Grape Growers Association, 'Abhinav Gaurav and Gold Medal' and 'Shri A. Dabholkar 'Utkrisht Shashtrgya' award for contributions in Indian viticulture. He has received scholarship from MASHAV (Israel) and has five years editorial experience for Potato Journal of Indian Potato Association.

**Dr. Vijay Kumar Dua**, Principal Scientist and Head, ICAR-Central Potato Research Institute, Shimla has specialisation in climate change studies, crop modelling, potato agronomy, remote sensing and GIS techniques. He has more than 27 years of research experience and has made significant contributions in climate change impact and adaptations studies, resource management in potato based cropping systems and development of several Decision Support Systems. He is recipient of the IPA-Kaushalya Sikka Memorial Award for his work on decision support tools. He is the Fallow of the Indian Society of Agronomy, New Delhi and India Potato Association, Shimla.

Potato Science and Technology for Sub-Tropics

Anand Kumar Singh
Deputy Director General (Horticultural Science)
Indian Council of Agricultural Research
New Delhi-110012

Swarup Kumar Chakrabarti
Director
ICAR-Central Potato Research Institute
Shimla-171 001

Brajesh Singh
Principal Scientist and Head
ICAR-Central Potato Research Institute
Shimla-171 001

Jagdev Sharma
Principal Scientist
ICAR-Central Potato Research Institute
Shimla-171 001

Vijay Kumar Dua
Principal Scientist & Head
ICAR-Central Potato Research Institute
Shimla-171 001

CRC Press
Taylor & Francis Group
Boca Raton London New York

CRC Press is an imprint of the
Taylor & Francis Group, an **informa** business

NEW INDIA PUBLISHING AGENCY
New Delhi – 110 034

First published 2021
by CRC Press
2 Park Square, Milton Park, Abingdon, Oxon, OX14 4RN

and by CRC Press
6000 Broken Sound Parkway NW, Suite 300, Boca Raton, FL 33487-2742

CRC Press is an imprint of Informa UK Limited

British Library Cataloguing-in-Publication Data
A catalogue record for this book is available from the British Library

Library of Congress Cataloging-in-Publication Data
A catalog record has been requested

ISBN: 978-1-032-00556-0 (hbk)

त्रिलोचन महापात्र, पीएच.डी.
एफ एन ए, एफ एन ए एस सी, एफ एन ए ए एस
सचिव एवं महानिदेशक

TRILOCHAN MOHAPATRA, Ph.D.
FNA, FNASc, FNAAS
SECRETARY & DIRECTOR GENERAL

भारत सरकार
कृषि अनुसंधान और शिक्षा विभाग एवं
भारतीय कृषि अनुसंधान परिषद
कृषि एवं किसान कल्याण मंत्रालय, कृषि भवन, नई दिल्ली 110 001

GOVERNMENT OF INDIA
DEPARTMENT OF AGRICULTURAL RESEARCH & EDUCATION
AND
INDIAN COUNCIL OF AGRICULTURAL RESEARCH
MINISTRY OF AGRICULTURE AND FARMERS WELFARE
KRISHI BHAVAN, NEW DELHI 110 001
Tel.: 23382629; 23386711 Fax: 91-11-23384773
E-mail: dg.icar@nic.in

Foreword

Potato is a staple food that contributes to the energy and nutritional needs of more than a billion people worldwide but has often been under-appreciated as far as its role in the global food system is concerned. It has a wide range of uses: as a staple food, animal feed, and as a source of starch for many industrial uses. The crop is ideally suited to places where land is limited and labour is abundant, conditions that characterize much of the developing world. Moreover, potatoes are a highly productive crop and produce more food per unit area and per unit time than field crops like wheat, rice and maize. Besides, potato cultivation and post-harvest activities constitute an important source of employment and income in rural areas, especially in countries like India. FAO has recommended that sincere effort should be made to realize the full agricultural potential of this crop in the Asian region. In recent past, remarkable progress has been made in potato production in countries like China (99 mmt) and India (52 mmt). The opportunities for further development of the potato industry appear to be very good in these countries. However, at the same time the problems to be addressed are substantial.

ICAR-Central Potato Research Institute, Shimla is internationally acclaimed for developing suitable varieties and technologies that virtually transformed the temperate potato crop to a sub-tropical one enabling its spread from cooler hill regions to the vast Indo- Gangetic plains as a *rabi* crop. It triggered a revolution in potato production causing very fast growth in area, production and productivity. However, the impact of global warming started manifesting during 1990s and it became imperative that further adaptation of potato from sub-tropical to more warmer growing condition would be necessary in near future to sustain its cultivation in the plains. The target of producing 125 million metric tonnes of potatoes in India by 2050 may appear to be unrealistic at first stance; however, analysis of factors and facts responsible for future potato demand validate this big target. Nevertheless, this target is full of challenges that need to be addressed with focused research and development solutions during next three decades. This book on Potato Science and Technology for Sub-tropics deals with issues and strategies for varietal development, preparedness for climate change, bio-security & disease management, enhanced availability of quality seed and post-harvest management, besides emphasizing on the vision for potato crop in the country. I appreciate the efforts made by ICAR-Central Potato Research Institute in compiling this book, which shall disseminate the latest knowledge on potato crop to scientists, students and all the stakeholders so as to prepare them for the new challenges associated with this important crop.

Dated the 7th January, 2020   (T. MOHAPATRA)
New Delhi

Preface

Both potato production and consumption are accelerating in most of the developing countries including India and it is expected that the trend will continue for years to come. The two emerging Asian economies, viz. China and India together contribute nearly 1/3rd of the global potato production at present. Potato is preferred in these densely populated countries largely because of its high productivity, flexibility in terms of fitting into many prevailing cropping systems, and stable yields under conditions in which other crops may fail. Potato consumption in this region is increasing due to increasing industrialization and participation of women in the job market that created demand for processed, ready-to-eat convenience food, particularly in urban areas. There is a perceptible shift in food preference from cereals to vegetables and fruits. As per the projection made by ICAR-CPRI, Shimla, India would require about 125 million tonnes of potato annually by 2050. This enormous jump in production has to come from productivity enhancement, since availability of additional cultivable land for potato cultivation would be virtually nil due to unfavorable changes in land utilization pattern. On the contrary, plateauing of yield gain in potato has emerged as a roadblock for achieving productivity enhancement in a sustainable manner. Innovative technologies are immediately required for breaking this yield barrier.

Potato is a predominant vegetable in India. At present most of the domestic supply of potatoes is consumed as fresh (68%) followed by processing (7.5%) and seed (8.5%). The rest 16% potatoes are wasted due to post harvest losses. However, the proportion of potato used/ wasted due to various reasons is expected to change in the medium and long term scenario. In future, potato has to emerge from just a vegetable to a serious food security option. Considering limited availability of cultivable land in the country higher potato production has to be led by growth in productivity. Future roadmap of potato R&D would be primarily focused on enhancing potato productivity to 35 tonnes/ha by the year 2050. The second focus will be to improve quality of potato as desired by the industry as well as potato consumers in the era of economic development, higher purchasing power and willingness to pay more for the desired quality. Research on improved post-harvest practices will be targeted as another vital component.

For addressing the set goals on potato production and productivity enhancement, use of technological advancements is unavoidable. This book in its 20 chapters elaborates the latest scientific knowledge and technological achievements for development of potato in sub-tropics and also suggests the future strategies for likely adoption. It is our sincere belief that it would act as a compendium of potato research in the country and similar regions and researchers, students and other stakeholders will benefit from the compiled information in a big way.

Anand Kumar Singh
Swarup Kumar Chakrabarti
Brajesh Singh
Jagdev Sharma
Vijay Kumar Dua

Contents

Foreword .... *vii*

Preface .... *ix*

List of Contributors .... *xiii*

1. **Potato in India: Present Status and Future Scenario** 1
   SK Chakrabarti, AK Singh and Brajesh Singh

2. **Impact of Climate Change on Potato Cultivation** 19
   VK Dua, Jagdev Sharma, PM Govindakrishnan, RK Arora and Sushil Kumar

3. **Genetic Resources of Potato and Its Utilization** 43
   Vinod Kumar, Jagesh Kumar Tiwari, SK Luthra, Dalamu, Vanishree G and Vinay Bhardwaj

4. **Breeding for Table Potatoes** 55
   RP Kaur, SK Luthra, Salej Sood, KN Chourasia and Vinay Bhardwaj

5. **Potato Breeding for Processing** 75
   VK Gupta, SK Luthra and Vinay Bhardwaj

6. **Potato: Genome Sequencing and Applications** 97
   Virupaksh U Patil, Ayyanagouda Patil, Jagesh K Tiwari, Vanishree G and SK Chakrabarti

7. **True Potato Seed: Achievements and Opportunities** 119
   Jagesh Kumar Tiwari, Salej Sood, Vinay Bhardwaj, SK Luthra, Dalamu, Sundaresha S, Hemant B Kardile, VU Patil, Vanishree G, Vinod Kumar, Shambhu Kumar and SK Chakrabarti

8. **Potato Physiology for Crop Improvement** 135
   Bandana, Brajesh Singh, Devendra Kumar, Som Dutt Milan K Lal, Sushil S Changan and N Sailo

9. **Precision Agriculture in Potato Production** 155

Manoj Kumar, Preeti Singh, Brajesh Nare and Santosh Kumar

10. **Natural, Zero Budget, Organic Agriculture for Sustainability and Cost Effectiveness** 169

SP Singh, Sanjay Rawal, VK Dua, Jagdev Sharma, YP Singh MJ Sadawarti and S Katare

11. **Weed Management in Potato Crop** 187

Sanjay Rawal, Pooja Mankar, SP Singh and VK Dua

12. **ICT Applications in Potato Cultivation** 207

Shashi Rawat, VK Dua and PM Govindakrishnan

13. **Role of Mechanization in Potato Crop Management** 223

Brajesh Nare and Sukhwinder Singh

14. **Potato Late Blight and Its Management** 233

Sanjeev Sharma, Mehi Lal and Sundaresha Sidappa

15. **Soil and Tuber Borne Diseases of Potato and Their Management** 247

Vinay Sagar and Sanjeev

16. **Viral and Viroid Diseases of Potato and Their Management** .. 267

Ravinder Kumar, A. Jeevalatha, Baswaraj R and Rahul Kumar Tiwari

17. **Important Potato Pests and Their Management** 295

Mohd Abas Shah, Aarti Bairwa, Kailash C Naga, Subhash S, Raghavendra KV, Priyank H Mhatre, Anuj Bhatnagar, Kamlesh Malik, Venkatasalam, EP and Sanjeev Sharma

18. **Potato Seed Production: Present Scenario and Future Planning** 327

Rajesh K Singh, Tanuja Buckseth, Ashwani K Sharma, Jagesh K Tiwari and SK Chakarbarti

19. **Potato Post-Harvest Management Strategies** 343

Brajesh Singh, Pinky Raigond, Arvind Kumar Jaiswal and Dharmendra Kumar

20. **Issues, Strategies and Options for Doubling the Income of Potato Producers in India** 361

Pynbianglang Kharumnuid and NK Pandey

List of Contributors

A Jeevalatha, Scientist
(SS), ICAR-Indian Institute of Spices Research, Kozhikode-675 012 (India)

Aarti Bairwa
Scientist, ICAR-Central Potato Research Institute, Shimla-171 001 (India)

Anuj Bhatnagar
Principal Scientist, ICAR-Central Potato Research Institute Regional Station Modipuram-250 110 (India)

Arvind Kumar Jaiswal
Scientist, ICAR-Central Potato Research Institute Regional Station, Jalandhar-144 003 (India)

Ashwani Kumar Sharma
Principal Scientist, ICAR-Central Potato Research Institute Regional Station, Kufri-171 012 (India)

AK Singh
Deputy Director General, Horticultural Science, ICAR, New Delhi-110012 (India)

Ayyanagouda Patil
Assistant Professor (Biotechnology), University of Agriculture Sciences, Raichur (India)

Bandana
Scientist (SS), ICAR-Central Potato Research Institute Regional Station Modipuram-250 110 (India)

Basawaraj R., Scientist
(SS), ICAR-Central Potato Research Institute, Shimla-171 001 (India)

Brajesh Nare
Scientist, ICAR-Central Potato Research Institute Regional Station, Jalandhar-144 003 (India)

Brajesh Singh
Principal Scientist and Head, ICAR-Central Potato Research Institute, Shimla-171 001 (India)

Dalamu, Scientist
(SS), ICAR-Central Potato Research Institute Regional Station, Kufri-171 012 (India)

Devendra Kumar
Principal Scientist, ICAR-Central Potato Research Institute Regional Station
Modipuram-250 110 (India)

Dharmendra Kumar
Scientist, ICAR- Central Potato Research Institute, Shimla-171 001 (India)

Hemant Balasaheb Kardile
Scientist, ICAR-Central Potato Research Institute, Shimla-171 001 (India)

Jagdev Sharma
Principal Scientist, ICAR-Central Potato Research Institute, Shimla-171 001 (India)

Jagesh Kumar Tiwari
Senior Scientist, ICAR-Central Potato Research Institute, Shimla-171 001 (India)

Kailash Chandra Naga
Scientist, ICAR-Central Potato Research Institute, Shimla-171 001 (India)

Kamlesh Malik
Principal Scientist, ICAR-Central Potato Research Institute Regional Station
Modipuram-250 110 (India)

Kumar Nishant Chourasia
Scientist, ICAR-Central Potato Research Institute, Shimla-171 001 (India)

Manoj Kumar
Principal Scientist & Head, ICAR-Central Potato Research Institute Regional Station
Modipuram-250 110 (India)

Mehi Lal
Senior Scientist, ICAR-Central Potato Research Institute Regional Station
Modipuram- 250 110 (India)

Milan Kumar Lal
Scientist, ICAR-Central Potato Research Institute, Shimla-171 001 (India)

MJ Sadawarti
Senior Scientist, ICAR-Central Potato Research Institute Regional Station
Gwalior-474 006 (India)

Mohd. Abas Shah
Scientist, ICAR-Central Potato Research Institute Regional Station, Jalandhar-144 003 (India)

N. Sailo, Scientist
(SS), ICAR-Central Potato Research Institute Regional Station, Shillong-793 009 (India)

NK Pandey
Principal Scientist & Head, ICAR-Central Potato Research Institute, Shimla-171 001 (India)

Pinky Raigond
Scientist (SS), ICAR-Central Potato Research Institute, Shimla-171 001 (India)

PM Govindakrishnan
Principal Scientist (Retd), ICAR-Central Potato Research Institute, Shimla-171 001 (India)

Pooja Mankar
Scientist, ICAR- Central Potato Research Institute, Shimla-171 001 (India)

Preeti Singh
Scientist, ICAR-Central Potato Research Institute, Shimla-171 001 (India)

Priyank Hanuman Mhatre
Scientist, ICAR-Central Potato Research Institute Regional Station, Ooty-643004 (India)

Pynbianglang Kharumnuid
Scientist, ICAR-Central Potato Research Institute, Shimla-171 001 (India)

Raghavendra KV
Scientist, ICAR-Central Potato Research Institute Regional Station, Modipuram-250110 (India)

Rahul Kumar Tiwari
Scientist, ICAR-Central Potato Research Institute, Shimla-171 001 (India)

Rajesh Kumar Singh
Principal Scientist & Head, ICAR-Central Potato Research Institute, Shimla-171 001 (India)

Ratna Preeti Kaur
Scientist (SS), ICAR-Central Potato Research Institute Regional Station
Jalandhar-144 003 (India)

Ravinder Kumar
Scientist (SS), ICAR-Central Potato Research Institute, Shimla-171 001 (India)

RK Arora
Principal Scientist (Retd), ICAR-Central Potato Research Institute Regional Station
Jalandhar-144 003 (India)

Salej Sood
Senior Scientist, ICAR-Central Potato Research Institute, Shimla-171 001 (India)

Sanjay Rawal
Principal Scientist, ICAR-Central Potato Research Institute Regional Station
Modipuram-250 110 (India)

Sanjeev Sharma
Principal Scientist, ICAR-Central Potato Research Institute, Shimla-171 001 (India)

Santosh Kumar
Scientist, ICAR-RMRSPC (ICAR-IIMR), Begusarai-851129, Bihar, (India)

Shambhu Kumar
Principal Scientist, ICAR-Central Potato Research Institute Regional Station
Patna-801 506 (India)

Shashi Rawat
Principal Scientist, ICAR-Central Potato Research Institute, Shimla-171 001 (India)

SK Chakrabarti
Director, ICAR-Central Potato Research Institute, Shimla-171 001 (India)

SK Luthra
Principal Scientist
ICAR-Central Potato Research Institute Regional Station, Modipuram-250 110 (India)

Som Dutt
Principal Scientist, ICAR-Central Potato Research Institute, Shimla-171 001 (India)

SP Singh
Principal Scientist, ICAR-Central Potato Research Institute Regional Station
Gwalior-474 006 (India)

Subash Katare
Senior Scientist, ICAR-Central Potato Research Institute Regional Station, Gwalior-474 006 (India)

Subhash S.
Scientist, ICAR-Central Potato Research Institute, Shimla-171 001 (India)

Sukhwinder Singh
Scientist (SG), ICAR-Central Potato Research Institute Regional Station
Jalandhar-144 003 (India)

Sundaresha Sidappa
Scientist (SS), ICAR-Central Potato Research Institute, Shimla-171 001 (India)

Sushil Kumar
ACTO, ICAR-Central Potato Research Institute, Shimla-171 001 (India)

Sushil Sudhakar Changan
Scientist, ICAR-Central Potato Research Institute, Shimla-171 001 (India)

Tanuja Buckseth
Scientist (SS), ICAR-Central Potato Research Institute, Shimla-171 001 (India)

Vanishree G.
Scientist (SS), ICAR-Central Potato Research Institute, Shimla-171 001 (India)

Venkatasalam EP
Principal Scientist, ICAR-Central Potato Research Institute Regional Station, Ooty-643 004 (India)

Vinay Bhardwaj
Principal Scientist, ICAR-Central Potato Research Institute, Shimla-171 001 (India)

Vinay Sagar
Principal Scientist, ICAR-Central Potato Research Institute, Shimla-171 001 (India)

Vinod Kumar
Principal Scientist, ICAR-Central Potato Research Institute, Shimla-171 001 (India)

VK Dua
Principal Scientist & Head, ICAR-Central Potato Research Institute, Shimla-171 001 (India)

VK Gupta
Principal Scientist, ICAR-Central Potato Research Institute Regional Station, Modipuram-250 110 (India)

VU Patil
Scientist (SS), ICAR-Central Potato Research Institute, Shimla-171 001 (India)

YP Singh
STO, ICAR-Central Potato Research Institute Regional Station, Gwalior-474 006 (India)

1

Potato in India: Present Status and Future Scenario

SK Chakrabarti[1] AK Singh[2] and Brajesh Singh[1]

[1]ICAR-Central Potato Research Institute, Shimla-171 001
Himachal Pradesh, India
[2]Division of Horticultural Science, ICAR, New Delhi-110012, India

Introduction

At the time of inception of CPRI, in the year 1949, India used to produce 1.54 million tonnes potatoes from 0.234 million ha area at an average productivity level of 6.58 tonne/ ha. As per the first advance estimate of the NHB, the potato production in India during 2018-19 was 52.5 million tonnes from 2.18 million ha area at 24 tonnes/ ha productivity. The potato production, area and productivity increased over 7 decades by 34, 9.3 and 3.7 times, respectively. ICAR-CPRI has been adequately recognised by the nation on several occasions for this stupendous contribution. However, potato in India still has to transform from simply a vegetable supplement to serious food security option. Ability of potato to produce highest nutrition and dry matter on per unit area and time basis, among major food crops, made FAO to declare it the crop to address future global food security and poverty alleviation during 2008. Rising number of working couples, rapid rate of urbanization, enhanced tendency of eating out of home, higher disposable income levels of people and important place of potato in fast food items, create an ideal situation for expansion of potato consumption in the near and distant future.

A perusal of various R&D efforts and outcomes in the field of agriculture in general and, more specifically the potato, reveals that "Business as Usual" scenario would not hold much longer. We have to anticipate and get ready to tackle much more complex and diverse future challenges in our respective fields. As the time lag between research and development efforts and the final application/ adoption of the output may stretch over decades, a long term vision and blueprint of action plan is highly important. This chapter attempts to envisage long term state of affairs of potato industry in India and formulate a strategy to fulfil national needs through a well-documented plan to tackle anticipated challenges.

Global and Indian Scenario

Potato is the third most important food crop in the world after rice and wheat in terms of human consumption. Global annual potato production during the triennium ending (TE) 2013 was 370 million tonnes resulting in per capita availability of over 50 kg. As per FAOSTAT, India is the second largest annual producer of potato after China, leaving the Russian Federation far behind (43.1, 88.2 and 30.8 million tonne, respectively, during TE 2013). Developed countries were the major potato producers as well as consumers till the last millennium. A comparison of potato production growth during TE 2003 and 2013 showed that Africa (97%) experienced the highest proportionate growth followed by Asia (Figure 1). India and China were not only the major contributors to the Asian growth of potato production but being producer of one third global potato, contributed significantly to world potato production. Potato consumption in India and China is accelerating due to increasing industrialization and participation of women in the job market that created demand for processed, ready-to-eat convenience food, particularly in urban areas.

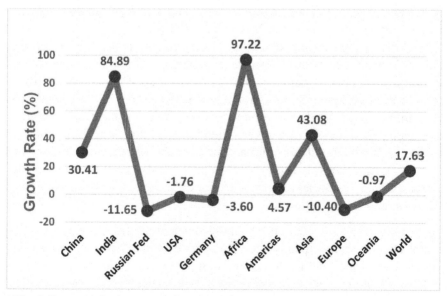

Fig. 1: Potato production growth (%) over major potato producing nations and continents during TE 2003 and TE 2013 (Data source: FAOSTAT, 2015).

During the last decade, developed world has experienced fall in per capita potato consumption (Americas, Europe, Oceania, Russian Federation having - 8.8, -9.4, -8.3 and -2.4% growth in per capita potato consumption, respectively), while at the same time per capita potato consumption in the developing world showed increasing trend (Africa, Asia, India, and China showed 40.6, 25.6, 37.1 and 28.8% growth in per capita potato demand, respectively) during the TE 2001 and 2011 (FAOSTAT, 2015). In absolute terms Asia is the biggest gainer in per capita as well as total potato consumption during this period. However, the productivity in most of the developing countries continues to be very low.

CONTRIBUTION TO AGRICULTURAL ECONOMY

Agriculture, including allied activities, contributed 13.9% of the GDP at constant prices (2004-05) in 2013-14 (Anonymous, 2014) while this sector still accounts for 54.6% of total employment in the country (Anonymous, 2011). Current share of potato to agricultural GDP is 2.86% out of 1.32% cultivable area. On the contrary, the two principal food crops, rice and wheat, contribute 18.25% and 8.22% of agricultural GDP, respectively from 31.19 and 20.56% cultivable area, respectively (FAOSTAT, 2015). It indicated that contribution of potato in agricultural GDP from unit area of cultivable land is about 3.7 times higher than rice and 5.4 times higher than wheat.

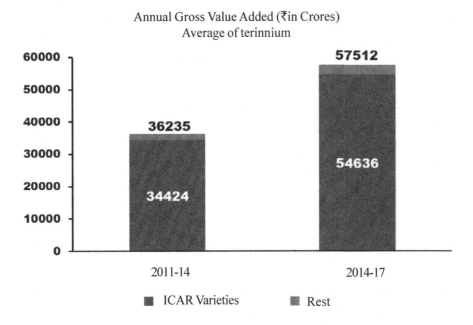

Fig. 2: Annual Gross Value Addition by Potato

Potato varieties developed by ICAR-Central Potato Research Institute are very popular among farmers and cover nearly 95% of total area under potato. India produced ca 45.87 million tonnes of potato annually during the triennium 2014-17 and contributed Rs. 57,512 crore annually to the Gross Value Added (GVA) at current price (Figure 2). The varieties developed by ICAR-CPRI contributed Rs. 54,636 crore annually during this period. Four varieties, viz. Kufri Jyoti, Kufri Bahar, Kufri Pukhraj, and Kufri Chipsona 1 together contributed around 75% of total area under potato (Anonymous, 2018).

Potato being a labour-intensive crop requires about 145 man days for cultivation of one ha of land. Thus nearly 293 million man-days of employment have been generated only for potato cultivation during 2013-14. Besides, large number of semi-skilled labour is required for carrying out post-harvest operations like transportation, storage, processing, marketing etc. Moreover, about 75% of the total labour force employed in potato cultivation is constituted by the women. Therefore, potato encourages gender equality in agricultural labour market. Input-intensive nature of potato crop helps in overall economic development of the country by supporting other sectors of the economy like industry, finance and services. For example, relatively higher demand of fertilizer, pesticide, farm machineries, cold storage equipment and structures, packaging materials, etc. for potato cultivation enables healthy industrial growth. Similarly, the crop supports services sectors through agricultural loans, insurance, marketing and technical consultancy etc.

DRIVERS OF GROWTH

Demand outlook

Potato is a predominant vegetable in India. At present most of the domestic supply of potatoes is consumed as fresh (68%) followed by processing (7.5%) and seed (8.5%). The rest 16% potatoes are wasted due to post harvest losses. However, the proportion of potato used/ wasted due to various reasons is expected to change in the medium and long term scenario.

Fresh potatoes: Per capita consumption of fresh potatoes (FAOSTAT) increased from 1991 to 2010 at an ACGR of 2.34%. Will this consumption rise in the future at the same rate? The stagnating growth rates of cereals' productivity, large scale diversion of food grains to feed & bio-fuel and expected steep rise in per capita consumption of pulses, edible oil, fruits, vegetables, milk, sugar and non-vegetarian food in the regime of steadily rising population is bound to put pressure on existing cultivable land. Since, cultivable land is expected to remain more or less constant in the next 40 years, the role of crops like potato having higher production potential per unit land and time will become imperative. In this context potato crop has very high probability of making crucial contribution to the future national food security agenda.

The perceived changes in Indian socio-economics in the medium and long term are expected to enhance per capita food consumption of fresh potatoes. Potato is an important ingredient of most of the fast foods in organised as well as unorganised sector. Rapid urban population growth from 375 to 840 million over next 40 years at an ACGR of 2.04% is expected against the overall national ACGR of population at 0.78%. Faster rise of number of nuclear families, higher disposable incomes on account of fast economic growth resulting into higher tendency of out-of-home eating and rapid increase in the number of working women in the medium and long run are expected to maintain the ACGR of 2.34% in per capita consumption of fresh potatoes. Per capita food demand of fresh potatoes at this ACGR will be 48.5 kg in the long term (Table 1). The corresponding national food demand for fresh potatoes will be 78.5 million tonne in 2050.

Table 1: Per capita and total national food demand for fresh potatoes

Items	2010	2050
Per capita fresh (kg)	19.784	48.470
National demand (million tonne)	23.94	78.47

ACGR of per capita demand (2.34) is based only on fresh consumption of potato

Processing quality potatoes: Agri-processing sector experiences very fast growth rate when an economy transforms from developing to developed economy. The rise of Indian economy from $ 1.57 to between 13 and 34 trillion (under varied scenarios; NCAP estimates) is not possible without corresponding rise in agri-processing industry. Further, potato is always the front-runner when we take processing of agri-commodities into consideration. Analysis of past experience and pattern of Indian processing industry suggests that demand for processing quality potatoes over next 40 years will rise at the fastest pace for French fries (11.6% ACGR) followed by potato flakes/ powder (7.6%) and potato chips (4.5%). The actual demand for processing potatoes will rise from 2.8 million tonne in 2010 to 25 million tonne during the year 2050 at an ACGR of 5.61% (Table 2).

Table 2: Raw material demand of potato processing industry (million tonne)

Product(s)	2014	ACGRs	2050
Potato chips	2.92	4.5	14.22
Potato flakes/ powder	0.39	7.6	5.44
Frozen potato products	0.09	11.6	5.40
Total	3.40	5.61	25.06

Potato starch making industry doesn't exist in India and its future is difficult to predict on account of short duration domestic potato cultivation, unknown future developments in potato starch industry of China and European Union (predominant global players) and possible allocation of more and more cultivable land to food security options rather than industrial products. However, if some quantity of potatoes is used for making potato starch or other industrial products, then this demand is not quantified and will be over and above the estimated one.

The corresponding total and per capita demand of actual processed potato products at current level and in the future is depicted in Table 3. The total demand of processed potato products is estimated to rise from 0.7 million tonne in 2010 to 7.3 million tonne in 2050. The corresponding per capita demand for processed potato products will increase from 0.6 to 4.5 kg (6.03% ACGR) over the period of 40 years.

Seed potatoes: As per FAOSTAT, India used 2.96 million tonne potato tubers (8.5% of national potato production) as seed during triennium ending year 2010. The absolute quantity of potatoes used as seed is estimated to increase to 6.1 million tonne during 2050. However the proportion of tubers used as seed will fall to 5% of national potato production during 2050, on account of anticipated higher potato productivity.

Post-harvest losses: At present a higher proportion of potatoes (16%) is wasted as post-harvest losses (PHLs) than that used as seed (8.5%) or processing (7.5%). This is an unfortunate situation that nearly 1 million tonne in excess to the total potato production in Canada during the triennium ending 2010 (FAOSTAT) was wasted in India. Although, due to hot summer temperatures, lack of state of the art cold storage facilities and massive transportation of potatoes from northern to southern states are the causes of high wastage of potato in absolute terms, the proportion of PHLs to the total potato production in the country are targeted to be lower. The estimated PHLs are targeted to lower from the current 15.75 to 10% in the year 2050, while the corresponding quantity of PHLs is estimated at about 5.5 (current), and 12 million tonne (in 2050).

Export: India contributes to about 13% of the total world potato production, but our contribution in world potato export is around 1.6% only, which is not even 1% of the total in house production. Potato being semi-perishable and bulky agri-commodity, its export from India is not guided by a long term policy support. As potato is a politically sensitive crop, targeted steps are taken to keep its retail prices at affordable level for the ordinary consumer. Export is generally a crisis management tool during the years of oversupply. However, in international markets exports can only be increased through building credibility and making long term contracts. It is anticipated that the future food policy will concentrate more and more on ensuring national food security and low value export of agri-commodities may be discouraged. Under these circumstances steep rise in export of fresh potatoes from the country is unlikely. India being the massive producer of potatoes a healthy growth in processed potato products is anticipated. However, unknown and complex future developments at international level are difficult to be assessed. Moreover, in the absence of a clear cut support through a robust export policy it is not possible to estimate and assign future targets for the export of processed potato products.

Total potato demand: The demand for potatoes in India is estimated to rise to 122 million tonne in the year 2050. Demand for processing quality potatoes will rise at the highest rate and their proportion in the net domestic supply will rise from 7.5% in the triennium ending 2010 average to 20.5% in 2050 (Table 3). Considering net domestic supply the proportion of potatoes put to seed and wastage due to post-harvest losses will gradually decrease in future. Sharp increase in potato utilization for processing purpose will slightly lower down the proportion of fresh potato consumption as food by 2050.

Table 3: Component wise potato demand (million tonne)

Year	Demand$	Fresh	Processing	Seed	Waste
2010	35.10(100.00)	23.94(68.21)	2.67(7.61)	2.96(8.43)	5.53(15.75)
2050	121.81(100.00)	78.47(64.42)	25.06(20.57)	6.10(5.01)	12.18(10.00)

Note: Figures in parenthesis are percentages; $: net domestic demand

Operating environment

After putting 46% of total area under agriculture, India is seriously constrained as far as bringing more land under agriculture is concerned. On the contrary we are already doing agriculture on marginal and degraded lands. Rising population and resulting fragmentation of land holdings has created a situation where farming doesn't generate complete livelihood for majority of the farmers, including the potato farmers. Resultantly, the farmers are not practicing agriculture professionally following the recommended scientific package of practices. This peculiar situation creates task of technology transfer quite tough and challenging.

Climate change scenario is supposed to adversely affect potato production and productivity in India. Modelling research at CPRI suggests that by the year 2020 potato yield is estimated to fall by 19.65% in the state of Karnataka followed by Gujarat (18.23% fall) and Maharashtra (13.02% fall) with an overall fall of

9.56% at national level if needed steps are not taken to mitigate the effects of climate change. The situation is expected to further worsen by the year 2050 when the national level potato production is expected to fall by 16% in the absence of needed steps. However, the potato production fall may be much severer in the states of Karnataka (45% fall), Gujarat (32% fall), Maharashtra (24.5% fall) and Madhya Pradesh (16.5% fall) if preventive steps are not taken.

Potato growth in India has largely been uneven as nearly 85% potato in the country is produced in north Indian plains. Average potato productivity in India is more than 20 tonne/ ha in states like Gujarat, Punjab, Uttar Pradesh, Haryana and West Bengal while several states in hills and plateau region have yield less than 10 tonne/ ha (Maharashtra, all North Eastern states, Sikkim and Karnataka). About 95% of the national potato is harvested in *rabi* season. The regional and seasonal production of potato is largely responsible for price fluctuation in this agri-commodity. About 90% of disease free seed potato is produced in Punjab, specially the Jalandhar district, which makes several potato producing areas in the country quite distant, from seed potato marketing angle.

Crop husbandry is affected by the latest developments in other related fields to a great extent. Such developments have taken place at a fast pace during the recent past and consequently the farming conditions are also changing to an equal extent. Contract farming is one of the new concepts of current nature in Indian context which is extensively used in potato farming. This concept is much preferred by the potato processing industry in order to have higher certainty and control over raw material supply chain *i.e.* processing grade potatoes. The concept is equally favoured by seed potato producing companies. The contract farming route to procure desirable produce is also exploited by retail chains for specific quality potato material. Crop insurance is another area that affects potato farmers to a great extent. However, this concept is still to attain the desired maturity. Future trading of agri-commodities is another development of recent past in India and it is still ill-understood by most of our farmers. However, its effective utilization provides a potent price risk management tool to the producers. Need of mechanization due to the shortage of farm labour and higher importance of efficient inputs delivery system are gaining increased attention of potato farmers in India.

Challenges

Future challenges in potato R&D being much more complex and difficult to be addressed, our preparedness has to be much more rigorous, precise and should extend over longer period of time. With every new addition to our knowledge on impending issues like climate change and their implications on potato industry we need to envisage more robust and superior strategies to deal with them.

Every new document on this policy issue provides an opportunity to assess the relevance of strategies designed and put forth in the light of unanticipated recent developments having bearing on the future of Indian agriculture as well as potato industry. Based on past experience we can visualise that several of the developments were not possible to be envisioned. Therefore, there is a need to make midterm course correction and embark on a better strategy with higher probability of success in the future.

Adapting potato to warmer growing condition: ICAR-Central Potato Research Institute, Shimla is internationally acclaimed for developing suitable varieties and technologies that virtually transformed the temperate potato crop to a sub-tropical one enabling its spread from cooler hill regions to the vast Indo-Gangetic plains as a *rabi* crop. It triggered a revolution in potato production causing very fast growth in area, production and productivity. However, the impact of global warming started manifesting during 1990s and it became imperative that further adaptation of potato from sub-tropical to more warmer growing condition would be necessary in near future to sustain its cultivation in the plains. In fact, the Intergovernmental Panel on Climate Change (IPCC) in its Fourth Assessment Report predicted that the potato growing season in 2055 is likely to be warmer by 2.41–3.16°C. Suitable growing period for potato in eastern UP, Bihar, and West Bengal will be not more than 75-80 days. Potato production may decline by 15.32% in the year 2050. To overcome this situation and to satisfy the projected demand, it is necessary to strengthen research work on developing varieties and production technologies for cultivating potato under warmer condition. Research emphasis should be on developing short duration, early-maturing varieties with heat tolerance for both fresh consumption and processing. Concomitantly, safe and sustainable technologies should be developed for heat and water stress management, nutrient management, management of invasive and range-expanding pests and diseases, and cold chain management.

Productivity enhancement: The food basket of the country is undergoing a drastic change due to economic growth, life style change and dietary preference. There is a perceptible shift in food preference from cereals to vegetables and fruits. As per the projection made by ICAR-CPRI, Shimla, India would require about 124.88 million tonnes of potato annually by 2050. This enormous jump in production has to come from productivity enhancement, since availability of additional cultivable land for potato cultivation by the year 2050 would be virtually nil due to unfavourable changes in land utilization pattern. On the contrary, plateauing of yield gain in potato has emerged as a roadblock for achieving productivity enhancement in a sustainable manner. Innovative technologies are immediately required for breaking this yield barrier. Research emphasis should be targeted for harnessing maximum yield potential of potato by broadening genetic base of varieties, improving photosynthetic energy conversion efficiency, conferring atmospheric nitrogen fixation ability, root biology & architecture for input use efficiency and improving sink strength. Besides, emphasis should be given on diploid breeding to exploit hybrid vigour in potato, which is yet to be exploited. Precision breeding tools such as genomic selection, genome editing, and marker assisted selection should be fully integrated with conventional breeding to break yield barrier in potato.

Sustainable production system: The green revolution during 1960s took the country out of the "ship-to-mouth" existence to a status of food surplus nation. On the other hand, all the natural resources including soil, water, and energy are under severe constraint now. Inputs for agricultural production will also become scarcer and dearer with time. It is also imperative that future food production technologies should be carbon neutral and sustainable. Moreover, income of farmers in real term has not appreciated adequately during recent years causing wide spread agrarian distress. Therefore, a paradigm shift is necessary now from the policy of mere food production to income security of farmers. Agriculture is the largest private enterprise in India, consisting of more than 138 million holdings, ~ 85% of which are < 2 ha in size. Most of these family farms are engaged in multiple agricultural activities like agri/horticulture, poultry and livestock rearing, fishery, beekeeping, sericulture, and agroforestry. All our future technologies should aim at addressing the farming system in its entirety instead of particular mandate crop of any institute. Under this background, we have to strike a balance between cutting edge technologies and their environmental cost for sustainable production enhancement of potato. Research emphasis should be given on integrated farming system (IFS) approach for technology development, water use efficiency, nutrient responsive technologies, conservation agriculture, and bio-intensive crop management. Use of ICT-enabled advisories and artificial intelligence (AI) should be encouraged for technology dissemination.

Post-harvest management: It is estimated that ~ 2.8-10% in non-perishable, 6.8-12.5% in semi-perishables and 5.8-18% in perishable agricultural products are lost after harvesting in India. It may be much higher (~ 20%) for a perishable commodity like potato that is harvested at the onset of summer season. About 50% of these losses can be prevented using appropriate post-harvest measures. Establishing on-farm primary processing facilities would capacitate small farmers in a big way. The family farmers can be trained to undertake post-harvest processing and packaging of farm produce, preferably on-farm or near to the production site. Such technologies would promote entrepreneurship in rural areas by strengthening forward linkage in agriculture. This would generate additional working days to farm family members, add value to harvest and generate additional income. The following areas should be given thrust for lowering postharvest losses: development of processing varieties and technologies, on-farm storage and primary processing units, energy-efficient storage structure, technologies for cold chipping, managing bruising injuries, and technologies for export facilitation.

Integrated pest management and bio-security: Potato is affected by late blight, viruses, bacterial wilt, aphids, and other common soil and tuber borne pests and diseases. Movement of plant pests, pathogens and invasive weeds does not recognize physical and political boundaries. Globalization of commodity and food trade has increased the bio-security risk, which threatens food security of the nation. Pests and pathogens have their strength in number and are capable of adapting to changing climate much faster than our effort to breed new varieties. Keeping in view the future challenges, emphasis on the following areas should be given for effective plant health management: robust diagnostic tools for effective interception of alien invasive pests and pathogens, application of pathogenomics for understanding epidemiology and management, breeding for multiple stress tolerance, use of info-chemicals for integrated pest management (IPM), emphasis on biological control of pests and diseases, and DSS & forecasting for environmental safety.

Opportunities

Varied climatic conditions across the country provide opportunity of round the year potato production in India. The advantage has been thoroughly exploited by potato processing industry for round the year supply of raw material and seed potato producers for taking double crop to ensure faster multiplication. However, the country still has tremendous scope of further progress in this direction.

At present potato is being consumed mainly as a vegetable by the entire population (~1.30 billion) of India. It is projected to increase to ~1.62 billion during 2050, out of which more than 840 million would be living in cities (against the current 375 million). Besides, much higher proportion of working women and nuclear families would totally transform the demographic structure of India. Incidentally this change will be conducive for higher demand for potato (as vegetables and fast food ingredients) and processed potato products in the future. Fast increment in per capita as well as total household consumption of potato in India during recent past is expected to sustain in the foreseeable future; hence, there is great potential of enhancing potato production in the country. The gradual shift of potato cultivation from developed nations to the developing ones provides great scope of potato exports in the future for countries like India.

Strong existing potato research and development base and adequate preparedness for future challenges is expected to provide requisite support to the future needs of potato industry in India. Application of new science in the fields of bio-technology, nanotechnology, genomics, phenomics, diagnostics, precision farming, aeroponics ICT, GIS and remote sensing has already been successfully demonstrated by ICAR-CPRI. We expect these areas to serve the cause of future potato R&D in a big way. The available physical and scientific infrastructure at ICAR-CPRI is enough for not only further accelerating potato R&D momentum at national level but also for emerging as a leading global player, especially in the tropics and subtropics. However, sustained efforts will be made to upgrade these facilities and pursue rigorous human resource development at the institute level.

TARGETS

Production target

At present in India, about 68% of potato production is consumed as fresh while the rest is utilized as seed (8.5%) and processing purposes (7.5%) and the remaining 16% goes as waste due to various reasons that includes rottage and wastage during the entire potato supply chain. After accounting for rapid growth of potato processing due to fast economic development of the country, lower proportionate growth rate of seed demand (due to higher productivity), rise in per capita consumption of fresh potato [due to rapid urbanisation, future role of potato in food security (Thiele *et al.*, 2010; Singh and Rana, 2013), and fast economic development], and as a result of ongoing efforts to lower post-harvest losses, the estimated demand of potatoes in 2050 would be about 122 million tonne.

In addition to the estimated demand it is likely that potato tubers will be used as animal feed and for other industrial uses such as potato starch manufacturing in the future. The net export of potato tubers from India is also likely to increase in the form of seed potato and processed potato products, from the current average of a meagre 0.1 million tonnes per year. However, due to non-existing potato starch industry, non-utilization of potato as animal feed, and numerous international factors as detriments of Indian export of seed potato and processed potato products, it is difficult to estimate future targets for the said uses of potatoes. However, using professional judgment a target of 3 million tonnes during 2050 is assigned to all these three uses collectively. Hence, we would require potato production of about 125 million tonnes in 2050 at an ACGRs of 3.2%.

Productivity goal

Productivity and profitability of potato will determine future growth of this crop in India. Considering more or less stagnant cultivable land and impending food insecurity threat in the country we will have to strive very hard in the direction of increasing productivity of the crop. The WOFOST model estimates indicate that at present level of technological intervention the achievable yield of potato in India would be 35.95 tonnes/ha during the year 2050. However, due to the technological advancement at national level the achievable yield during 2050 would be 43.14 tonnes/ha.

At present level of farm management practices we are actually able to harvest only 60.8% of the achievable yield. However, an enhanced emphasis on efficient dissemination of farm technologies and consequent improvement in farm management practices in the country, it is estimated that we would be able to harvest 80% of achievable yield in 2050 which would result in estimated yield of 34.51 tonnes/ha. Potato productivity in India is required to rise at an ACGR of 1.46% up to the year 2050 in order to meet this yield target.

Area requirement

The projected potato area based on the ACGRs of last 40 years is not possible to achieve as country has already about 46% (141.4 million ha) of the total geographical part as net sown area. Further increase in cultivable area in the country is likely to cause huge environmental damage. Fast increment in land used for non-agricultural purposes makes this expansion still more difficult. ICAR-National Institute of Agricultural Economics and Policy Research has forecasted net sown area in the country during 2050 at 142.6 million ha which is more or less same as during 2010. Therefore, primary goal of future potato R&D in India in general, and at ICAR-CPRI in particular, would be increasing average potato productivity in the country to 34.51 tonnes/ha during the year 2050. The potato production targets as dictated by the corresponding demand of potatoes for various uses would be met with the adjustment in the area. Higher or lower demand will be reflected in prices which will consequently affect farmers' profitability and ultimately the area under potatoes will automatically be adjusted.

At estimated production and yield targets, we would require 3.62 million ha of area under potato. However, availability of this enhanced area under potato cultivation in future needs to be critically analysed. Higher per unit (area and time) production potential of potato crop compared to other food crops will help getting higher allocation of area under this crop in the future. Rice and wheat have 41.85 and 27.75 million ha area under cultivation in India and potato is generally taken as a sequence crop with both these crops. With the ongoing efforts of developing short duration potato varieties and tropicalization of the crop, faster mobility of perishable crops on account of upcoming dedicated freight corridors of Indian Railways, potato emerging as an important food security option in India, and elevated demand will bring needed area under potato through relative improvement in crop profitability.

WAY FORWARD

In future, potato has to emerge from just a vegetable to a serious food security option. Considering limited availability of cultivable land in the country higher potato production has to be led by growth in productivity. Future roadmap of potato R&D would be primarily focused on enhancing potato productivity to 34.51 tonne/ ha by the year 2050. The second focus will be to improve quality of potato as desired by the industry as well as potato consumers in the era of economic development, higher purchasing power and willingness to pay more for the desired quality. Research on improved potato storage will be targeted as another vital component in order to lower post-harvest losses during the next 30 years.

Strategies for achieving targets

In order to meet targets and tackle anticipated challenges, following seven-pronged strategy need to be employed to accomplish the set goals.

1. Effective exploitation of genetic resources for varietal improvement

- Molecular characterization and development of core collection of the germplasm.

- Development of mapping population and pre-breeding including somatic hybrids for exploiting wider gene pool.

- Heterosis and hybrid vigour leading to enhancement in production potential of potato.

- Development of potato varieties and populations for short duration, processing, starch making, heat & drought tolerance, biotic stress tolerance, nutrient use efficiency, *kharif* season, exports, early bulking, and TPS populations.

2. Safe application of biotechnology for potato improvement

- Structural genomics and bioinformatics for developing robust molecular markers for qualitative and quantitative traits.

- Functional genomics for gene discovery for targeted traits like late blight durable resistance, heat tolerance, high temperature tuberization, better water and nutrient use efficiency.

- Proteomics and metabolomics for basic studies on tuberization, photosynthesis, partitioning of photo-assimilates, starch metabolism, carotenoid and flavonoid synthesis, storage protein quality, processing quality.
- Technology development for marker-free and site-specific integration of transgenes.
- Development of transgenic potato with improved resistance/ tolerance to biotic/ abiotic stresses and to improve nutritional and processing quality.

3. Encouraging production of quality planting material

- Development & standardization of low cost and efficient mass propagation methods – aeroponics, bio-reactor technology.
- Vector dynamics and its implications on seed quality.
- Development of homozygous TPS populations using apomixes and monohaploidy.

4. Resource based planning and crop management

- Development of IT based Decision Support Systems/ tools for crop scheduling and management of weeds, nutrients, water, diseases and pests under climate change scenario.
- Standardization of technologies leading to improved carbon sequestration and soil health.
- Development of technologies for enhancing inputs use efficiency through precision farming and micro-irrigation.

5. Eco-friendly crop protection

- Cataloguing genome variability and dynamics of new pathogen/ pests populations (Pathogenomics).
- Development of diagnostics for detection of pathogens both at laboratory & field level using micro-array and nano-technologies.
- Ecology and management of beneficial microorganisms for enhancing crop productivity and disease management.

6. Encouraging energy efficient storage and diversified utilization of potato

- Technology refinement for elevated temperature storage for both on- and off-farm situation.

- Development of new processes, products and utilization technologies for diversified use of potatoes including waste utilization.

- Food fortification to enhance nutritional quality of processed foods.

- Technologies for lowering glycemic index.

7. Strengthening institute-farmer interface for technology dissemination

- Comparative farm profitability studies vis-a-vis ability to contribute to GDP by various crops, for providing efficient policy input.

- Proficient technical dissemination through an optimal mix of traditional and modern extension tools.

- Besides, following cutting edge research themes need to be focussed as future R&D agenda for making India a leading international hub on potato research.

- Development of transgenic potatoes to address high risk areas *viz.* biotic and abiotic stresses, quality enhancement and wider adaptation.

- Development of cold chipping varieties.

- Development of potatoes with low glycaemic index and high antioxidant contents.

- Identification of new genes and markers for important traits.

- Studies on potato proteomics and phenomics with reference to tuberization.

- Development of next generation molecular marker, SNP, with reference to disease resistance and quality traits using allele mining and resequencing.

- Development of bio-risk intelligent system (surveillance of racial pattern of different pathogens and pests and early warning systems) for taking informed decision at the local, regional and national levels.

- Application of ICT, GIS and remote sensing to understand and mitigate ill effects of climate change & global warming, identify new potato growing areas and to develop decision support systems to meet impending complex challenges.

CONCLUSION

With 52.5 million t potato production India (2019) is the second largest potato producer after China (99 million t in 2017, FAOSTAT). The target of producing 125 million t potatoes in India by 2050 may appear to be unrealistic at first stance, however, analysis of factors and facts responsible for future potato demand validate this big target. Nevertheless, this target is full of challenges that need to be addressed with planned potato research and development solutions during next three decades.

REFERENCES

Anonymous (2011) Census of India. Ministry of Home Affairs, Government of India, New Delhi

Anonymous (2014) Economic Survey 2014-15. Ministry of Finance, Government of India, New Delhi

Anonymous (2015) Vision 2050. ICAR-Central Potato Research Institute, Shimla, India 33p

Anonymous (2018) CPRI varieties brought potato revolution in India. Extension Folder, ICAR-Central Potato Research Institute, Shimla.

FAOSTAT (2015) Statistical database. http://faostat.fao.org/

Singh BP and Rana Rajesh K (2013) Potato for food and nutritional security in India. *Indian Farming* **63**: 37-43

Thiele G Theisen K Bonierbale M and Walker T (2010) Targeting the poor and hungry with potato science. *Potato Journal* **37**(3-4): 75-86

2

Impact of Climate Change on Potato Cultivation

VK Dua[1], Jagdev Sharma[1], PM Govindakrishnan[1], RK Arora[2] and Sushil Kumar[1]

[1]ICAR-Central Potato Research Institute, Shimla-171001, Himachal Pradesh India
[2]ICAR-Central Potato Research Station Regional Station, Jalandhar-144003 Punjab, India

INTRODUCTION

Potato is an important crop for food and nutritional security of India. It is mainly grown as *rabi* crop and 90 per cent of its cultivation is confined to northern Indo-Gangetic plains comprising the states of Uttar Pradesh, West Bengal, Bihar, Punjab and Haryana. Due to climate change, the main potato season is likely to be warmer than the rest of the seasons of the year and will affect seriously the potato production in India. Climate change is defined by Intergovernmental Panel on Climate Change (IPCC) as a plausible future climate that has been constructed for explicit use in investigating the potential consequences of climate change. It is the difference in climate over a period of time (with respect to a base line or a reference period) and corresponds to a statistically significant trend of mean climate or its variability persistent over a long period of time. Climate change may be due to both natural as well as anthropogenic factors. Reference periods are typically of three decades used as climatological baseline period in impacts and adaptation assessments to Global climate model. The future climate minus baseline climate are used for model scenario comparison of most climate variables. Typically a number of fixed time horizons in the future are produced from model output. For example 2020 (2010-2039), 2050 (2040-2069), 2080 (2070-2099). Future climates are estimated through models, called global circulation models, or global climate models (GCM). GCM simulations are carried out by means of a large set of heavy computations. The

IPCC TAR reported 7 GCMs for developing climate scenarios while IPCC FAR reported 21 different GCMs.

Anthropogenic activities are the major factors contributing to climate change. They cause changes in atmospheric concentration of greenhouse gases and aerosols, land use, land cover change and solar radiations. The human activities result in emissions of four long lasting greenhouse gases: carbon dioxide, methane, nitrous oxide and halocarbons. The human activities have led to higher production of these gases than their removal. In the Fifth Assessment Report of the IPCC , the scientific community has defined a set of four new scenarios, denoted Representative Concentration Pathways (RCPs), which are identified by their approximate total radiative forcing in year 2100 relative to 1750, i.e. 2.6 W/m^{-2} for RCP 2.6, 4.5 W/m^{-2} for RCP 4.5, 6.0 W/m^{-2} for RCP 6.0, and 8.5 W/m^{-2} for RCP 8.5. These four RCPs include one mitigation scenario leading to a very low forcing level (RCP 2.6), two stabilization scenarios (RCP 4.5 and RCP 6), and one scenario with very high greenhouse gas emissions (RCP 8.5). For RCP 8.5, missions continue to rise throughout the 21st century; for RCP 2.6 emissions peak around 2080, then decline; and for RCP 4.5 it stabilizes by 2100. It is further stated in the report that global surface temperature change for the end of the 21st century is likely to exceed 1.5°C relative to 1850 to 1900 for all RCP scenarios except RCP 2.6. It is likely to exceed 2°C for RCP6.0 and RCP8.5, and more likely than not to exceed 2°C for RCP 4.5. Warming will continue beyond 2100 under all RCP scenarios except RCP2.6. Further, warming will continue to exhibit inter-annual to decadal variability and will not be regionally uniform. Land is already under growing human pressure and climate change is adding to these pressures. Global warming to well below 2°C can be achieved only by reducing greenhouse gas emissions from all sectors including land and food (IPCC, 2019).

Assessing future projections of climates through a GCM requires large storage and processing capacity and due to those temporal and spatial resolutions of GCM outputs are still limited. Agricultural landscapes vary on a small scale and rough resolution results of GCMs do not provide the necessary spatial accuracy in order to assess the likely impacts of climate change on agricultural production. Downscaling is therefore required to provide a higher resolution in future climate forecast. Regional climate modelling (RCMs) is the most common approach for forecasting climate changes. The impact of climate change is generally studied in terms of the effect of predicting changes in temperature and CO_2 predicting shifts in rainfall patterns, changes in extreme events and sea level rise. In agriculture changes in temperature and CO_2 as well as changes in rainfall and their shifts are important and widely studied. The effect of these is studied empirically using past production data to develop regressions between climate conditions and agricultural production.

Another approach is the use of crop modelling. Many simulation models have been developed which can quantify the effect of changes in temperature, CO_2 and rainfall on crop phenology and yields. The third approach adopted to study the impact of climate change is the Niche based approach where environment niche models are used to quantify the effect of climate change in terms of changes in environmental niches for different crops or pest and diseases. The negative impact of climate change on agriculture is projected to have severe influence on food production and security.

The temperature and rainfall changes projected for the SE Asia due to climate change are given in Table 1. Perusal of the table shows that the increase in temperature is likely to be highest during December, January and February months and the least during June, July and August. The implication of this on potato in India is that in Indo-Gangetic plains, which accounts for about 80% of potato acreage in the country, the potato season which extends from October/ November to January/ February would be warmer by about 0.8–1.2 °C during 2010–2039 and by 1.5–3.2 and 2.2–5.4 °C by 2040-2069 and 2070-2099, respectively. In the plateau areas where the season extends from June to September, it is expected to be warmer by up to 3.1 °C and the rainfall is also likely to be higher by up to 26%, depending upon the time slice and scenario. The current level of CO_2 (396 ppm) in the atmosphere, the main GHG, is 35.4% more than the pre-industrial level and is rising. The CO_2 level is predicted to be 543 and 789 ppm in year 2050 and 2080, respectively and in the light of these expected future climates that assessment of potato production scenario is to be analysed.

IMPACT OF CLIMATE CHANGE ON CROP GROWTH AND QUALITY

Effect of elevated temperature

Temperature affects the growth and development of potato. The reported cardinal temperatures are minimum (0-7°C), optimum (16-25°C) and maximum (40°C) temperatures for net photosynthesis. The main effect of temperature is through the control of the duration of growing period. Higher temperature may shorten the duration of the growing season and the length of the growing cycle. A reduction in maturity period of potato in Punjab up to 10 and 25 days due to 1 and 3 °C increase in temperature, respectively.

Effect of temperature on potao growth and yield

Although photosynthesis in potato is suppressed by high temperature, process of tuberization and partitioning of photosynthates to tuber are highly sensitive to temperature and even moderately high temperature drastically reduces tuber

Table 1: Projected changes in surface air temperature and precipitation for sub-regions of Asia under SRES A1FI (highest future emission trajectory) and B1 (lowest future emission trajectory) pathways for three time slices, namely 2020s, 2050s and 2080s in sub region South Asia (5N-30N; 65E-100E).

Season	2010 to 2039				2040 to 2069				2070 to 2099			
	Temperature °C		Precipitation %		Temperature °C		Precipitation %		Temperature °C		Precipitation %	
	A1FI	B1	A1FI	B1	A1FI	B1	A1FI	B1	A1FI	B1	A1FI	B1
DJF	1.17	1.11	-3	4	3.16	1.97	0	0	5.44	2.93	-16	-6
MAM	1.18	1.07	7	8	2.97	1.81	26	24	5.22	2.71	31	20
JJA	0.54	0.55	5	7	1.71	0.88	13	11	3.14	1.56	26	15
SON	0.78	0.83	1	3	2.41	1.49	8	6	4.19	2.17	26	10

DJF= December, January, February; MAM= March, April, May; JJA= June, July, August; SON= September, October, November (Source: IPCC (2007).

yield with little effect on photosynthesis and total biomass production. The highest tuber yield of potato is obtained at 20°C under short day (12 hour photoperiod) condition. Temperature above 25°C not only reduce and delay tuberization and hence the yield. Being a cool weather crop potato grows well under moderate temperatures. The optimal temperature requirements for above ground plant growth and below ground tuber growth are different. For above ground growth the optimum temperature is 20–25°C and for tuber growth 15-20°C. Temperature above the optimal is likely to inhibit plant growth and development leading to reductions in tuber yield and productivity (Tang *et al.*, 2018) mainly by inhibiting tuberization and tuber yield through change in assimilate partitioning and impaired sucrose translocation to tubers.

High temperature decreases harvest index and thus, tuber weight by inhibiting the carbon export to tubers (sink) from the leaves (source) and the assimilated carbon is accumulated in source leaves or diverted to other leaves as indicated by increased leaf dry matter (Hancock *et al.*, 2014). [With increase in temperature total and tuber dry mass and canopy leaf area decreased, whereas stem dry mass increased and apical and basal stem leaf area covered the major component of the leaf area in every canopy layer]. The consequence of high temperature, however, largely depends upon cultivars, growth stage and duration of stress Temperature sensitivity of potato plant is growth stage dependent and the early stage high temperature stress impacts the plant growth and tuber yield more negatively. In potato, although few studies suggest that high temperature inhibits PSII activity (Havaux, 1993), its effect on RuBisCO activity and at present little is known about photorespiration. The delayed tuberization is associated with high temperature-induced inhibition of tuberization signal known as StSP6A.

Accumulation of carbon in the source leaves has multiple consequences and negatively impacts fresh market and processing quality. High temperature triggers skin russeting deteriorates secondary growth and cause hollow heart, tuber cracking, heat necrosis in the tuber flesh and other malformations. Under high temperature stress conversion of starch to reducing sugars is stimulated thereby affecting quality of French fries. Other consequences of high temperature stress include formation of chain and misshapen tubers, field sprouting and decreased dry matter content in response to high soil temperature.

It has been predicted that potato yield will decline substantially by 2055 as a consequence of global warming and drought. Hijmans (2003) anticipated that world potato production will decline by 18–32% in the projected period of 2040–2069. The climate change and global warming will have a profound effect on potato growth and development in India also, affecting not only production and profitability, but also seed multiplication and processing sectors. The productivity in Indian states is likely to decrease by 19% to 55 % in the year 2050 (Figure 1). All India estimates of production based on current relative contribution of different states in total production, showed decline in production from the current levels by 13.72 % in the 2050, respectively compared to base line scenarios.

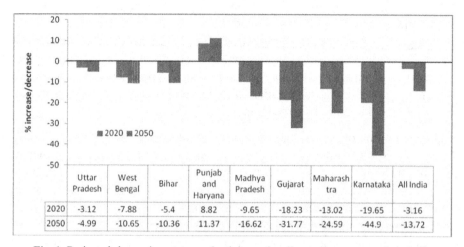

Fig. 1: Projected change in potato productivity under climate change scenario in India (Singh *et al.*, 2009)

Using simulation studies impact of climate change on the productivity of the major potato cultivars in India has been assessed (Dua *et al.,* 2013, 2015, 2016). Over all under A1FI scenario of climate change the impact analysis using WOFOST crop growth model has shown the decline in productivity in 2055 (Uttar Pradesh - 9.4 to 13.4%, Bihar - 8.7 to 12.7%, West Bengal – 8.8 to 12.0%, Madhya Pradesh – 10.9 to 14.3% and Gujarat – 19.5 to 25.2%). By adopting the simple and practical adaptation measures like change in date of planting and selection of suitable variety, the decline in productivity can be brought down to 5.2, 1.9, 4.1, +1.7 and 0.8% in 2020 and 9.2, 6.6, 8.1, 3.6 and 10.4% in 2055 in Uttar Pradesh, Bihar, West Bengal, Madhya Pradesh and Gujarat, respectively (Dua *et al.*, 2013, 2015, 2016). However, the response of potato plants to temperature varies across the cultivars and cultivars like Russet Burbank exhibited maximum rates of photosynthesis at 24 to 30°C with photosynthetic reduction observed only at or above 35°C. Similarly, growth at 30/20 °C day/night temperature reduced tuber yield whereas, the high temperature stimulated the rates of photosynthesis (Hancock *et al.*, 2014).

Effect of temperature on quality and tuber size

Dry matter content as well as number of tubers per plant in potato depends upon temperatures. Minimum temperature of more than 12°C during the last month of tuber maturation is desirable to obtain high dry matter. High temperature can affect tuber quality by causing heat sprouting, which is premature growth of stolons from immature tubers. Internal necrosis is another disorder which may occur at high temperature. Potato processing requires large size tubers with high dry matter. Since, warming may reduce proportion of marketable and processing grade tubers thus climate change may adversely impact processing quality.

Effect of temperature on seed production

In vegetatively propagated potato crop, the disease free quality seed tubers as planting material has a special significance. However, use of tubers as seed material is mainly responsible for rapid and drastic reduction in yield in successive generations of clonal multiplication since it is a carrier of a host of fungal, bacterial and viral diseases. Rapid degeneration of planting materials in potato crop is due to viral diseases transmitted by aphid and other vectors. The technology of 'seed plot technique' was developed for growing seed tubers in relatively low aphid periods in plains during winters and termination of vines by dehaulming before aphid population crosses a threshold to minimize infection of viral diseases. The appearance of vectors is dependent upon climate and the appearance of potato peach aphid (*Myzus persicae*) is reported to advance by two weeks for every 1°C rise in mean temperature. The population build up is also positively correlated with maximum temperature and minimum relative humidity. Thus, under the impact of climate change and global warming the early appearance coupled with increased population is likely to limit the aphid free period to the detriment of seed tuber quality and quantity.

Effect of elevated CO_2

Effect of elevated CO_2 on plant growth and yield

Carbon dioxide affects plants indirectly through climate change, by acting as a greenhouse gas, and directly as a substrate in photosynthesis and by causing partial closure of stomata since the photosynthetic apparatus of C_3 plants is not fully saturated with CO_2 at present day concentrations. Potato being indeterminate crop shows a larger response than determinate crops such as cereals. The saturation CO_2 concentration for C_3 species is approximately 650 μmol mol^{-1}. Consequently, any increase in atmospheric CO_2 up to that concentration should result in an increase in the photosynthetic rate. Thus elevated atmospheric CO_2 level tends to enhance photosynthesis, which results in to better growth and yield of majority crops In the case of potato, considerable response to elevated atmospheric CO_2 is due to its large below ground sinks for carbon and efficient phloem loading mechanism has been reported in many studies worldwide (Komor *et al.*, 1996). Since CO_2 concentration and assimilation are positively correlated. Approximately 10% increase in tuber yield for every 100 ppm increase in CO_2 concentration. These positive effects are attributed to increased photosynthesis from 10 to 40%. Elevated CO_2 cause an increase in potato tuber yield with a concomitant reduction in allocation to above ground biomass, which results in a large increase in root: shoot ratio.

Increased atmospheric CO_2 reduces photorespiration of C_3 plants by promoting carboxylation and diminishing oxygenation of the photosynthetic enzyme Rubisco. The enhancement of growth in C_3 species under elevated CO_2 may also be expected to increase demand for nutrients; however, as the photosynthetic apparatus may exhibit partial or complete acclimation to elevated CO_2, individual nutrients may be affected differently. This change, together with the increased CO_2 diffusion gradient into the leaves is expected to lead to enhanced photosynthetic rates. The increase in photosynthesis is most marked in young leaves of potato. This is attributed to the phenomenon of photosynthetic acclimation later in the growing season particularly in old leaves. Varietal differences in response to elevated CO_2 concentration exists. Further, simulation studies have also shown that the yield effect of increased CO_2 would have been almost twice if acclimation had not down-regulated the photosynthesis rate, indicating that acclimation may be responsible for a major reduction of the CO_2 effect on biomass and tuber production. Total tuber fresh weight was 36% higher, under high CO_2 treatments, as compared with that in the ambient and the response of Kufri Chipsona-3 was more pronounced, to elevated CO_2 concentration, as compared with Kufri Surya.

Studies have shown that elevated CO_2 increases tuber yield in potato reduces nitrogen content. Reduced nitrogen content implies a decrease in protein content thus increased applications of nitrogen fertilizer may be needed to derive maximum benefit from predicted future increases in atmospheric CO_2 concentration. Number of tubers is expected to remain unaffected under elevated CO_2, but mean tuber weight is likely to increase mainly through increase in number of cells in tubers without influencing the cell volume. However, an increase in tuber number has also been reported. Elevated CO_2 concentration advances the tuber initiation and flowering but also hastens senescence of leaves. The relationship between leaf senescence and atmospheric CO_2 levels was found to be linear up to 660 ppm. Further elevated CO_2 concentration has been reported to reduce chlorophyll content in leaves particularly during later growing season after tuber initiation. Effect of elevated CO_2 was studied in two potato varieties in transparent walled open top chambers (OTCs) with ambient and elevated CO_2 levels (550 and 700 ppm). Kufri Surya was found more responsive to elevated CO_2 than Kufri Pukhraj as showed up to 24 and 70 per cent increase in yield, respectively by increasing CO_2 level till 700 ppm (Kumar et al., 2018).

In other study elevated CO_2 increased the fresh tuber yield by 36%. The magnitude of the yield response is determined by interaction with a number of other factors including cultivar, agronomy and growing conditions. Tuber yield stimulation was observed in most cases, accompanied by an increase in the number of tubers. Simulation studies have shown that CO_2 levels of 700 ppm could stimulate tuber yield by 20-30% The few negative effects of elevated CO_2 concentration include reduction in chlorophyll content in leaves particularly during later growing season after tuber initiation. Later in the season, leaf photosynthesis is also likely to decrease progressively in the higher CO_2 environment due to senescence. Leaf N concentration is reported to decrease faster in the leaves grown under high CO_2 environment which further supports the conclusion that leaf senescence is accelerated in the plants grown under high CO_2. Varietal differences for response to elevated CO_2 has been observed and Kufri Chipsona-3 was found more responsive to elevated CO_2 concentration, compared to Kufri Surya. Overall rising atmospheric CO_2 in the future climatic change scenario may be beneficial to such tuber crops like potato to enhance growth as well as tuber number and weight.

Effect of temperature and CO_2 on water use efficiency, nutrient acquisition and tuber qulaity

Water use efficiency

Water use efficiency (WUE) is defined as the amount of carbon assimilated as biomass or yield produced per unit of water used by the crop. The response of WUE at the leaf level is directly related to the physiological processes controlling the gradients of CO_2 and water, which shows vapour pressure deficits, between the leaf and its surrounding air. From the leaf to the canopy, the dynamics of crop water use and biomass accumulation reflects soil water evaporation rate, transpiration from the leaves, and the growth pattern of the crop. Factors responsible for affecting water use by plants are increasing carbon dioxide (CO_2) level, temperatures, more variable precipitation and variations in humidity. Projections of climate change are the result of a combined set of simulation models using various scenarios of changes in carbon dioxide (CO_2) levels and the associated forcing functions .

Higher temperatures and heat waves are becoming more frequent, compounded by increasing episodes of drought, flood and land degradation and drought has been identified as the most detrimental environmental stress affecting agriculture worldwide causing greatest loss of yield to field crops like rice, chickpea and potato. Increase in WUE at the canopy level can be achieved by adopting practices that reduce the soil water evaporation component and divert more water into transpiration which can be through crop residue management, mulching, row spacing and irrigation. Climate change will affect plant growth, but there are opportunities to enhance WUE through crop selection and cultural practices.

Over the past 50 years WUE of potato has increased due to an increase in temperature and a decrease in precipitation. In China WUE increased in the 1990–2009 period in potatoes by 0.00045 mm m^{-2} yr^{-1} compared to the previous values. However, this response could be expected, if the temperatures during the growing season were not above the optimum for the specific crop. Guoju *et al.* (2013) found that under controlled conditions WUE increased as temperatures increased to 1.5°C above normal and then began to decrease. They also observed that WUE began to decrease linearly with increases in annual precipitation above 310 mm and proposed that the combination of increased temperatures and precipitation affected the respiration rate of potato, which directly influenced the productivity of the plant.

Use of plastic mulches has been found to increase the water use efficiency. Li *et al.* (2018) evaluated the effect of plastic and straw mulch on WUE of potato and found plastic mulch increased productivity by 24% and straw mulch by 16%. Water use efficiency, however, was affected by seasonal air temperature, precipitation, baseline soil fertility, fertilizer management and the effectiveness of the mulching practices on WUE were increased when precipitation was less than 400 mm and decreased when precipitation was above 400 mm.

Decreased stomatal conductance of potato leaves at elevated (up to concentrations of 700 μmol mol^{-1}) and super-elevated (1000-10,000 μmol mol^{-1}) CO_2 has been reported by many researchers in crops including potato. Decreased stomatal conductance in response to elevated CO_2 could be expected to improve the water use efficiency (WUE) of potato crops. Doubling CO_2 concentration is reported to reduce stomatal conductance in potato by about half although this reduction did not limit the net photosynthetic rate, which rather increased by approximately 50% in some studies. Reduced transpiration rate coupled with increased instantaneous transpiration efficiency resulted saving to the extent of 12 to 14 %.

Irrigation will play an important role under future climate scenarios. In the absence of irrigation, tuber yields are predicted to be much lower, declining slightly with rise in temperature and with increases in radiation, increasing considerably with increasing CO_2 while, declining slightly with increasing O_3 in Europe under two holistic climate change scenarios. However, irrigated tuber yields in northern Europe are predicted to increase by 1-4 tonnes dry matter per ha while tuber yield of un-irrigated crops would increase by 1-3 tonnes dry matter per ha. Yield increases were predicted to be smaller in southern Europe and (NPOTATO and POTATOS) predicted moderate to strong increases in tuber yield in northern Europe with little or no change in the British Isles, central and southern Europe in 2050. This modelling analysis also revealed advancing the planting dates as the best management strategy to minimize the effects of climate change. Changing planting dates has also been shown as best management strategy in Indian potato growing regions by simulation studies using WoFoSt model.

Effect of elevated CO_2 and temperature on nutrient content

Both temperature and CO_2 are known to influence the nutrient acquisition. Some studies have reported reduction in macro and micronutrient concentrations in plants grown under elevated CO_2. Decrease in concentrations of grain nutrients such as N, Ca, Zn and Fe were observed in soybean, sorghum, potatoes, wheat and barley grown in the free-air CO_2 enrichment (FACE) facilities (Myers *et al.*, 2014; Dietterich *et al.*, 2015). Concentrations of P, K, Mg, Mn and Zn s were significantly higher in the stems and leaves of OTC grown potato, whereas Ca and Fe concentrations were lower and in tubers potassium, calcium, magnesium and zinc concentrations were significantly higher in the OTC plots in response to elevated CO_2 (Fangmeier *et al.*, 2002).

The decrease in leaf protein content can lead to decreased seed protein concentration as the N supply to tubers is largely from translocation from catabolized proteins in senescing photosynthetic tissues in seeds and this mechanism may also contribute to the decreased protein concentrations in potato tuber. Since approximately 50% of the nitrogen supply to tubers is from translocation from leaves and stems (a decrease in N concentration of potato tuber to the tune of 24% has been reported under elevated CO_2). This effect of elevated CO_2 on protein nutrition might be mitigated by adopting agricultural practices that increase the protein content to a large extent. However, reductions in macro and micronutrient concentrations have been observed in plants grown under elevated CO_2. Studies conducted in wheat have demonstrated that reductions in macro and micronutrient concentrations have been observed in plants grown under elevated CO_2. Thus, nitrogen fertilizer can minimize, but not eliminate, the reduction in protein concentration associated with increased atmospheric CO_2. This will increase the pressure on resources since more nitrogenous fertilizers will be neded and may also cause potentially undesirable ecological effects across a wide variety of ecosystems.

Effect of elevated CO_2 on tuber nutritional quality

Nutrient supply and availability affect tuber quality aspects in potato, CO_2 induced changes in mineral use efficiency and thus mineral concentrations in tubers may also affect related quality parameters. It is established that excessive application of N decreases the dry matter content as well as the citric acid and malic acid concentrations, while it increases glycoalkaloid and ascorbic acid concentrations in potato tuber. As a consequence, elevated CO_2 may result in reverse effects because of increased NUE. Improvements with regard to potato tuber quality may also be expected as P concentration increases at elevated CO_2 because of the positive relationships with taste and concentrations of protein, starch and ascorbic acid. Higher application of N decrease the dry matter content as well as the citric acid and malic acid concentration, while glycoalkaloid and ascorbic acid concentration is increased in potato tuber (as a consequence, elevated CO_2 may result in reverse effects because of increased nitrogen used efficiency). Reduced protein content in response to elevated CO_2 is a matter of concern. Assuming today's diets and levels of income inequality, approximately 148 million of the world's population may be placed at risk of protein deficiency because of elevated CO_2 and 53 million people may become at risk of protein deficiency in India (Danielle *et al.*, 2017).

The mechanism(s) by which elevated CO_2 decreases tissue concentrations of N and protein are not thoroughly understood, as growth at elevated CO_2 can affect multiple processes involved in nitrogen uptake and metabolism. Dilution of nitrogen by increased concentrations of nonstructural carbohydrates is one mechanism that has often been suggested to account for decreased N and protein concentrations. The few studies that have measured both non-structural carbohydrates and protein (or N) in the edible portions of crop plants suggest that dilution can account for at best a small proportion of the observed decreases in protein concentrations. Fangmeier *et al.* (2002) reported on the effect of elevated CO_2 on a wide range of both macro and micronutrients in above and below ground tissue of potato in the CHIP project. They found a decrease in both phosphorous and potassium concentration in above-ground biomass in response to elevated CO_2 at an intermediate (maximum leaf area) harvest at five sites across Europe. A reduction in nitrogen, potassium and calcium concentration in tubers was also reported at this harvest. Nitrogen, manganese and iron concentration, ecological responses and adaptations of crops in above-ground biomass at final harvest were all significantly reduced in response to elevated CO_2 whereas tuber nitrogen, potassium and magnesium concentrations were also reduced (Fangmeier *et al.*, 2002).

Effect of climate change on carbon sequestration and soil quality

The great potential of C sequestration in cropland has provided a promising approach to reducing the atmospheric concentration of CO_2 for mitigating climate change. However, this approach depends on cropping systems, which may be defined as an operating system for growers to follow in their practices for crop production. An ideal cropping system for C sequestration should produce and remain the abundant quantity of biomass or organic C in the soil. The organic C concentration in the surface soil (0-15 cm) largely depends on the total input of crop residues remaining on the surface or incorporated into the soil. It decreases soil C greatly to remove crop top from the soil by cleaning up the land. Therefore, to improve C sequestration, it is critical to increase the input of plant biomass residues. Biomass accumulation can be enhanced by an increase in cultivation intensity, growing cover crops between main crop growing seasons, reducing fallow period of land, crop rotations and intercropping systems. Biomass return to the soil can be improved by elimination of summer or winter fallow and maintaining a dense vegetation cover on the soil surface, which can also prevent soil from erosion for SOC loss. In different agro-ecologies of Peru 55.6 t/ha carbon stock was quantified under potato (Monneveux *et al.*, 2012).

The impact of climate change on soil is a slow complex process as because soils not only be strongly affected by climate change directly (for example effect of temperature on soil organic matter decomposition) and indirectly for example changes in soil moisture via changes in plant related evapotranspiration) but also can act as a source of greenhouse gases and thus contribute to the gases responsible for climate change (Soil-climate models assuming constant inputs of carbon to soils from vegetation predicts the expected changes in temperature, rainfall and evaporation with associated increase in organic matter turnover facilitating increased losses of CO_2 in mineral and organic soils.) The losses of soil carbon, affect soil functions like poorer soil structure, stability, topsoil water holding capacity, nutrient availability and erosion. Soil organic matter is undoubtedly the most important soil component as it improves soil quality though the influences in soil structure, water holding capacity, soil stability, nutrient storage, turnover and oxygen-holding capacity. Soil organic matter is highly susceptible to changes in land use and management, soil temperature and moisture. In the last decades changes in land use and management have already led to a significant decline in organic matter levels in many soils which increase the susceptibility to soil erosion.

Increase in temperature coupled with decline in precipitation may affect water availability indirectly by increasing salinity. The water stress caused by high evaporative demand due to increase in temperature may supersede the beneficial effects of increased CO_2 in the atmosphere unless irrigation can be stepped up to compensate the water requirement of potato. In conclusion, increased productivity would generally lead to greater inputs of carbon to soil, thus increasing organics. This may be true in other crops but not much is expected from potato where more than 80% of the biomass is harvested and field preparation involves intesive ploughing.

IMPACT OF CLIMATE CHANGE ON PESTS AND DISEASES

Potato is affected by number of pests and diseases which limit its production. Climate change involves changes in temperature, moisture and CO_2 levels which influences survival, multiplication and vigour of many insect and pathogens and can result in loss in production and supply chain. Besides this, other greenhouse gases comprising of water vapour, methane, nitrous oxide, hydrofluorocarbons and ozone directly or indirectly influence crop productivity.

Global warming brings an increase in ambient temperature. Temperature is the single most important factor which influences the behaviour, distribution, development, survival and reproduction of pests and diseases. Insects especially aphids and whiteflies are vector of many viral and mycoplasmal diseases of potato. Any changes either in host plant or insect vector population due to climate change has potential to spread plant viruses. A 2°C increase in temperature in temperate climate zones could result in one to five additional life cycles of certain insects per season and thus the insect population is likely to increase by the global warming. Warmer conditions may also promote a range of insect species to new geographical zones. The insects may survive in new zones which are currently limited by low temperature or the altitude at which they can survive. A 2°C rise in temperature may result in a shift of certain insect population by as much as 600 km of latitude or 330 m in elevation thus bringing an increase in the geographical area in the range of their infestation. Higher average temperature might also result in some crops being able to be grown further north in cooler regions where these are not grown currently and some insect pests of those crops shall follow the expanded crop areas. Temperature also plays a vital role for the occurrence of bacterial diseases such as *Ralstonia solanacearum* in the areas where it has not been previously observed (Kudela, 2009).

The elevated CO_2 can lead to changes in physiology and anatomy of plants such as formation of extra layers of epidermal cells, greater number of mesophyll cells, greater accumulation of carbohydrates, more wax deposition and increased fibre content in leaves and thus have an impact on plant disease interaction. This can indirectly influence conditions for initial establishment of a pathogen and its build up. The initial establishment of some pathogens may be delayed because of modifications in host susceptibility and an increase in fecundity of pathogens. The combination of increased fecundity and a more humid microclimate within dense crop canopies associated with increased CO_2 concentrations might provide more opportunities for severe infection for the foliar pathogens such as *Phytophthora infestans* causing late blight. The higher CO_2 concentrations may also result in greater spore production by some pathogens.

Frequency of extreme events such as droughts and floods are increasing due to climate change. Changes in rainfall pattern have implications on life cycle and survival of certain insect pests and pathogens. It is predicated that frequency of rainfall will decrease but its intensity will increase which might lead to excess water resulting into floods on one hand and long dry spells on the other. Small insects like sucking pests' *viz.* aphids, mites, psylla, whitefly *etc.* might be affected adversely by such patterns of rainfall. More intense rains might lead to outbreaks of late blight disease, pests like armyworms and beetles. This can result in a 2.4 to 2.7 fold increase in use of pesticide by 2050. Influence of climate change on major pests and diseases of potato is discussed here.

Impact on climate change on insects

Myzus persicae **and** *Aphis gossipii*

Aphid species like *Aphis gossypii, Myzus persicae etc.* are worldwide in distribution. They play an important role as vectors to spread many viral diseases. An increase in their population and activity can increase spread of many potato viruses. Change in climate may result in change in their geographical distribution, changes in population growth rates, increases in their number of generations, changes in crop-pest synchrony and increased risk of invasion by migrant pests. Temperature is the single most important environmental factor which influences the behaviour, distribution, development and survival of aphids. In *M. persicae* an increase of mean temperature by 1°C could advance the time of progeny migration by two weeks. Population of *M. persicae* could get enhanced by higher levels of CO_2. Thus an increase in temperature and CO_2 could result in an increase in number of generations of *M. persicae* in a year resulting in more infestation and loss in yield. Results tend to suggest that in sub-tropical plains of India, *M. persicae* population is on the rise. An analysis of aphid data for 20 years (1984-85 to 2003-04) have revealed that appearance of aphids has advanced by 5 days resulting in reduction of the seed production window from 80 to 75 days. Any further reduction in the seed production period may adversely affect the seed production programme in the subtropical plains in India. Population of *Aphis gossypii,* another aphid affecting potato in the region, has already increased three fold during the last 20 years. Although *A. gossypii* has low vector efficiency but its appearance right from the emergence of the crop and prevalence throughout the crop season could pose serious problems to seed production in sub-tropical plains of India.

White flies

White flies (*Bemisia tabaci*) affects potato besides many crops such as cotton, chilly, capsicum, okra, crucifers, cucurbits, tobacco and many others. The nymphs and adult suck cell sap resulting in lower vitality of plants. The whiteflies are vector for many viral diseases of potato including Potato Apical Leaf Curl Virus (PALCV) the incidence of which has already increased in recent years. In India, the whiteflies are distributed mostly in northern and western region of the subcontinent. Whiteflies can transmit a number of Gemini viruses and have been responsible for the spread of apical leaf curl disease in potato in northern India in the last two decades. Increase in temperature is likely to increase their population and their distribution further in the northern states thus bringing more geographical area under its attack. Findings from CIP Peru suggest that global warming is already affecting the distribution of white flies. Experiences from Canete Valley in Peru has revealed that whiteflies can develop resistance to pesticides and multiply faster at high temperature at higher population densities white flies has potential to devastate whole crop stands. Earlier, only biotype A of *Bemisia tabaci*, which causes only moderate yield loss was found in Canete Valley. But a more aggressive whitefly biotype B, was identified in 2001. In India, *B. tabaci* was a minor pest in the early eighties but data on population build up recorded during the past 20 years has revealed that the average population of *B. tabaci* has increased more than two folds from 1984 to 2004. During this period, average ambient temperature has increased by 1.07 °C. This indicates that warming induced by climate change could increase whitefly infestation. The outbreak of a new viral disease caused by a gemini viruses has added a new challenge to seed potato production in sub-tropics of India. It requires a new approach to manage such viruses through discarding of the old stocks, their replacement with disease-free stocks produced through tissue culture and through management of the whitefly population in an integrated manner. PALCV is likely to flare up in new areas with the increase in ambient temperature and requires special attention in the seed producing areas of the country (Singh and Bhat, 2010).

Leaf hoppers

Empoasca fabae (leaf hopper) is widely distributed in all the potato growing areas of the country. The Nymphs and adults suck sap from the leaves and their prolonged feeding can cause hopper burn. The leaf hoppers while feeding on potato also inject a toxin with saliva which cause browning and drying of the leaves resulting in drying up of the foliage. The leaf hoppers has become an important pest especially in early planted potato crop in sub-tropical plains of India. Its population has increased over the years. The average population during 1984 was 16.6 which rose to 23.8 in 2004. Consequently, the hopper burn damage has also increased from 45 to 68% during this intervening period. Further warming up the atmosphere could flare up the pest population. Severe attack of leaf hoppers were observed during 2006-07 in Gujarat when the increase in temperature was up to 5.8°C during fourth week of December, 2006. This led to faster multiplication of the pest resulting in very heavy (up to 100 %) hopper burn damage This pest could pose a problem in new areas with the rise of temperature due to global warming

Mites

Mites (*Tetranychus urticae* Koch and *Polyphagotarsonemus latus* Banks) are widely distributed in India. They suck cell sap from young plants and leaves and cause bronzing, curling and discolouration of leaves. Severe infestations can cause 12 to 60% loss in yield depending upon the stage of the crop. Early planted crop which is exposed to high temperature in Punjab and western UP is severely affected by mites during some years. In Maharashtra and Karnataka the *rabi* crop can be seriously affected toward its maturity in February-March. Mite infestation has steadily increased in early planted crop in recent years. According to an estimate its damage has increased from 86% during 1984-85 to 100 % in 2004. An increase in temperature and humidity can further result in increase in crop damage due to mites (Singh and Bhat, 2010).

Tuber moth

Potato tuber moth (*Phthorimaea operculella* Zeller) is a major pest of potato. The larvae of the pest cause more damage to potato. The larvae cause mines in leaves, petioles and terminal shoots causing wilt of the affected foliage. The larvae enter the tubers and feed on them. They leave black excretory mass inside the tunnels formed in the affected tubers and make these unfit for human consumption or their use as seed. Population of tuber moth is expected to rise with increase in temperature and the pest is predicted to spread further into areas like Gujarat and Rajasthan, which are currently too cold for them. The potato tuber moth (*Phthorimaea operculella*) damage potential will progressively increase in all regions where the pest already prevails today in cropping regions of the tropics and subtropics. It will increase until the year by 2050 to 42.4% from the present infestation which is equal to an increase of 2,409,974 ha of potato under new infestation. It is likely to increase in many areas in Asia, North America, Europe and particularly in Bolivia, Ecuador and Peru.

Insect vectors and viral diseases of potato

Apart from vectors, global warming can change the entire scenario of viruses *per se* in temperate regions and sub-tropics. The rate of multiplication of most of the potato viruses is expected to rise with the increase in temperature. By the end of this century when average temperatures are expected to increase by about 3°C there is a possibility of flaring up of the viral and mycoplasmal diseases. Studies carried out in Holland has revealed that during 12 years (1994-2008), new viral strains (PVYntn, PVYnw) have been detected. Similar results have been obtained in India. This will also result in an increase in the number of insecticide sprays required to keep the vector population in check. Increased use of insecticides can lead to development of insecticide resistant aphids, which shall be difficult to manage through normal use of pesticides. Clones of *M. persicae* resistant to three major groups of insecticides have already been reported. An increase in temperature may also alter the physiology of the host and its resistance mechanism. The rate of multiplication of viruses in host tissue can increase substantially leading to early expression of the disease symptoms and increased reduction in crop yields.

Impact on fungal and bacterial diseases

An overall temperature increase could also influence crop pathogen interactions by speeding up growth rates of the fungal and bacterial pathogens resulting in their additional generations per crop cycle and simultaneously a decrease in pathogen mortality due to warmer winter temperatures which can make the crop more vulnerable to such diseases. A more humid microclimate within dense crop canopies associated with increased CO_2 concentrations could provide more opportunities for the foliar pathogens of potato to severe infection. Higher CO_2 concentrations may also result in greater fungal spore production. Higher rainfall coupled with rise in temperature can result in spread of certain bacterial diseases such as *Ralstonia solnacearum* causing bacterial wilt of potato.

Late blight

Several potato growing regions of the world are likely to face a further increase in temperature and precipitation thus making them warmer and wetter. This will make conditions favourable for late blight especially in the temperate regions. Work in Finland, has predicted that for each 1°C rise in temperature, late blight would occur 4 to 7 days earlier, and the crop susceptibility period will be extended by 10 to 20 days. Early onset of warm temperatures could also result in an early threat from late blight in some regions. Late blight is also expected to expand into new areas that have previously been fairly safe from this disease. According to an estimate the yield loss would be 2 t/ ha per 1°C increase in temperature. Consequently, additional 1 to 4 fungicide applications could be required to manage the disease there by increasing both input costs, environmental risk and there is a likelihood of the development of fungicide resistance which has been reported in other crops. Early onset of warm temperatures could result in an early appearance of late blight disease in temperate regions resulting in more severe epidemics and increased number of fungicide applications to manage the disease in this region. Late blight scenario can change drastically with climate change in India. Currently, late blight is not a very serious problem in autumn during most of the years in the state of Punjab, Haryana and parts of Uttar Pradesh, primarily due to sub-optimal temperature prevailing during December-January. However, the disease outbreaks can become more intense with increase in ambient temperature during December-January coupled with high RH.

WAY FORWARD

Potato crop, like most other crop species, is highly prone to abiotic stress particularly, high temperature, drought and soil salinity. In order to maintain a sustainable potato production under climate change, we must identify best cultivation practices and develop heat, drought, insect and pathogen tolerant cultivars that can best adapt to the changing environment. It is one of the challenges for crop scientists to pinpoint means of improving crop yield and quality at high CO_2, high temperature and drought stress as well as biotic stress. One important approach is to understand stress-related molecular, biochemical and physiological markers that can be used to develop screening procedures for selection of crop cultivars that can better adapt to sub-optimal growth conditions. The mechanism by which potato plants initially sense the changes in their surrounding CO_2, temperature, water status, soil salinity and biotic stress and consequently respond to these changes at the molecular, biochemical and physiological levels is still under infancy. Future research needs be concentrated on the identification of signaling molecules, target genes and their networks that govern plant phenotypic and metabolic plasticity in response to elevated CO_2, high temperature and water deficit. Elucidation of such mechanism may provide new insights into the identification of specific characteristics that may be useful in breeding programs aimed at (i) developing new potato cultivars with enhanced yield potential and (ii) developing potato germplasm with adaptation to climate change. This will help in sustaining or even enhancing potato productivity under predicted climate change.

CONCLUSIONS

The effect of high temperature associated with climate change on potato crop yield is still a major area of research and still little information is available about the most critical growth stage in relation to temperature stress. Understanding and elucidation of mechanism by which potatoes respond to changes in surrounding temperature at the physiological, biochemical and molecular levels may offer new insights into the identification of specific characteristics that will be useful in breeding new cultivars aimed at sustaining or even enhancing potato crop productivity and quality in response to climate changes.

Although crop dependent the predicted rise in atmospheric CO_2 level is expected to increase crop yield and biomass through stimulation of photosynthesis and suppression of photorespiration an important concern with increase in atmospheric CO_2 is concomitant increase in heat stress and associated water deficit and biotic stress. Thus, the yield deterioration due to climate change associated heat stress, drought stress and biotic stress can outpace the benefit achieved by any increase in atmospheric CO_2. It is one of the challenges for plant breeders and agriculturists to maintain yield stability through improved tolerance to climate change associated stresses such as drought, low and high temperature in future.

Enhancing potato productivity is important to meet the global food demand of an increasing population. However, potato plant growth and tuber yield is constrained by high temperature, water limitation, soli salinity, and insect and pathogen threats. Climate change will likely further aggravate tuber yield losses by intensifying potato plant's exposure to these stress conditions. Hence, there is an urgent need to adapt to the new cropping challenges by developing heat, drought, insect and pathogen tolerant crop cultivars that are appropriately engineered for the changing environment. Improving potato plant adaptability to environmental stresses under increasing CO_2 is one of the most important and challenging targets. This challenge can be approached through the identification of stress-related traits at the physiological, biochemical and molecular levels and their deployment in new cultivars. In addition, the development of new knowledge and techniques on "omics-driven" high-throughput approaches is crucial to advance screening procedures for breeding potato cultivars aimed at improving plant adaptability and tuber productivity in response to climate change. In order to meet the challenges posed by the climate change, there is an imminent need to develop a multipronged strategies which may include:

- Development of early bulking, heat tolerant and virus resistant potato varieties which can be planted 7-10 days in advance of their normal planting. These can help to neutralize early appearance of aphids and their exposure to high population. Besides, efforts need to be made to breed drought, salinity tolerant and disease resistant cultivars.

- Refining and development of improved agronomic practices which include: Use of crop residue mulches after planting, use of drip and sprinkler irrigation, conservation tillage, on farm crop residue management, development of agro-techniques for warm weather cultivation and potato based cropping systems *etc.*

- Identification of new areas for potato cultivation and advance planning for its possible relocation.

- Improvement and augmentation of cold storage facilities and air conditioned transportation chain from producing to consumption centers.

- Adjustment in the time of planting of potato according to the specific plant protection problem prevalent for that particular region. Leaf hoppers, mites and whiteflies usually affect the early planted crop and a delay in planting by 10-15 days along with Integrated Pest Management (IPM) practices can help to manage these pests in the specified areas.

- Regular monitoring of pest population, pest forecasting, developing diagnostic tools and following IPM measures can help in formulate strategies to cope with the effects of climate change.

- Study of virus-vector relationship involving aphids and potato virus Y, potato leaf roll virus. Study of ecology and biology of whiteflies in spread of potato apical leaf curl virus for a greater understanding of the virus-vector interactions. Such studies will help us in devising new measures to protect our crop from these important vectors.

- Shifting of cultivation of nucleus seed in net houses and replacement of conventional seed production by tissue culture based seed production system are some of the measures which can help to reduce virus incidence to prevent degeneration in our seed stocks.

- Identification and incorporation of resistance against late blight disease in early maturing potato varieties. This can help the crop to escape the damage from late blight disease.

- Development of new and modification of the existing late blight forecasting systems in tune with the climate change.

- Monitoring changes in pathogen population and rescheduling of sprays based on distribution of new pathogen population.

- Enhanced research on production of potato under soil less and multi-tier system under protected cultivation.

REFERENCES

Danielle EM, Joel S and Samuel SM (2017) Estimated effects of future atmospheric CO_2 concentrations on protein intake and the risk of protein deficiency by country and region. *Environmental Health Perspectives* **125**(8): 087002-1-8

Dietterich LH, Zanobetti A, Huybers P, Leakey ADB, Bloom AJ, Carlisle E, Fernando N, Fitzgerald G, Hasegawa T and Holbrook NM (2015) Impacts of elevated atmospheric CO_2 on nutrient content of important food crops. *Scientific Data* **2**: 150036

Dua VK, Singh BP, Govindakrishnan PM, Kumar Sushil and Lal SS (2013) Impact of climate change on potato in Punjab–a simulation study. *Current Science* **105**: 787–794

Dua VK, Singh BP, Kumar Sushil and Lal SS (2015) Impact of climate change on potato productivity in Uttar Pradesh and adaptation strategies. *Potato Journal* **42**(2): 95-110

Dua VK, Kumar Sushil, Chaukhande Pooja and Singh BP (2016) Impact of climate change on potato (*Solanum tuberosum*) productivity in Bihar and relative adaptation strategies. *Indian Journal of Agronomy* **61**(1): 79-88

Fangmeier A, De Temmerman L, Black C, Persson K and Vorne V (2002) Effects of elevated CO_2 and/or ozone on nutrient concentrations and nutrient uptake of potatoes. *European Journal of Agronomy* **17**: 353–368

Guoju X, Fengju Z, Zhengji Q, Yubi Y, Runyuan W and Juying H (2013) Response to climate change for potato water use efficiency in semi-arid areas of China. *Agricultural Water Management* **127**: 119–123

Hancock RD, Morris WL, Ducreux LJ, Morris JA, Usman M and Verrall SR (2014) Physiological, biochemical and molecular responses of the potato (*Solanum tuberosum* L.) plant to moderately elevated temperature. *Plant, Cell and Environment* **37**: 439–450

Havaux M (1993) Rapid photosynthetic adaptation to heat stress triggered in potato leaves by moderately elevated temperatures. *Plant, Cell and Environment* **16**: 461–467

Hijmans RJ (2003) The effect of climate change on global potato production. *American Journal of Potato Research* **80**(4): 271-279

IPCC (2019) Expert meeting on mitigation, sustainability and climate stabilization senarios, report, Priyadarshi R. Shukla, Jim Skea, Renée van Diemen, Elizabeth Huntley, Minal Pathak, Joana Portugal-Pereira, Juliette Scull and Raphael Slade. Published by the IPCC Working Group III Technical Support Unit, Imperial College London, the United Kingdom. Available on http://ipcc.ch/ and the IPCC WGIII website http://www.ipcc-wg3.ac.uk/ 41p

Komor E, Orlich G, Weig A and Kockenberger W (1996) Phloem loading - not metaphysical, only complex: towards a unified model of phloem loading. *Journal of Experimantal Botany* **47**: 1155-1164

Kudela V (2009) Potential impact of climate change on geographic distribution of plant pathogenic bacteria in Central Europe. *Plant Protection Science* **45**: 527-532

Kumar Prince, Minhas JS, Sharma Jagdev, Dua VK, Kumar D, Saha Sunayan and Gupa YK (2018) Impact of elevated CO_2 level on growth, tuber yield and mineral content of Indian potato cultivars. *Potato Journal* **45**(2):123-130

Li Q, Li H, Zhang L, Zhang S and Chen Y (2018) Mulching improves yield and water-use efficiency of potato cropping in China: a meta-analysis. *Field Crops Research* **221**: 50–60

Monneveux P, Quiroz R, Posadas A and Kleinwechter U (2012) Facing climate change effects on potato cultivation: an integrative approach. In, International Conference FAO and IAEA on Managing Soils for Food Security and Climate Change Adaptation and Mitigation, Vienna, Austria, 23-27 July 2012

Myers SS, Zanobetti A, Kloog I, Huybers P, Leakey ADB and Bloom AJ (2014) Increasing CO_2 threatens human nutrition. *Nature* **510**: 139–142

Singh BP and Bhat NM (2010) Impact assessment of climate change on potato disease and insect pests. In, Challenges of climate change: Indian Horticulture. Singh HP, Singh JP and Lal SS (eds). Westville Publishing House, New Delhi 178-184

Singh JP, Lal SS and Pandey SK (2009) Effect of climate change on potato production in India. ICAR-Central Potato Research Institute, Shimla, Newsletter **40**: 17-18

Tang R, Niu S, Zhang G, Chen G, Haroon M and Yang Q (2018) Physiological and growth responses of potato cultivars to heat stress. *Botany* **96**: 897–912

3

Genetic Resources of Potato and Its Utilization

Vinod Kumar[1], Jagesh Kumar Tiwari[1], SK Luthra[2], Dalamu[3], Vanishree G[1] and Vinay Bhardwaj[1]

[1]*ICAR-Central Potato Research Institute, Shimla-171 001, HP, India*
[2]*ICAR-Central Potato Research Institute, Regional Station Modipuram-250 110, UP, India*
[3]*ICAR-Central Potato Research Institute, Regional Station, Kufri Shimla-171 012, HP, India*

INTRODUCTION

Potato belongs to a genetically diverse genus *Solanum* having nearly 235 tuber bearing species. These species form a polyploid series ranging from diploids (2n=2x=24) to hexaploids (2n=2x=72), with 12 as the basic chromosome number. There is also a rich diversity within the cultivated species of potato and they were initially grouped under six different species. Later studies, however, recognized those species as land race populations of *S. tuberosum*, with the eight groups: Ajanhuiri, Andigenum, Chaucha, Chilotanum, Curtilobum, Juzepczukii, Phureja and Stenotomum (Spooner & Salas, 2006). This rich genetic diversity of the species is available in the Andean mountain where potato originated. In India, however, it was quite natural that genetic variability within the locally available material will be meager since, it was introduced here only about 400 years ago. On the contrary, more than 500 different varieties existed in India by 1940. This paradoxical situation was primarily due to assigning different local names to a single variety in different parts of the country. The available local varieties in India consisted of both early introductions belonging to either pure *Solanum tuberosum* ssp. *andigenum* or their hybrids with *S. tuberosum* ssp. *tuberosum* as well as later European introductions belonging to *S. tuberosum* ssp. *tuberosum*.

GENETIC RESOURCES OF POTATO

The sum total of genes in a crop species is referred to as genetic resources or gene pool or genetic stock or germplasm. Genetic resources are a strategic resource for sustainable crop production. Their efficient conservation and use are critical to keep feeding increasing world populations. Gene banks play a key role in the conservation and distribution of germplasm for crop improvement and research for sustainable food production.

GERMPLASM COLLECTION

Up to early 20th century, *S. tuberosum* L. was the only potato species known outside South America. It was only after the first expedition carried out in center of origin by Prof. S.M. Bukasov (1933) and his co-workers in 1925-26 that knowledge about the wealth of genetic diversity among the tuber bearing *Solanum* started accumulating. Thereafter a number of N.I. Vavilov expeditions largely led by S.M. Bukasov and S.W. Juzepczuk were held in the Americas. This laid the foundation of the first gene bank in Leningrad starting in 1927. This gene bank still survives as N.I. Vavilov Institute of Plant Industry, Russia. Taking advantage of the pioneering Russian work, British Empire expedition held in 1939, collected materials in Mexico and the Andean countries of South America. As a result, Commonwealth Potato Collection was established in Cambridge and is now situated at Scottish Crop Research Institute, Pentlandfield, Scotland. Lately, the major bulk of collection has been mediated by the International Potato Centre (CIP), Lima, Peru, a CGIAR organization with the primary mandate of collection and conservation of vast potato genetic resources. Presently CIP is the holder of the largest collection of the potato germplasm. In addition, large potato collections also exist at Inter-Regional Potato Introduction Station (IR-1), Sturgeon-Bay, Wisconsin, USA; Instituto Nacional de Tecnologia Agropecuaria (INTA), Balcarce, Argentina; Chilean Potato Gene bank in Valdivia, Chile; Duch-German Potato Collection, Braunschweig, Germany, and Institut fur Kartoffelforschung, Gross-Lusewitz, Germany. Some national potato programmes also maintain potato genetic resources e.g. in India, ICAR-Central Potato Research Institute, Shimla holds a modest collection of more than 4,500 accessions of elite potato varieties, parental lines as well as wild species imported from 30 countries. This is the largest potato collection in Southwest Asia. In spite of concerted efforts to collect the genetic variability, only about 130 out of the total known number of 235 potato species exist in gene banks. Much more efforts are needed to collect the others, which are known only as descriptors or dried specimens. The Food and Agriculture Organization of the United Nations (FAO, 2010) reported there were approximately 98,000 accessions currently conserved ex situ and 80% of them are maintained in 30 key collections (Table 1).

Table 1: List of holders of ex situ collections of potato germplasm (Solanum sp.) (FAO 2010)

Name of institute	Accessions		Types of accession (%)				
	Number	%	WS	LR	BL	AC	OT
INRA-RENNES, France	10,461	11	6	2	84	8	
VIR, Russian Federation	8,889	9		46	3	26	25
CIP, Peru	7,450	8	2	69	2	<1	27
IPK, Germany	5,392	5	18	37	7	32	6
NR6, USA	5,277	5	65	21	9	5	<1
NIAS, Japan	3,408	3	3	1	31		65
CORPOICA, Colombia	3,043	3					100
CPRI, India	2,710	3	15		85		
BNGTRA-PROINPA, Bolivia	2,393	2	26	74			
HBROD, Czech Republic	2,207	2	5	1	29	52	13
BAL, Argentine	1,739	2	85	15			
CNPH, Brazil	1,735	2					100
SASA, UK	1,671	2					100
ROPTA, Netherland	1,610	2	3	1		1	95
PNP-INIFAP, Mexico	1,500	2					100
TARI, Taiwan	1,282	1					100
SamAI, Uzbekistan	1,223	1					100
IPRBON, Poland	1,182	1			8	92	
RIPV, Kazakhstan	1,117	1	26	2	15	57	
SVKLOMNICA, Slovakia	1,080	1	1	2	47	41	9
Others (154)	32,916	33	19	15	3	16	46
Total	98,285	100	15	20	16	14	35

*WS: wild species, LR: landraces/old cultivars, BL: research materials/breeding lines, AC: advanced cultivars, OT: (others) the types are unknown or a mixture of two or more types. *INRA-RENNES: d'Amélioration des Plantes Institut national de la recherche agronomique/ Station, VIR: N.I. Vavilov All-Russian Scientific Research Institute of Plant Industry, CIP: Centro Internacional de la Papa; IPK: External Branch North of the De- partment Genebank, Leibniz Institute of Plant Genetics and Crop Plant Research, Potato, NR6: Potato Germplasm Introduction Station, United States Department of Agriculture, Agricultural Research Services, NIAS: National Institute of Agrobiological Sciences, CORPOICA: Centro de Investigación La Selva, Corporación Colombiana de Investigación Agropecuaria, CPRI: Central Potato Research Institute, BNGTRA- PROINPA: Banco Nacional de Germoplasma de Tubérculos y Raíces Andinas, Fundación para la Promoción e Investigación de Productos Andinos, HBROD: Potato Research Institute Havlickuv Brod Ltd., BAL: Banco Activo de Germoplasma de Papa, Forrajeras y Girasol Silvestre, CNPH: Embrapa Hortaliças, SASA: Science and Advice for Scottish Agriculture, Scottish Government, ROPTA: Plant Breeding Station Ropta, PNP-INIFAP: Pro- grama Nacional de la Papa, Instituto Nacional de Investigaciones Forestales, Agrícolas y Pecuarias, TARI: Taiwan Agricultural Research Institute, SamAI: Samarkand Agricultural Institute named F. Khodjaev, IPRBON: Institute for Potato Research, Bonin, RIPV: Research Institute of Potato and Vegetables, SVKLOMNICA: Potato Research and Breeding Institute.*

GERMPLASM CONSERVATION

Conservation refers to protection of genetic diversity of crop plants from genetic erosion. There are two important methods of genetic resources conservation: *in-situ* conservation and *ex-situ* conservation.

In-situ conservation

Conservation of germplasm under natural habitat is referred to as in situ conservation. *In-situ* conservation is based on the concept that it allows natural evolution to continue. Through the interplay of evolutionary processes in natural habitats, *in-situ* conservation maintains genetic diversity in a dynamic state. *In situ* conservation is generally suggested only for wild relatives because they alone live in natural communities. It is carried out under the auspices of national governments through biospheres reserves, national parks, world heritage sites, and other protected areas. It can also be carried out through farmers who manage land races and wild relatives in eco-geographic pockets of genetic diversity. In potato, there is no specific programme to follow this method of conservation because of the practical problems of maintenance and economic viability etc. Further, in this population size should be large enough to avoid risk of inbreeding and of genetic drift as these lead to the decay of genetic diversity. It is not realistic to expect that all species and their component populations can be covered by natural reserves, national parks and other protected areas. Further this method can't safeguard the species in face of unforeseen natural calamities.

Ex-situ conservation

Due to the practical limitations of *in-situ* conservation, it is preferred to maintain variability under managed conditions (*ex-situ*) in gene banks. This method, however, freezes the evolutionary process of species though it ensures conservation of its existing genetic variability. Potato genetic resources can be conserved as vegetative propagules or as true (botanical seeds). Potato is a highly heterozygous crop due to which sexual reproduction results in segregating populations. So when the objective is to maintain the exact genotype of an accession, vegetative propagation (*in-vitro* or *in-vivo*) is the only option. However, if the objective is not to maintain the exact genotype, but total gene pool, germplasm can be conserved as true seeds.

In vivo propagation

In vivo propagation is done through tubers in glass house, as well as in fields. Genetic identity is maintained by rouging mixtures and clones are protected from diseases and pests by using various protective measures. Still there is risk of exposure of germplasm to viral and mycoplasmal diseases from year to year, resulting in degeneration of the germplasm stocks. Loss of material due to natural calamities and loosing identity due to mechanical mixtures or wrong labeling are the other risks associated with this method. Instances are known of entire genetic stocks having been lost. Labour and maintenance costs are also high. This is the traditional method of conservation. This method, however, provides a continuous opportunity to evaluate and compare the characteristics of different genotypes. Further, seed/propagules can be readily made available to the users.

In vitro **propagation:** The ability to grow plants under aseptic conditions has allowed the development of *in-vitro* preservation techniques for germplasm conservation and exchange. *In vitro* maintenance of germplasm has several advantages. A large number of accessions can be conserved in a small space under disease free conditions irrespective of the crop season. The risk of loss due to biotic and abiotic factors is minimized and the possibility of cross infection between accessions is eliminated. *In vitro* materials can be made free of systemic bacteria, fungi and mycoplasmas and maintained as pathogen-free stocks. This eases the quarantine regulations for international distribution of genetic materials. The main disadvantage of an in vitro propagation is the induction of genetic variation or somatic mutations during subculturing. Maki *et al*. (2015) pointed out that DNA mutation occurs due to the long period of subcultures of the plants, while stored shoot tips in super-low temperature retained their original genetic structure. For this reason, a minimal growth method is desirable for preservation of in vitro materials in order to reduce the subculture numbers. The frequency of sub culturing can be reduced by incubating them at low temperature, under low light intensity and varied photoperiods, and growing the micro plants on the Murashige and Skoog medium (MS medium) supplemented with growth retardants or osmotic stress-inducing polyols (Gopal and Chauhan 2010).

Slow-growth conservation

This is based on the micropropagation of apical or axillary buds (nodal cuttings) on Murashige and Skoog medium (for potato tissue culture techniques consult Bajaj, 1987). In order to avoid frequent sub-culturing of micro propagated plants, sub-culture period is enhanced by following the slow-growth conservation strategy. For this a number of approaches like low temperature storage, reduced light intensity, high sucrose concentration, use of osmoticums or growth retardants in the medium, increase in volume of medium, sealing of the culture vessels, mineral-oil layer on the medium, etc. are used. A number of protocols have been developed for *in vitro* conservation of potato germplasm under minimal growth conditions. The most commonly used are based on combination of low temperature, reduced light intensity and use of osmoticums. By this method, *in-vitro* plantlets can be conserved for 2-3 years depending on the genotype. In sub-tropics, where maintenance of low temperature is problematic and expensive due to high demand on energy, plantlets could be conserved at normal propagation temperatures using MS medium with osmoticum (4% sorbitol, 2% sucrose) for 12 months without sub-culture.

Cryo-preservation

By this method plant material is frozen at ultra low temperature around -196°C of liquid nitrogen. This technique was derived from the observations that in temperate areas plant species could survive below freezing temperatures. At ultra low temperatures, the cells are in state of metabolic inactivity. Due to inhibition of cell division this method allows storage of material with minimal risk of genetic instability. Tissues can be stored virtually indefinitely with low labour costs. The other advantage of this method is that it requires highly reduced space. This technique involves several steps.

i.   **Selection and isolation of the material**: Keeping in view the intrinsic genetic stability of the material, meristems are preferred. Meristematic tissue is also suitable because it has compact and dense cytoplasmic cells with a lower water content making them less prone to ice damage during cryo-preservation. The material can be sampled from *in-vivo* or *in-vitro* plants. *In-vitro* material is generally preferred, since the explants are already miniaturized, free of superficial contamination and may also be pathogen free. The technique can also be used for preserving embryos, callus, pollen, cell suspension etc. These are, however, rarely used as potato germplasm due to their intrinsic genetic variability.

ii. **Pretreatment and cryo-protection**: This step is important to reduce the problem of cryo-injury. Cryo-injury often occurs when intracellular water freezes and forms ice crystals which rupture the internal membranes. It can also lead to adverse effects of the intracellular solutes accumulating to a toxic level, or leakage of vital solutes during freezing. The use of cryo-protectants with low molecular weight such as glycerol and dimethy sulphoxide which penetrate the cell with ease, and high molecular weight compounds such as poly vinyl pyrrolidone and dextran, which penetrates slowly, can reduce cryo damage significantly. They protect surface membranes by reducing growing rate and size of ice crystals, and by lowering the effective concentration of solutes in equilibrium with ice inside and outside the cell. They also help to increase membrane permeability which aids removal of water from the cell and facilitates protective dehydration in the early stages of freezing. By pretreatment with cryo-protectants some degree of dehydration is induced into cells and tissues thereby avoiding the damage caused by the formation of ice crystals during the freezing and defrosting process.

iii. **Freezing**: This is the most crucial step in the whole process of cryo-preservation. This can be done in any of the following ways.

    a. **Slow-Freezing:** Cultures are frozen by slow cooling at freezing rate between $0.5°$ C to $4°$ C per minute, starting from $0°C$ until the temperature reaches $-100°$ C, and finally transferred to liquid nitrogen. It requires programmable freezer in order to obtain accurate and reproducible results.

    b. **Rapid-Freezing:** Materials contained in vials are lowered directly into a tank filled with liquid nitrogen. The temperature decrease rapidly at the rate of $300°C$ to $1,000°$ C per minute.

    c. **Stepwise Freezing Method:** This method combines both the procedures of slow and rapid freezing. Initially plant material is cooled slowly and stepwise (ca $1-5°C$ per minute) to an intermediate temperature, maintained at that temperature for 30 min, and then rapidly cooled by plunging it into liquid nitrogen. In the initial slow freezing, ice is formed outside the cells and the unfrozen protoplasm losses water due to the vapour pressure deficit between the super cooled protoplasm and the external ice.

    d. **Vitrification:** Vitrification is a process by which water undergoes a phase transition from a liquid to an amorphous 'glassy state'; in this form, water does not possess a crystalline structure. Vitrification occurs when the solute concentration becomes so high that ice formation is prevented and the water molecules form a glass. For this, tissue is sufficiently dehydrated with a highly concentrated vitrification solution at 25 or $0°$ C without causing injury prior to immersion in liquid nitrogen. Vitrification solution PVS2 and PVS3 have been developed which are glycerol based and less toxic.

    e. **Encapsulation/Dehydration:** The cells/tissues are trapped into calcium alginate beads followed by incubation in 0.85 M sucrose (as sole source of cryo-protection) for 14-16 h, air-drying for 3-4 h in a laminar flow chamber and rapid freezing in liquid nitrogen.

The above alternatives are also combined e.g. encapsulation-vitrification, pre growth-desiccation, droplet freezing etc. to achieve the desired results.

iv. **Defrosting:** Thawing of the frozen material is achieved by transfer to warm water at 37-40°C. The optimal thawing rate is one that prevents ice formation by re-crystallization in the process of warming.

v. **Recovery:** The recovery of thawed cultures can be improved by nursing them through an initial recovery period involving gradual dilution of cryo-protectants through several steps of washing and by keeping osmotic disruption to the minimum. Plants are regenerated from the recovered tissue by culturing on suitable tissue culture media. To avoid somaclonal variations, shoot tips are directly regenerated into plantlets, without any adventitious growth or callus formation. This is ensured by developing suitable protocols of cryo-conservation and regrowth media for different cultivars/accessions.

Though different protocols based on above basic concepts are being developed, yet this technique is not being routinely used for germplasm conservation. Cumbersome procedures and low survival rate of the material conserved by this method appears to be the reasons for this. However, interest in this innovative germplasm method is increasing rapidly.

Sexual propagation

Potato true seed (botanical seed) is orthodox type, meaning it is able to tolerate a high degree of desiccation and storage at low temperature i.e. between 5°C and –20°C without loss of viability. Due to this, preservation of potato germplasm through sexual propagation as true seed is less laborious and much cheaper than the preservation of vegetatively propagated material. In addition, it is easy to maintain the material free of pathogens this way, as only a few virus diseases are known to be seed transmitted. Seeds occupy relatively small space and their transport is also economical. However, majority of the wild species of potatoes are diploid and self-incompatible. In such species true seeds are required to be produced by sib-mating. The sample size should be large enough to maintain the heterozygosity and to preserve all alleles at the loci of interest. Sample size depends on the nature of genetic control of the character(s) concerned.

True seeds with low moisture content can be kept at low temperatures for many years. Successful seed storage depends on effective control of several factors including physiological maturity of the seed, temperature, seed moisture content, storage atmosphere etc. It is desirable to harvest seeds at physiological maturity as germination, and vigour are maximum in fully mature seeds and longevity of mature seeds is more than in those harvested at other stages. The seeds extracted from the mature berries should be dried in shade or in seed dryers at temperature < 30°C. Drying in direct sunlight or at temperatures above 30°C can adversely affect the viability of the seed. The moisture content of the seeds can be further reduced by keeping them in Silica gel for a week or so. With in certain limits, viability of the seeds increases by drying and storing them at low temperature. For long-term storage, potato seeds should have moisture content 5± 1% and stored at -10 to -20°C. For short to medium term storage samples should be stored at 0 to 10°C after drying to 8± 1% moisture content. Before storage, the samples should be hermetically sealed in airtight containers. The methods described above are complimentary to each other. Based on the requirements and nature of the material to be conserved all these methods are being followed in various potato gene banks.

GERMPLASM EVALUATION

Evaluation refers to screening of germplasm in respect of morphological, genetical, economical, biochemical, physiological, pathological and entomological attributes. Evaluation of germplasm is essential to identify gene sources for resistance to biotic and abiotic stresses, earliness, productivity and quality characters and also to get clear picture about significance of individual germplasm line. The species evaluated for various characters of resistance or adaptation are even fewer than those collected. Important sources of resistance to the major potato diseases and pests, and adaptation to environmental extremes are listed in Table 2. Information on usefulness of various accessions and species is getting enriched and published as more and more germplasm is evaluated. Where such information is not available, passport data on characteristics of the natural habitats of species are of great importance. For example late blight resistant species are found in Mexican gene pool where late blight fungus has been found to reproduce sexually. Similarly frost resistance is found in species capable of growing at altitudes above 3,500 meters. Species from dry and warm climate areas are tolerant to these stresses.

Table 2: Wild species as sources of resistance to various diseases & pest and adaptation to environmental extremes

Diseases	Sources
Viruses -PVX	*S. acaule, S. berthaultii, S. brevicaule, S. chacoense, S. commersonii, S. curtilobum, S. phureja, S. sparsipilum, S. sucrense, S. tarijense and S. tuberosum* ssp. *Andigena*
PVY	*S. acaule, S. chacoense, S. demissum, S. gourlayi, S. phureja, S. rybinii, S. stoloniferum, S. tuberosum* ssp. *Andigena*
PLRV	*S. acaule, S. brevidens, S. chacoense, S. demissum, S. etuberosum, S. raphanifolium, S. stolonifrum* and *S. tuberosum* ssp. *Andigena*
Spindle tuber viroid	*S. acaule, S. berthaultii, S. gurreroense, S. hjertingii* and *S. multidissectum*

(Contd.)

Late blight vertical	*S. cardiophyllum, S.demissum, S. ediense, S. stoloniferum* and *S. verrucosum*
Horizontal	*S. berthaultii, S. bulbocastanum, S. chacoense, S. circaeifolium, S. demissum, S. microdontum, S. phureja, S. pinnatisectum, S. polyadenium, S. stoloniferum, S. tarijense, S. tuberosum* ssp *andigena, S. vernei* and *S. verrucosum*
Wart	*S. acaule, S. berthaultii.*
Common scab	*S. chacoense, S. tuberosum* ssp. *Andigena*
Bacterial wilt	*S. chacoense, S. microdontum, S. phureja, S. sparsipilum* and *S. stenotomum*
Cyst nematodes	*S. acaule, S. berthaulti , S. boliviense, S. bulbocastanum, S. capsicibaccatum, S. cardiophyllum, S. demissum, S. gourlayi, S. kurtzianum, S. leptophyes, S. multidissectum, S. oplocense, S. sparsipilum, S. spegazzinii, S. sucrense, S. tuberosum* ssp. *andigena* and *S. vernei*
Root knot nematode	*S. bulbocastanum, S. cardiophyllum, S. chacoense, S. curtilobum, S. hjertingii, S. kurtzianum, S. microdontum, S. phureja, S. sparsipilum* and *S. tuberosum* ssp. *Andigena*
Aphids	*S. berthaultii, S. bukasovii. S. bulbocastanum, S. chomatophilum, S. infundibuliforme, S. lignicaule, S. marinasense, S. medians, S. multidissectum, S. neocardenasii, S. stoloniferum.*
Colorado beetle	*S. berthaultii, S. chacoense*, S. commersonii, S. demissum, S. jamesii, S. pinnatisectum, S. polyadenium* and *S. tarijense*
Tuber moth	*S. chacoense, S. stenotomum* and *S. tuberosum* ssp. *Andigena*
Frost	*S. acaule, S. ajanhuiri*
Heat and drought	*S. acaule, S. bulbocastanum, S. chacoense, S. commersonii, S. gourlayi, S. megistacrolobum, S .microdontum, S. ochoae, S. papita, S. pinnatisectum, S. spegazzinii* and *S. tarijense*
High protein content	*S. phureja* and *S. vernei.*
High starch content	*S. vernei*

UTILIZATION

Utilization refers to the use of germplasm in crop improvement programmes. Vast amount of genetic variability exists in potato, but very little has been actually used in the improvement of cultivated potato. In spite of the involvement of some wild species in potato breeding programs, the genetic base of Indian potato varieties is quite narrow. The Pedigree analysis of 77 advance potato selections developed for Indian plains since 1990 showed that their origin could be traced to only 49 ancestors out of which 29 were exotic which accounted for 69.52% of the total genetic constitution and maximum contribution (40.65%) was by 10 ancestors from U.K. (Gopal and Oyama, 2005). The genetic base of Indian potato selections is narrow owning to the tendency to involve in hybridization clones that carry specific gene complexes necessary in successful varieties. The most frequent ancestors in Indian selections were two clones 2814(a)1 and 3069(d)4. These were selection from the cross between *Solanum rybinii* (a variant of *Solanum phureja*) and *Solanum demissum*, which was made by Dr Black in 1937 to introgress late blight resistance from *S. demissum*. This situation is similar in several other countries. In USA, parentage of as many as 171 American varieties could be traced to a single variety Rough Purple Chili introduced from Chiloe region of southern Chile. Only 10 primitive cultivars or species had been used in the pedigrees of 62 per cent of the 627 cultivars listed in the Index of European Potato Varieties (1985). Remaining 38 per cent cultivars involved only subspecies *tuberosum*. Ross (1986) has estimated that only 13 species have been used so far in the variety improvement programmes of the world. Though, tetraploid and diploid crosses are not very difficult to make, in one direction at least, the viability of interspecific hybrids is not very high (Hawkes and Hjerting, 1969). The use of bridging species belonging to other series in the interspecific crosses can be exploited. Furthermore, somatic hybridization with conventionally non-crossable species can be other alternative (Helgeson *et al.*, 1986). Crossability of different species of potatoes is dependent on the endosperm balance number (EBN) and not the ploidy status. Species with same EBN are cross-compatible. The EBN in fact is a number initially assigned to cross-compatible species and assignment of an EBN number to species is determined by its cross-compatability with a species of known EBN. Molecular basis of EBN is not known and studies on this aspect may facilitate the use of various species in all desired combinations. The wild species which occur in the pedigree of several cultivars are briefly described below.

S. demissum: A hexaploid, self-fertile sps. from Mexico. It is resistant to PVY and PLRV, hypersensitive and resistant to late blight, resistant to wart and both *Globodera* species; alkaloids in this confer resistance to Colorado bettle and other insects. It is moderately resistant to frost also. Genes of *S. demissum* have been incorporated into more than 50% of the world's cultivars mainly for resistance to late blight and PLRV.

S. acaule: A tetraploid and also hexapoid, self fertile sps. from Andes of Argentina to Peru. It is extremely resistant to PVX and resistant to PLRV, PSTVd, wart, both sps. of *Globodera* and frost.

S. chacoense: A diploid sps. from Argentina, Southern Brazil, Bolivia, Paraguay and Uruguay. It is extremely resistant to PVA and PVY, resistant to late blight, colorado bettle, tuber moth and other insetcs. *S. chacoense* has been used mainly with *S. phurja* to produce a hybrid which served as a bridge between *S. demissum* and *S. tuberosum* to create tetraploid hybrids by the smooth euploid transfer of gene into ssp. *tuberosum*.

S. spegazzinii Bitt: A diploid sps, from Northwest Argentina. It is resistant to *Fusarium,* wart and *Globodera* sps and has low content of reducing sugars. This has been mainly used as a source of resistance to nematodes.

***S. stoloniferum*:** Schlechtd. et Beche: A tetraploid, self fertile sps. from Mexico. It is extremely resistant to PVA and PVY, hypersensitive and field resistant to late blight. About 20 cultivars carry the gene Ry for extreme resistance to PVA and PVY derived from *S. stoloniferum*.

***S. vernei* Bitt. et Wittm:** A diploid sps. from Northwest of Argentina. It is outstanding for protein and starch content and resistant to late blight and both sps. of *Globodera*. It has low content of reducing sugars. It has been used in breeding mainly for resistance to nematodes.

In the utilisation of gene pool, it is just as important to be aware of possible undesirable characters, as it is to know of desirable characters in a specific source. Undesirable tuber traits of the wild species and crossability problem of certain species has acted as deterrents for their use by the breeders. Pre-breeding of the wild species for combining resistance to various diseases and insect pests with agronomic characters requires special attention. This will undoubtedly lead to a closer association between gene banks and gene bank users and help in turning "gene bank collections" into 'working collections".

REFERENCES

Bajaj YPS (1987) Biotechnology in Agriculture and Forestry, Vol. III: Potato, Springer-Verlag, Berlin.

Bukasow SM (1933) The potatoes of South America and their breeding possibilities. Bulletin of Applied Botany, Genetics and Plant Breeding (Leningrad)- 58. 192p

FAO (2010) The second report on the state of the world's plant genetic resources for food and agriculture. Italy, Rome: 370p

Gopal J and Chauhan NS (2010) Slow growth *in vitro* conservation of potato germplasm at low temperature. *Potato Research* **53**:141–149

Gopal J and Oyama K (2005) Genetic base of Indian potato selections as revealed by pedigree analysis. *Euphytica* **142**: 23-31

Hawkes JG and Hjerting JP (1969) The potatoes of Argentina, Brazil, Paraguay and Uruguay - A Biosystematic Study. Oxford University Press: 459-460p

Helgeson JP, Hunt GJ, Haberlach GT and Austin S (1986) Expression of a late blight resistance gene and potato leaf roll resistance. *Plant Cell Reports* **5**: 212-214

Maki S, Hirai Y, Niino T and Matsumoto T (2015) Assessment of molecular genetic stability between long-term cryopreserved and tissue cultured wasabi (*Wasabia japonica*) plants. *Cryo Letters* **36**:318–324

Ross H (1986) Potato breeding-problems and perspectives. Springer-Verlag, Berlin: 132p.

Spooner, DM and Salas A (2006) Structure, biosystematics and genetic resources. In, Handbook of potato production, improvement and post-harvest management (J. Gopal and S.M. Paul Khurana, Eds.), Food Product Press, New York, USA:1-39p

4

Breeding for Table Potatoes

*RP Kaur[1], SK Luthra[2], Salej Sood[3], KN Chourasia[3]
and Vinay Bhardwaj[3]*

[1]*ICAR- Central Potato Research Institute, Regional Station, Jalandhar-144 003
Punjab, India*
[2]*ICAR- Central Potato Research Institute, Regional Station
Modipuram- 250 110, UP, India*
[3]*ICAR-Central Potato Research Institute, Shimla - 171 001
Himachal Pradesh, India*

INTRODUCTION

India has emerged as second highest producer of potato in the world, after China. The crop has been identified as the most important crop for ensuring food security for the developing countries, and a highly nutritious food providing more calories, vitamins and nutrients per unit area of cropped land than any other staple food. The crop has witnessed phenomenal increase over the seven decades of potato research in India. For a crop which was introduced in the country only in the early 17th century, and was neither adapted, nor its staple food, the remarkable increase in potato production and productivity is an exceptional achievement of breeding this temperate introduction into the present day potato cultivars.

BREEDING REQUIREMENTS

The table potatoes introduced in India were adapted to long photoperiods of up to 14 hours and crop durations of 140-180 days allowing them to be cultivated only in the hilly regions of India. The unique and diverse agro-climatic condition of the country, comprising of contrasting conditions in the Indian plains, where potato could be grown in the winter months necessitated breeding efforts to develop the potato varieties suitable for the sub-tropical plains and the other potato growing regions. The winters in sub-tropical plains have short photoperiod (with about 10-11 hours sunshine) and the crop duration is limited to 90-100 days because of short and mild winter. Further constraints on photosynthetic activity due to early morning fog, posing a strong limitation on tuber bulking. This is followed immediately by a long hot summer period, which coincides with the post-harvest storage period of potato, creating challenging storage problems. The rapid degeneration of seed stocks, with concomitant decrease in yield due to accumulation of viral diseases and tuber storage and utilization in hot and humid climate thus needed to be addressed, in priority. Besides, these southern, north Bengal and Sikkim hills and the plateau offer agro climatic area specific problems related to the biotic and abiotic stresses prevalent in the regions. Moreover, the unique reproductive biology of the plant made it possible to flower only under the long day conditions of the hills, where hybridization was initiated at Kufri (Shimla), Himachal Pradesh. However, during the initial years, it was not possible to evaluate the breeding progenies in the warmer sub-tropical plains due to the quick degeneration of hill bred progenies and dormancy of hill potatoes.

EARLY HISTORY OF POTATO BREEDING IN INDIA

The initial stage of potato crop improvement was at its centre of origin itself, where the Andean farmers carried out selection from naturally occurring variation, which was mainly responsible for evolution of many Andean varieties. There was a drastic reduction in toxicity/ bitterness in potatoes during this course of domestication mainly due to gradual reduction in glycoalkaloids. Even to this day an enormous variation in tuber shape and color of skin and flesh has been maintained by Andean farmers. With the establishment of the science of crop improvement, the breeding for potato in the twentieth century remained largely empirical and genetically unsophisticated, based on mainly targeting yield and late blight resistance as the major characters of interest to ensure food security. With gradual shifting of focus to breeding and selection of potato varieties for early maturity, dormancy and resistance to different abiotic and biotic stresses for adaptation of potato to the different agro-climatic conditions. In recent years varieties are being developed for multi traits and specialized breeding purposes.

In India, foundation of ICAR-Central Potato Research Institute (CPRI) in 1949 at Patna, marked the beginning of the era for oriented research towards development of technologies and varieties for growing potato under sub-tropical conditions, for increasing the overall potato production of the country. Subsequently, shifting of ICAR-Central Potato Research Institute (CPRI) to Shimla in 1956, advent and perfection of seed plot technique in 1963 and initiation of All India Coordinated Research Project (AICRP) on potato in 1971 marked the major milestones for potato breeding in India. These allowed planned and well-coordinated research on potato in the country, making it possible to raise, maintain and evaluate segregating populations of progenies (developed in hybridization carried out at Kufri), in the plains under disease-free low aphid periods, allowing development of high yielding varieties for the plains, which were the main contributors for the potato revolution witnessed by the country. The identified insect-free zones further allowed rapid multiplication of disease free seed which could be supplied to the other parts of the country resulting in high productivity.

Early potato introductions in India were mostly *S. tuberosum* ssp. *andigena*, bearing small, misshapen tubers with deep eyes, having low yield, and most importantly restricted to the cool long day climate of the mountains. A lot of misperceptions regarding the identity and nomenclature of these introductions initially existed, which varied with dialects and regions, hence. A potato synonym committee was thus formed from National Institute of Agricultural Botany which led to identification and characterization of 16 non-European varieties, which came to be known as desi or indigenous samples or varieties, of these Phulwa, Darjeeling Red Round and Gola were most popular. Besides, these 38 European varieties were also identified, which were referred to as the exotic varieties having limited adaptation to the Indian climate. The major breeding objectives of potato in the country have been summarized in the subsequent sections.

BREEDING OBJECTIVES FOR TABLE POTATOES AND PROGRESS

Based on climate and photoperiod of growing region

India is a country of great geographical diversity. The variations in its terrain, temperature, rainfall and soils have closely influenced the cropping patterns and other agricultural activities. With the objective to increase the area under potato cultivation and overall production of potato in the country, it is important to breed suitable potato varieties. Based on the breeding objectives, areas have been determined for the cultivation of potato in the country, each having its unique biotic and abiotic stresses afflicting the crop (Table 1).

The breeding for early bulking potatoes has gained a major perspective in relation to its cultivation in the Indo-Gangetic plains of India, where the crop is grown as an intercrop between the predominant rice and wheat, staple crops. The early bulking potato is planted after the harvesting of rice in September end to first fortnight of October and is harvested just before the planting of wheat in December. Since, potato is not present in the market at this time of the year, the early harvested crop ensures good profitability to the growers.

The main potato growing region in India falls in the Indo Gangetic plains where 80% of the potato crop is grown under the short day conditions of winters. The early (60 to 75 days) and medium maturing (75 to 90 days) varieties are most suited to this environment. As the initial introduction of potato varieties in India

Table 1: Potato growing regions in India and their salient features and relevant breeding objectives

Agro-climatic zone	Areas		Potato growing period of the year		Breeding objectives
			Planting	Harvesting	
North-western plains	Punjab, Haryana and Rajasthan	Early autumn Autumn Spring	September October January	November January/ February April-May	Early bulking, short day photoperiod, tolerance for high temperatures, tolerance to frost, slow rate of degeneration, late blight resistance, good keeping quality
West-central plains	West-central Uttar Pradesh and north-western districts of Gujrat and Madhya Pradesh	Autumn	October to	January/February	Medium maturing early bulking varieties, short day adapted, moderate resistance to late blight, slow rate of degeneration and good keeping quality.
North-eastern plains	Assam, Bihar, Jharkhand, West Bengal, Orrisa, eastern Uttar Pradesh and north-eastern and eastern districts of the states of Madhya Pradesh and Chhattisgarh	Winter	November	February/ March	Early bulking, short day adapted varieties, resistance to biotic stresses (late blight and viruses), good keeping quality, regional preference to red and round potatoes
Plateau region	Maharashtra, Karnataka and parts of Gujarat, Madhya Pradesh and Orissa	Kharif	June-July	September-October	Early bulking, ability to tuberise under high temperature, resistance to bacterial wilt, mites and potato tuber moth, slow rate of degeneration
		Rabi	November	January-February	

(Contd.)

Region	States	Season	Planting time	Harvesting time	Breeding objectives
North Western and Central Hills	Jammu and Kashmir, Himachal Pradesh and Uttaranchal. Potato is cultivated in from to	Summer	April	September	Resistance to Late blight and bacterial wilt. Adapted to long day conditions
North eastern hills	Meghalaya, Manipur, Mizoram, Tripura and Arunachal Pradesh	Spring Autumn	January-February August	May-June November-December	Adaptability for long days with resistance to late blight and bacterial wilt
Southern hills	Hilly regions of Tamil Nadu mainly the Nilgiris and Palani Hills	Summer Autumn Spring	March-April to August-September January-February	August-September May	Early bulking, adaptability to long days and resistance to late blight and cyst nematodes December-January
Sikkim and North Bengal hills	Sikkim and hills of West Bengal	Spring Autumn	January-February September	July-August December	Resistant to late blight and immunity to wart, preference to red skin tubers

comprised of the long day adapted genotypes, these over the period of several years were bred and selected under the short day conditions of the plains to develop short day adapted varieties. *S. andigena* which is short day adapted has been reported to be an advantageous source for exploiting heterosis for early bulking in the potato breeding programmes under short day subtropical environments (Gopal *et al.* 2000). The progenies from crosses involving tuberosum × andigena were reported to depict significantly higher heterosis for yield and tuber number as compared to tuberosum × tuberosum progenies. Several varieties have been developed utilizing the andigena parents which are well adapted to the sub-tropical plains of the country and give high yields. The physiological mechanism for earliness is governed by early tuberization and subsequent early dry matter partitioning, affected by both temperature and photoperiod. To date, 60 improved potato varieties have been released by the ICAR-Central Potato Research Institute for different agro-climatic zones of the country. Among them, the most popular table potato varieties are Kufri Bahar, Kufri Pukhraj, Kufri Jyoti and Kufri Himalini (Figure 1).

Fig. 1: Popular table potato varieties

Breeding for external quality parameters: Shape, size, colour

The external tuber quality or morphology is the one of the most important criterion for the acceptability of a potato variety for table purpose comprising of skin colour, tuber size, tuber shape and eye depth, physical injuries, scab etc (Table 2). The tuber shape and colour preference has been observed to vary not only in the different parts of the world, but within country *per se*. In India the white, white-cream and yellow skin varieties are preferred in the North-western parts, whereas the red skinned varieties are preferred in the eastern parts of the country. For tuber colour R locus causes production of red anthocyanins, while P is required for the synthesis of purple pigments. Whereas, the synthesis of red or purple anthocyanins in tuber skin is mainly due to I locus. These three loci have been mapped in the potato genome. Related studies for candidate gene analysis of anthocynin pigmentation in tuber skin reported three genes R (dfr), P (f3´5´h) and D (f3h) expressed in the periderm of red and purple skinned clones, while dfr and f3´5´h were not expressed, and f3h was only weakly expressed, in white skinned clones. Similarly, potato tuber flesh color shows a large variation from white to purple mainly due to two naturally occurring pigments, anthocyanin and carotenoids. Red, blue and purple color flesh is due to accumulation of anthocyanin pigment. Red and purple fleshed potatoes have acylated glucosides of pelargonidin while purple potatoes have, in addition, acylated glucosides of malvidin, petunidin, peonidin and delphinidin. Natural anthocyanin pigments are the economic constituent of colored potato adding nutritional value to potatoes. Antioxidant values also depend on color of flesh of tuber. Red fleshed potato genotypes have the high antioxidant value (300%) as compare to white fleshed, while purple fleshed tubers have antioxidant value of 250% than white fleshed. The single dominant allele at the Y locus on potato chromosome III controls the white, yellow or orange flesh of tuber which determines by the carotenoid level. A combination of the dominant â-carotene hydroxilase 2 (Chy2) allele and homozygous recessive Zep allele controls the yellow flesh (accumulation of high levels of zeaxanthin) of potato tubers.

Table 2: External and internal quality parameters in Potato

External quality traits	Internal quality traits
Tuber skin characteristics (russet, smooth)	Dry matter
Tuber shape	Reducing sugar
Tubers size uniformity	Texture
Tuber skin colour	Nutritional value
Tuber eye depth	Glycoalkaloid content
Tuber flesh colour	Internal growth defects
External cracking	Keeping quality

Similarly, the preferable shape of round and oval tubers are preferred over the misshapen, pear etc. shaped tubers. The round shape of tuber mainly governed by a single locus on chromosome X with a dominant allele Ro, while other reports mention quantitative trait loci (QTLs) on chromosome II, V and XI, II and XI and VII and XII for tuber shape. Other than this, there are several factors which control this trait, most probably it depends on the genetic background of the respective populations used in the researches.

Eye depth of shallow to medium are more acceptable which reduces peeling losses. The breeding for external quality parameters indirectly becomes an integral part of all breeding programmes, where the genotypes with external quality possessing desirable characters are selected and continued for release as a variety. A single locus Eyd/eyd controls the appearance of eye depth, this locus is closely linked with Ro locus at chromosome X at a distance of 4 centi Morgan (cM), whereas, at least four QTLs on chromosomes III, V and X were identified.

Breeding for internal quality traits like starch quality, phytonutrients and Fe and Zn rich potatoes

The rising awareness among the population has led to a rising interest among the masses towards the overall nutritional and culinary value of potato, which comprise its internal quality. The internal quality characteristics, include nutritional properties, cooking/after cooking properties, and processing quality etc. encompassing traits such as dry matter content, flavor, sugar and protein content, starch quality and type and amount of glycoalkaloids. However, low glycoalkaloids (having teratogenic, embryotoxic, genotoxic and carcinogenic effects) and after cooking blackening are quite relevant to development of varieties for fresh table consumption.

The internal quality parameters are affected by the genetic makeup, crop maturity, agronomic practices, environmental conditions, storage temperatures and presence of pests and diseases.

Tuber dormancy is the ability of tubers to stay in a physiological dormant state after harvesting, when they do not sprout. It is directly correlated to weight loss and keeping quality of the potato tubers. Molecular mapping studies for dormancy detected a number of QTLs for tuber dormancy and demonstrated its complex character. There are several categories of genes which are involved in breaking dormancy and sprouting action. Among these, the first group represents the genes coding for homeotic proteins and transcription factors involved in dormancy breaking and sprouting action. The second class regulates metabolism and hormones mainly abscisic acid and ethylene, where the former maintains dormancy and latter is involved in loss of dormancy. The third group of genes is involved in metabolism of reserve storage while fourth gene category was involved in DNA replication.

Glycoalkaloid mainly α-solanine and α-chaconine are naturally occurring bitter tasting secondary metabolites present in all parts of potato plant and also in green tubers. Various factors influence the quantity of glycoalkaloid, that is, it increases during storage and transportation and under the influence of light, heat, cutting, slicing, sprouting and exposure to phyto-pathogens. The quantity of these chemicals decides the quality of tubers, that is, when this compound is present in low concentration, it may attribute to flavor of processed potato, but when its level crosses 15 mg/100 g fresh weight it cause bitterness of tubers. The consumption of large amounts of glycoalkaloids by humans could produce toxication symptoms ranging in severity from nausea to, in extreme cases, death.

The quality of potato tubers also depends on sugar content of the harvested crop. For the fresh market, sucrose levels above 1% fresh weight (FW) are reported to give an unacceptably sweet taste to the boiled potatoes.

The nutritional profile of potatoes has garnered a lot of interest in recent years. It has been identified as a balanced and complete source of human nutrition with 75% water, 21% carbohydrate, 2.5% protein and less than 1% fat. It is also a good source of vitamin C, B6, minerals, essential amino acids and fiber. The nutritional aspects of potato include breeding for enhanced iron and zinc content. The bioavailability of these important nutritional supplements during consumption is also most crucial and is affected by ascorbic acid and phytic acid present in potato, where the former enhances while the latter reduces its absorption. Amino acid profile of potato has also garnered much interest in the present day.

The remaining 20% dry matter of potato is mostly made up of starch. The processing of tubers with high levels of protein content is economically relevant for the potato starch industry as these compounds render high economic value. Therefore, improving protein content in starch potato varieties has emerged as a topic for innovation amongst starch potato breeders (Klaassen *et al.,* 2019). In diploid populations the broad sense heritability of protein content has been estimated between 56 and 66%, while in tetraploid biparental population upto 40-74 % (Klaassen *et al.,* 2019), indicating a moderate proportion of this trait variance can be ascribed to genetic factors within a particular experimental setup. The Genetic loci of QTL affecting the level of soluble protein content in potato tubers was illustrated to be located on chromosome 1, 3, 5 by Acharjee *et al.* (2018). The protein content was reported to be quantitative and complex and can be improved by fixating alleles with positive effects that underlie these QTLs in gene pools for breeding (Klaassen *et al.*, 2019).

The organoleptic properties of potato are much important, in the present day world, where some varieties appeal more to consumers as compared to the others. Similarly, Phureja potatoes have been reported to be tastier, than the tuberosum potatoes. Studies have been undertaken to analyse the compounds providing the taste preference in potatoes. European (*Solanum tuberosum* Group Tuberosum) and Phureja potatoes (*S. tuberosum* Group Phureja, also known as *S. phureja*) were observed to show distinction in sensory panels with respect to aroma and taste. The Phureja group was adjudged to be the "better tasting". There was a clear relationship between the elevated abundance of certain branched amino acids in tubers of raw Phureja relative to tuberosum and similarly elevated levels of branched short-chain aldehydes in the volatile proûle from cooked Phureja.

The quality of starch is of as much importance in this regard as out of the two type amylose and amylopectin, the amylose is an undesirable component in potato starch industry it gelatinizes to form crystals which reduce paste clarity. The starch is mainly used as a feeder to the paper industry, food additive as an agent for thickening, stability, shelf-life and texture, in textile industry for warp sizing and fabric printing. Besides these a number of other fields where starch products have a special role include fluid loss control during deep-well drilling for oil and gas, and flocculation in the purification process for drinking-water. Potato starch is a 100% biodegradable substitute for polystyrene and other plastics and used, for example, in disposable plates, dishes, and knives. Breeding for varieties specifically starch industries would require varieties having not only higher starch content but also starch of amylopectin type.

Similarly, the higher glycemic index of potato is another major concern for diabetics, since it gets rapidly broken down in the intestine, causing a sharp increase in the blood sugar levels. The amylose fraction of the starch is much important as this gets converted to resistant starch imparting health benefits to the human body. With the existence of wide variation in potato germplasm for this character, it can be an important breeding objective for potato in future. Low glycemic index potatoes with higher amylose would therefore be preferable having slower conversion of starch to sugar and thereby preventing sharp increase in blood sugar levels

Breeding for resistance to biotic stresses

The most important diseases of potato include late blight and bacterial wilt, while the major pests affecting potato crop are root and cyst nematodes. The viruses are the most important organisms causing potato degeneration especially in the warmer climate, spread by vectors like white fly and aphids. In recent year use of disease-free seed and use of chemicals has limited the losses incurred due to the diseases and pests. However due to breakdown of resistance to the chemicals and their associated toxicity in the environment, the development of resistant varieties is the most ecologically sustainable strategy in countering the disease and pests. The basic requisite for resistance breeding includes a) a durable source of resistance, b) a reliable screening method using the specific isolate and c) understanding the inheritance of resistance. Strategies for durable resistance are not singularly dependent on the resistant source *per se* but largely on the epidemiology and population genetics of the specific pathogen/ pest and its ability to overcome the resistance. Marker assisted selection for the genes of interest in screening of populations has met with success, in breeding for resistance where, several resistance genes have been identified.

In case of late blight monoculture of susceptible potato variety had proved to be devastating in the famous Irish famine of 1844-45. The resistance to late blight, initially identified in the wild species *S. demissum*, began to breakdown leading to the exploration of new sources of resistance. Race non-specific resistance imparted by the quantitative genes has been projected to be more relevant, for countering late blight resistance. Several Rpi-genes have been identified from the wild species like *S. stoloniferum, S. phureja, S. bulbocastanum, S. verrucossum, S. schenckii* and *S. venturi* imparting resistance to late blight. The severe constraint of break-down of single R-gene based resistance due to evolution of pathogen has led to development of strategies like R-gene pyramiding using multiple genes from diverse sources (Kim et al. 2012). In case of viruses' durable resistance to PVY and PVX has been reported and also H1 gene has remained effective against *Globodera rostochiensis*. The pyramiding of quantitative resistance genes with the important commercial traits like early maturity, high yield is presently the most effective strategy available.

Transgenic technology involving different genes encoding proteins for conferring resistance using various strategies like response to infection, host encoded genes for viral translation etc. against major diseases, viruses and pests have been reviewed by Halterman *et al.* 2016.

Breeding for abiotic stress tolerance: The cultivation of potato in the Indian conditions imposes several abiotic stresses on the crop including drought, frost and heat. In India, high temperature during crop growth and tuberization restricts adoption of potatoes in early planting conditions of north-western plains and peninsular India. Since a major proportion of the crop is grown in the sub-tropical climate, the crop witnesses low yields under the short day and early to medium maturing conditions as compared to its temperate counterpart. Besides, high rates of degeneration under the sub-tropical conditions, limited moisture and heat imposes severe physiological and hormonal imbalance in the crop, affecting the efficient partitioning of the carbon assimilates to the above ground vegetative parts at the cost of the tubers resulting in proficient vegetative growth. Minimum night temperature is critical to tuberization at potato, with reduction in tuberization at night temperatures of 20°C and complete inhibition above 25°C. Leaf bud cutting grown under heat conditions have been used to screen for heat tolerance, similarly seedlings are selected under the heat stress conditions for their ability to tuberize.

The drought conditions may arise due to erratic rainfall, inadequate irrigation, lack of water supply or due to high respiration rates. Like any other crop, drought causes reduction in productive foliage, rates of photosynthesis and shortening of vegetative period. Emergence and tuberization are the two most critical periods when the water stress affects the potato plant. Potato crop is highly sensitive to drought responding rapidly to low water potential of -3 to -5 bars, eliciting various responses like drought avoidance or drought tolerance. Breeding for drought therefore depends on screening of genotypes with respect to effect of drought on productivity and tuber yield, survival and recovery of plants after imposed stress and water use efficiency. Drought tolerance with high irrigated yields have been reported from Andean landraces belonging to the species *S. curtilobum, S. chillonanum, S. jamesii, S. okadae,* S. *tuberosum* L cultivar group Stenotomum, Andigenum and Chaucha. Transgenic technology involving manipulation of abscisic acid signal transduction through loss of function of the cap-binding protein 20 (CBP20) leading to increased drought tolerance in potato has been reported. Projected damages, caused on potato production, due to global warming and climate change, impose a new trend in breeding programme – creation of breeding lines and cultivars, tolerant to high temperature and drought.

The abiotic stress imposed by frost in potato crop relates to the severe losses in the crop at low temperatures under at -2°C, when white frost occurs under high relative humidity causing crystal formation on shoots, and black frost occurs under drier conditions, where the plant parts turn dark. In India, the winter potato crop grown in the plains get affected by frosts during the months of December and January. Excised leaflet test, freezing test and field test may be used for the screening for frost. The existence of variability in the germplasm for frost resistance has been reported, especially in the wild species *S. demissum, S. acuale* and *S. juzepczukii* which may be used in effective breeding of varieties for frost resistance. Species *S. spegazzinii* (diploid) *S. demissum* (hexaploid) and *S. tuberosum* ssp. *andigena* (tetraploid) were also found promising for frost tolerance.

Thus the breeding strategy for important crop growth nutrients and water should focus on the improvement of resource-use efficiency. Farmers need varieties that can give reasonable yield under stress conditions and respond to ideal conditions with yield increment. Phenotypic plasticity has a high negative association with yield stability. Thus, low plasticity (less yield difference between most environments) or high stability is not always a desirable characteristic. Tolerant genotypes usually give moderate yield whether it is under ideal growing conditions or marginal conditions (under permanent stress). A wide range of phenotypic variation has been reported for nitrogen use efficiency (NUE) among wild accessions and commercial cultivars; however, efforts to understand the physiological and genetic basis for differences in NUE among cultivars are generally limited mainly due to the high G × E interaction of this complex trait. NUE was observed to show consistently high heritability (h^2) and genetic advance as percent of mean (GA%) across treatments and locations, which indicates the possibility of direct selection for NUE (Baye 2017). However, selection for correlated traits like N-uptake efficiency and N-utilization efficiency have been suggested as a selection criterion for the improvement of NUE rather than direct selection.

Specialty potato breeding (Breeding for low cold induced sweetening, organic potatoes, baby potatoes etc.)

In relevance to the sub-tropical countries like India, where the 95% of table potatoes are produced in the plains and cold stores for storage of potatoes at 2-4°C or 10°C with CIPC treatment is employed to save the perishable produce from the hot summer conditions. This also helps in management of the glut due accumulation of large quantity of produce during the harvesting period and ensuring good profitability to the farmers. However, it causes low temperature sweetening due to the breakdown of starch in the tubers to reducing sugars, which makes the tubers sweet in taste and somewhat less organoleptically unacceptable for the consumers. This is of much more relevance to the processing varieties, where the product develops browning, due to high sugar content. Thus, development of varieties having resistance to the cold induced sweetening is of much importance in the Indian scenario. Many wild & semi-cultivated sources have been identified possessing minimal cold induced sweetening after cold storages (Bhardwaj *et al.,* 2011) but important is to exploit these through un-conventional methods. Anti-sense RNA technology using antisense UDPG-PPase RNA has been used to specifically inhibit UDPG-PPase enzyme activity in tubers, restricting the starch sugar pathways, however limited success has been achieved. Similarly, down-regulating the production of the enzyme acid invertase, which cleaves sucrose into glucose and fructose and genes required for asparagine synthesis has been suggested to counter carcinogenic acrylamide in processes potato.

In future the breeding for table potatoes worldwide needs to consider two main types of potatoes for the fresh market. The first type includes the standard merchandise at a standard discount price where breeders need to provide cost efficient, high yielding, multi-use varieties. The second includes a high-end nice looking, high price top quality table potato segment. Other recent quality goals include breeding for varieties that do not accumulate asparagine, which is a precursor for carcinogenic and cytotoxic acrylamide, formed during cooking of potatoes through Maillard reaction. Organically grown food is safe and nutritious, and would remain in high demand in domestic and international market for which suitable varieties needs to be developed with high biotic resistance and nutrient use efficiency, to enable organic cultivation without the use of chemicals. Kufri Khyati, Kufri Mohan, Kufri Garima (table potato varieties) Kufri Himsona and Kufri FryoM (processing varieties) are identified Indian potato varieties with sustainable production under organic farming system.

BREEDING METHODS

Major objective of potato breeding is to develop varieties with higher yields by overcoming various biotic and abiotic constraints, which limit the yield potential of cultivars. In India, varietal improvement programme for sub-tropical plains is being undertaken at Jalandhar, Modipuram and Patna centre of the institute. The choice of the parents in a hybridization programme is done carefully keeping in mind the objectives of the programme. The general methodology for potato improvement involves making several crosses among the selected parents with complementary traits so as to generate large range of genetic variation. Phenotypic selection is then carried out over a number of vegetative generations, where progenies are evaluated on the basis of external quality parameters and yield. The inheritance of characters is studied using statistically signiûcant offspring on mid-parent regressions for estimating heritable variation, and the slopes of the regression lines for measures of heritabilities. Assessment of offspring over vegetative generations is used to derive estimates of the genetic variation within progenies and superior crosses. Various hybridization methods like distant crossing, followed by back-crossing, bi-parental cross, multiple cross and poly-cross are usually utilized in potato breeding. Besides the above traditional approaches, non-conventional methods are also used in potato breeding and germplasm improvement programmes taking advantages of diversity in reproductive biology like synaptic mutants, unilateral sexual polyploidization, haploidy, stylar barriers, endosperm barriers, endosperm balance number (EBN) etc. Many a times, dihaploids are evolved for production of homozygous lines or for pre-breeding at diploid level for transferring desired traits through transgression.

However, in order to design more efficient breeding programmes, knowledge is required of the number of genes segregating, their chromosomal locations, and the magnitudes of their effects (Bradshaw *et al.,* 2008). The quantitative nature of the economically important characters and strong G × E, affects the phenotyping of the cultivars and the selections tend to perform differently under different agro-climatic conditions.

It has been emphasized by several workers that much importance should be given on parent selection for achieving an efficient breeding programme. Selection of parents based on genetic divergence followed by progeny test is suggested for effective exploitation of heterosis in potato breeding programmes. Superior parents and crosses for tuber yield and its components should be identified based on combining ability analysis in potato. Further the strategy of carrying out initial selection between crosses followed by within cross selection has been suggested to result in increased efficiency. This implies that the whole cross may be rejected on the basis of the poor performance of sampled progenies based on visual evaluation as they are less likely than others to contain progenies of commercial worth and vice versa. This will the allow a larger numbers of seeds of beneficial crosses to be sown and evaluated, for larger number of traits in these progenies, giving a rapid progress in a breeding programme. This genotypic recurrent selection, allows for identification of superior parents with good GCA to be recognized shortly after each round of hybridization.

The strategy has also been extended to develop durable quantitative resistance to insects and pests and fry colour. Multitrait (full sib recurrent selection) utilizing new cultivars from within re-sowings of the best crosses within each cycle has also been reported.

Strategies to improve breeding efficiency requires more detailed knowledge on several factors/aspects such as inheritance patterns and genetic architecture of important traits, number of genes segregating, their chromosomal locations and the amount of observed variation accounted for individual traits (Dale *et al.,* 2016). Theory of linkage analysis and quantitative trait locus (QTL) mapping in autotetraploid species have been reported in potato (Hackett *et al.,* 2013) considering random pairing and segregation of four homologous chromosomes. However, in practice, the complex potato genetics causes several departures for this proposed theory. There is limited cytological evidence regarding predominance of bivalents in potato with low frequencies of quadrivalents, trivalents and univalents are also observed, besides the effect of location of gene relative to the centromere affecting the double cross over (Bradshaw *et al.,* 2008).

The application of molecular technologies has huge potential for speeding up the process of conventional plant breeding. The identification of naturally existing allelic variation at the molecular level as well as the phenotypic level has the potential to accelerate the process of breeding for improved cultivars. Strong association of a genotype with a phenotype of interest allows genetically superior plants to be identified and selected early in plant growth before the phenotype can be observed (Dale *et al.*, 2016). A number of molecular maps of potato have led to the identification of genetic loci underlying many important agronomic, resistance and quality traits of potato. The mapping of the reference potato genome, including the annotation of around 39,000 protein-coding genes, has facilitated the identification of candidate genes in regions associated with important traits of interest such as the identification of the StSP6A gene for tuber initiation. Numerous molecular markers based on proteins and DNA sequences (RFLPs, RAPD, SSRs, AFLPs and SNPs). However, the tetraploid outbreeding background of potato, aggravated by the polgenic inheritance pattern of the economically important characters severely hinders the potential progress expected from marker assisted selection. There is further limitation of development of near isogenic lines and decreased robustness of MAS for small effect QTLs. In a practical example the potential of MAS for the introgression of *Globodera pallida* resistance genes derived from *S. sparsipilum* in a potato breeding program has been shown. Similarly, identification and utilization of two genetic markers flanking the H1 locus for association with PCN (*G. rostochiensis* RO1 & 4) resistant phenotypes has been demonstrated. The marker was found to be robust and has been used routinely within breeding programs. Further for identification of triplex and quadruplex H1 progeny imparting 100% resistance has been reported to be possible by sequencing of PCR products from different H1 resistant and susceptible genotypes and further screening of sequence polymorphism, to develop SNP markers whose dosage can be measured. Hackett *et al.* (2013) described the application of SNP dosage data to linkage analysis and QTL mapping in tetraploids and its potential application to population studies. There are more than 25 single dominant genes present on potato map. Among these, most of them show resistance to pest and disease while some of genes related to yield and quality traits together.

Interspecific hybridization between wild and cultivated genotype is a valuable approach used to transfer the useful genes and in this case the use of species-specific molecular markers would allow the wild genomic content to be reduced in few backcross generations (negative-assisted selection). Somatic hybridization provides a means to bypass the sexual incompatibility barriers between wild and cultivated Solanum species, leading to pollen fertile plants that can be used directly in potato breeding. Somatic hybridization bypasses gene segregation and enables to transfer both mono- and polygenic traits. Luthra *et al.* 2016 investigated the breeding potential of somatic hybrids developed through protoplast fusion between the dihaploid *Solanum tuberosum* L. 'C-13' and the diploid wild species *Solanum pinnatisectum* Dun.

Potato map is one of the most highly saturated maps with different molecular markers and there are more than 350 markers which covers approximately 90% of the potato genome which provides an extensive opportunity for optimal use of DNA analysis for MAS and making it valuable tool for fixing the genes that controls the expression of quality traits (Gebhardt *et al.*, 2001a). The important tuber traits such as skin color, flesh color, tuber shape and leptin content are controlled by single loci. The full deployment of MAS into practical potato breeding programs is now a reality and will further increase in the near future.

Association mapping of starch and yield related traits also have been performed in potato (D'hoop *et al.*, 2014, Schreiber *et al.*, 2014). The most prominent hot spot comprises the distal segment of approximately 5 Mbp on the north arm of chromosome V, which harbors among other candidate genes (Schreiber *et al.*, 2014), the StCDF1 locus that controls photoperiod dependent tuberization. Microarray technique helps to identify consistent differences in gene expression profiles between *S. phureja* and *S. tuberosum* cultivars, including genes likely to impact on flavor, texture, carotenoid content and tuber life-cycle. Candidate genes for flavor and texture have been reported. The concept of association mapping was introduced in potato genetics 15 years ago (Gebhardt *et al.*, 2001b). Since then it has been suitably exploited in the analysis of several potato traits (Baldwin *et al.*, 2011) using both more traditional (e.g. AFLPs and SSRs) as well as contemporary, sequence based (e.g. SNPs) marker types. Association mapping allows simultaneous analysis of a larger number of traits than possible in any one bi-parental population and is highly robust depicting statistically more significant marker-trait associations. In addition to this, association mapping offers significant advantages such as increased mapping resolution, reduced research time, and evaluation of marker-trait associations across a greater pool of alleles displayed in the whole set of selected diverse range of clones. As larger numbers of molecular markers (e.g. SNPs) become available, association mapping across the entire potato genome is increasingly being investigated (Genome-wide Association Studies, GWAS). While many QTLs have been identified in potato, only a handful, mainly confined to major R genes, have proven to be truly diagnostic and are deployed by breeders. In many cases the QTLs hold relevance to only those cultivars that have the same cultivars as parents as the ones used for those particular genetic mapping studies.

Genetic engineering is an additional tool to produce new genetic variability and to study important metabolic pathways. The rich diversity in potato germplasmcan contribute resistance to late blight, Verticillium wilt, potato virus Y, water stress, and cold-induced sweetening, from potato background thereby producing biotech varieties by inserting potato DNA. This is contrasted with traditional transgenic plants that use DNA derived from bacteria, viruses, or other organisms (Halterman *et al.,* 2016). In Netherlands, a variety Karnico1 has been developed by genetic modification of starch which leads to production of amylose free potatoes. Amylose production was completely suppressed by antisense RNA-mediated inhibition of granule bound starch synthase, an approach made possible by the identification of an amylose-free mutant produced by techniques associated with conventional breeding. Transgenic approaches have also provided new ways of understanding and manipulating carbohydrate metabolism aimed at developing genetically in-built resistance to low temperature sweetening caused by an accumulation of glucose and fructose.

Breeding for table potatoes in today's world has emerged as a highly complex objective, with multi trait breeding, for yield, quality as well as resistance. In his regard biotechnological tools are providing a strong background to conventional plant breeding, for identification of desirable genes and increasing the pace of genetic improvement. The genetic diversity of potato is immense, and its suitable utilization for improving the potato varieties, needs due consideration. A current major challenge in potato genetics remains the identification of further important gene combinations that lead to significant crop improvement. Increasingly the application of an ever increasing number of robust molecular tools will improve not only the quality of material progressing through programs, but also quality of the finished product.

REFERENCES

Acharjee A, Chibon PY, Kloosterman B, America T, Renaut J, Maliepaard C and Visser RG (2018) Genetical genomics of quality related traits in potato tubers using proteomics. *BMC Plant Biology* **18**:20-26

Baldwin SJ, Dodds KG, Auvray B, Genet RA, Macknight RC and Jacobs, JME (2011) Association mapping of cold-induced sweetening in potato using historical phenotypic data. *Annals of Applied Biology* **158**(3): 248-256

Baye BG (2017) Genetic diversity of potato for Nitrogen Use Efficiency under low input conditions in Ethiopia. Wageningen University, Wageningen, Ph.D. thesis

Bhardwaj V, Manivel P and Gopal J (2011) Screening potato (*Solanum spp*) for cold-induced sweetening. *Indian Journal of Agricultural Sciences* **81**(1): 20-24

Bradshaw JE, Hackett CA, Pande B, Waugh R and Bryan GJ (2008) QTL mapping of yield, agronomic and quality traits in tetraploid potato (*Solanum tuberosum* subsp. tuberosum). *Theoretical and Applied Genetics* **116**:193-211

D'hoop B, Keizer PC, Paulo MJ, Visser RF, Eeuwijk F and Eck H (2014) Identification of agronomically important QTL in tetraploid potato cultivars using a marker - trait association analysis. *Theoretical and Applied Genetics* **127**: 731-748

Dale MFB, Sharma SK and Bryan GJ (2016) Potato breeding now and into the genomics era. *Acta Horticultuare* **1118**:1-10

Gebhardt C, Ritter E and Salamini F (2001a) RFLP map of the potato. In, DNA-based markers in plants. Phih'pps RL and Vasfl IK (eds.), Vol. 6. Dordrecht: Kluwer Academic Publishers, Amsterdam: 319-336

Gebhardt C, Schuler K, Walkemeier B and Balllvora A (2001b) Association mapping in potato of a QTL for late blight resistance. Paper presented at: IX Plant Anim. Genome Conference (San Diego, CA, USA)

Gopal J, Chahal GS and Minocha JL (2000) Progeny mean, heterosis and heterobeltiosis in *Solanum tuberosum* X *S. tuberosum* and *tuberosum* X *andigena* families under a short day sub-tropic environment. *Potato Research* **43**(1): 61-70

Hackett CA, McLean K and Bryan GJ (2013) Linkage analysis and QTL mapping using SNP dosage data in a tetraploid potato mapping population. *Plos One* **8**(5) :e63939

Halterman D, Guenthner J, Collinge S, Butler N and Douches D (2016) Biotech Potatoes in the 21st Century: 20 Years Since the First Biotech Potato. *American Journal of Potato Research* **93**:1–20

Kim HJ, Lee HR, Jo KR, Mortazavian SM, Huigen DJ, Evenhuis B, Kessel G, Visser RG, Jacobsen E and Vossen JH (2012) Broad spectrum late blight resistance in potato differntial set plants MaR8 and MaR9 is conferred by multiple stacked R-genes. *Theoretical and Applied Genetics* **124**:923-935

Klaassen, MT, Bourke PM, Maliepaard C and Trindade LM (2019) Multi-allelic QTL analysis of protein content in a bi-parental population of cultivated tetraploid potato. *Euphytica* **215**(2):1-18

Luthra SK, JK Tiwari, Lal M, Chandel P, Kumar V and Singh BP (2016) Breeding potential of potato somatic hybrids: Evaluations for adaptability, tuber traits, late blight resistance, keeping quality and backcross (BC₁) progenies. *Potato Research* **59**:375–391. DOI: 10.1007/s11540-017-9336-1

Schreiber L, Nader-Nieto AC, Schönhals EM, Walkemeier B and Gebhardt C (2014) SNPs in genes functional in starch-sugar interconversion associate with natural variation of tuber starch and sugar content of potato (*Solanum tuberosum* L.). *G3: Genes, Genomes, Genetics* **4**: 1797-1811

5

Potato Breeding for Processing

VK Gupta[1], SK Luthra[1] and Vinay Bhardwaj[2]

[1]ICAR-Central Potato Research Institute Regional Station, Modipuram-250110 Uttar Pradesh, India
[2]ICAR-Central Potato Research Institute, Shimla-171 001, Himachal Pradesh India

INTRODUCTION

Potato (*Solanum tuberosum* L.) is 3rd most important non-cereal food crop in the world, after rice and wheat, and popularly known as 'the king of vegetables'. Potatoes are considered as a non-fattening, nutritious and wholesome food, which supply important nutrients to the human diet. Tubers contain significant concentrations of vitamin C and essential amino acids. Potatoes can be consumed in many ways, including baking, boiling, roasting, frying, steaming, and microwaving. It is integral part of food and traditional cousins and likely to find more importance in the dietary habit. Potatoes are either consumed directly or they are processed to give products such as chips and French fries, mashed and canned potatoes. Besides being important in human diet, potatoes are also used as animal feed and as raw material for starch and alcohol production. Potato was an occasional dish of the affluent class at the time of independence but now it is the most abundantly consumed horticulture based food by the poorest of the poor in the country. At present about 68% are utilized for table purpose, 7.5% for processing; 8.5% for seed and remaining 16% produce goes waste due to pre and post-harvest handling.

POTATO PROCESSING

Processing of potatoes appears to have originated in pre-Columbian South to extend their nutritional availability and portability. In high altitude zones, a product called *chuño* obtained following freezing, soaking, and drying was popular which could be easily transported, and can be stored for several years. Potato's popularity has nevertheless grown since the end of World War II, particularly in its forms of standardized industrially produced potato fries, chips, and other frozen and processed "convenience" foods. Acceptance of standard fries (with burgers) and packaged chips symbolizes the "globalization of diet". Processing or value addition refers to economically add value to a product and form characteristics more preferred in the market place. It increases the economic value of a commodity through production processes and has become the integral part of modern agriculture trade. The value added products fetches more profits and provides a choice to the consumers under changing food habits.

In North Indian Plains, potato harvesting is followed by hot summer, which makes it difficult to store potatoes under ordinary conditions. Present cold store capacity in country cold stores can accommodate only about 65% of the produce leaving a scope for further expansion in the cold store industry. In years of higher production, non-disposal of potatoes leads to distress sale and prices crash at the time of harvest. Lack of proper marketing avenues and low domestic utilization are the other major problems. Further, the glut years are normally followed by reduction in potato area and consequently lower potato production leading to higher prices in the following years. Keeping in view the ever-increasing trend of population in the country, it is essential that the production of potatoes keeps pace with the rising demands. Therefore, to bring stability in prices and avoid the periodic glut situations, it is essential to apportion a part of the acreage for production of raw material for value added products. Further, growing urbanization, rise in per-capita income, increase in number of working women's, changing food habits, preference for ready-to eat snacks and expanding tourism have made potato based processed products, the darling of all segment of the population. Value added processed products are opening up new market avenues in the national and international markets, and as a result, the farmers are finding it highly remunerative to grow processing potato varieties. Besides, dehydrated chips, cubes and other products can be easily prepared at the small scale industry level and can provide employment to the rural youth and village women. Value addition of potato not only contributes to crop diversification, improve the farm incomes and nutrition but also provide value export and additional employment.

In India, the major potato processing industries are located in West Bengal, Gujarat, Madhya Pradesh, Maharashtra, Karnataka, Tamil Nadu, Andhra, Assam, Punjab and around Delhi. Among these, potatoes produced in north-western and west-central plains contain low dry matter and high reducing sugars and not considered suitable. It is known that areas with minimum night temperatures above 10°C during the last 30 days of the crop growth produce potatoes with high dry matter (> 20%) and low reducing sugars (< 100 mg/ 100 g fresh wt.). Such areas include parts of Rajasthan and Madhya Pradesh, north-eastern plains and the entire plateau region. Potatoes for processing are, therefore, procured by the industry from west Bengal, Malwa region of Madhya Pradesh during March to June and from distant plateau region of Karnataka during August to December. This arrangement involves long distance transport of raw material adding to the cost of production and deterioration in raw material quality. Besides, this region is also vulnerable for many biotic and abiotic stresses (late blight in West Bengal and Bacterial wilt and late blight in Karnataka and Maharashtra), frosting (Malwa region of MP) and shorter window for growth and production of potato.

In India, presently about 7.5% of the total produce is being processed mainly as chips, flakes/powder and French fries, which is almost tenfold increase in comparison to less than 1% during late nineties. The major share of potato processing goes to potato chips which accounts nearly 85% of the total snack business in India.

QUALITY REQUIREMENT FOR PROCESSING

Quality of potato produce determines the marketability. Quality is related to visual appearance, culinary preference of the consumer, or ability to meet market specifications. Tuber size, shape, appearance, absence of diseases or defects, flavor and texture determine the quality of the produce for various purposes. On the basis of dry matter and texture, potatoes can be used for different purposes (Table 1). A mealy texture is associated with high solids and a waxy texture with low solids. Potatoes containing more than 20% dry matter content with mealy texture are preferred for fried and dehydrated products, while small size potatoes containing dry matter between 18- 20 % with waxy texture are preferred for salad making and canning. Specific characteristics in potato varieties are required for different purposes (Luthra *et al.*, 2004). Low glycoalkaloids content (< 15mg/100 gram fresh tuber weight) and ability to withstand cold induced sweetening are added advantages. The tuber quality requirements for different processed products are given in Table 2.

Table 1: Relationship between tuber dry matter and optimum use (Mosley and Chase, 1993)

Specific gravity	Dry matter %	Texture	Typical uses
Below 1.060 (very low)	Below 16.2	Very soggy	Pan frying, salads, canning
(1.060-1.069 (low)	16.2-18.1	Soggy	Pan frying, salads, boiling, canning
1.070-1.079 (medium)	18.2-20.2	Waxy	Boiling, mashing, fair to good for chip processing and canning
1.080-1.089 (high)	20.3-22.3	Mealy, dry	Baking, chips, frozen French fry, some cultivars tend to slough when boiled
Above 1.089 (very high)	Above 22.3	Very mealy or dry	Baking, frozen French fry, chip, tendency to produce brittle chips and to slough when boiled

Certain morphological and biochemical attributes are necessary in potato varieties to meet the requirement of processing product. Morphological attributes mainly includes size and shape of tubers, internal and external defects, whereas bio-chemical attributes includes dry matter, reducing sugars, free amino acids, phenol content etc. These attributes not only determine the quality and recovery of the finished products but also govern the production efficiency and operational costs of the processing industry. For making potato flakes, granules and dice/cubes, size and shape of tubers are not very important, however, they are inevitable for making chips and French fries. In general higher dry matter content in potato tubers results in higher recovery of processed products with lower energy and lesser oil consumption with better shelf life, while, low reducing sugars results in lighter and better colour of processed products. The ideal reducing sugar content for fried products is generally accepted to be < 0.1% tuber fresh weight.

MORPHOLOGICAL ATTRIBUTES

Tuber size, shape and eye depth

The tubers of 45 – 85 mm are considered ideal to obtain the desirable size of chips. Round shape is preferred to produce uniformly round chips; tubers with oval shape can also be used for making chips. For making good quality French fries, the oblong or long oval (>75 mm) tubers are desired. While for flakes, though round to oval shape (18-38mm) is desirable, the requirement of shape is not very strict. For canning, small tubers of round to oval shape are suitable. Processing varieties in general should possess tubers with shallow or fleet eyes so that peeling loss are minimum possible during slicing.

Table 2: Requirement of potato varieties for different purposes

Characters		Use requirements					
	Seed	Table potatoes			Processing		
		Boiled	Baking	French fries	Chips	Flakes
Tuber shape	Round/ovoid	Round/ovoid	Round/ovoid	Oblong/Long-oval (>75-110 mm)	Round (45-85mm)	Oval/Round (30-85mm)
Skin color	White/yellow/ red	White/yellow/red	White/yellow /red	White/yellow/red	White/yellow/red	White/ yellow/red
Eye depth	Shallow/ medium	Shallow/ medium	Shallow/medium	Shallow	Shallow	Shallow
Flesh color	White/yellow	White/yellow	White/yellow	White/yellow	White/yellow	Whit-cream
Texture	Waxy/mealy	Waxy	Mealy	Mealy	Mealy	Mealy
Uniformity	High	High	High	High	High	High
Defects	Minimum	Minimum	Minimum	Minimum	Minimum	Minimum
Dry matter (%)	18-22	18-20	>20	>20	>20	>20
Reducing sugars*	NA	NA	NA	<150mg	<100mg	<150mg
Phenols	NA	High	High	Less	Less	Less
Glycoalkaloids *	<15mg	<15mg	<15mg	<15mg	<15mg	<15mg
Keeping quality	Good	Good	Good	Good	Good	Good
Damage resistance	High	High	High	High	High	High

*mg/100g fresh tuber weight

External and Internal Defects

An internal or external defect of potato affects the quality of finished products. External defects may be due to unwanted shape or size, knobbiness, cracking, decay, greening *etc.* Internal defects are imperfections occurring within the tubers, such as hollow heart, brown centre, internal brown spots (IBS) etc. These may be caused by physiological or pathological reasons. Higher defects in the raw material increases the labour requirement during sorting of tubers and ultimately enhance the operational cost apart from reducing the quality of product.

BIOCHEMICAL ATTRIBUTES

Tuber dry matter and specific gravity

Potato contain on an average 80% of water. Tuber dry matter or solids is positively correlated with tuber specific gravity. The dry matter content of tubers is the most important character determining quality and yield of fried and dehydrated products. Potatoes with high dry matter content are preferred for preparation of fried and dehydrated products. Dry matter content of 18-20% is considered acceptable for canning but for chips, French fries and dehydrated products it should be more than 20% or > 1.080 specific gravity. Higher dry matter content or solids content result higher recovery of processed product, lower absorption of oil, lesser energy consumption, crispy texture of the product and ultimately lower the risk of fattening of consumers.

Reducing sugars

The reducing sugars (glucose and fructose) present in tubers play a critical role in determining colour of fried products like chips and French fries. The product become coloured during frying at high temperatures due to the *'Maillard reactions'* between reducing sugars and free amino acids present in tubers. Presence of excessive amounts of reducing sugars in potato tubers results in unacceptably dark colour and bitter taste in fried products. Beside colour and flavor of fried products, *'Maillard reactions'* is also related to formation of acrylamide, which is considered a potentially toxic compound. Reducing sugars content below 100 mg/100g fresh tuber weight is acceptable for producing chips. However, for French fries and dehydrated products reducing sugars content up to 150 mg/100g fresh tuber weight is acceptable.

Phenols

In addition to the discolouration of fried products, tubers show enzymatic discolouration and after cooking discolouration. Enzymatic discolouration occurs when the potatoes are peeled, cut or injured. Some of the constituents present in the tubers react with oxygen (air) and tuber flesh turns brown. This type of discolouration can be prevented if potatoes are not exposed to air and are immersed in water. After cooking discolouration develops in the tubers upon cooking and exposure to air. The canners often face this problem. However, this is not a major problem in our country as almost all the cultivated varieties are free from this defect.

BREEDING FOR PROCESSING

The cultivated potatoes (*Solanum tuberosum*) is tetraploid ($2n=4x=48$) and displays tetrasomic inheritance patterns that make genetic studies and breeding work very difficult. It is highly heterozygous and has a narrow genetic base. Unlike diploids, potato breeding is a cumbersome task due to inherent genetic and biological factors. Therefore, breeders need to screen large progenies to find noteworthy segregants. Many quality traits have a polygenic control, and as such they are not inherited in a simple Mendelian term, further phenotypes cannot be grouped into a small number of easily distinguished classes, and the genotype x environment interaction (GEI) is usually high.

In potato, high number of wild tuber bearing relatives that grow from the southern part of USA to southern Chile form a polyploid series with species having 24, 36, 48, 60, and 72 chromosomes and more than 70% of them are diploids ($2n=2x=24$). These species represent a rich source of valuable germplasm with traits of interest that can be transferred to the cultivated potato. This germplasm also possesses all the allelic diversity needed to broaden the genetic base of the cultivated potato and to produce highly heterotic genotypes. Potato breeding is positively influenced also by the fact that genomes (whole chromosome sets) can be easily manipulated through the functioning of gametes with an unreduced chromosome number ($2n$ gametes) and the use of *S. tuberosum* haploids. $2n$ gametes are the basis for sexual polyploidization events that allow increase in chromosome number through sexual hybridization. *S. tuberosum* haploids can be easily extracted following 4x X 2x crosses with pollinator clones of diploid *S. phureja* or through anther culture. Potato is a model crop for tissue culture approaches. In potatoes, genetic transformation offers a real alternative approach for new cultivar development. Considering the allelic diversity present in *Solanum* species, DNA markers open new doors for breeding based on marker-assisted selection. The production of starches with modified amylose to amilopectin ratio represents a good example of the possibilities offered by genetic engineering in improving potato quality traits.

A recent target of genetic engineering has been the production of potato plants with a higher nutritional value. Chakroborty *et al.* (2000) transformed a potato genotype with the gene *AmA1* from *Amaranthus hypocondriacus*, encoding a protein with a nutritionally balanced amino acid composition. Genetic engineering has been recently used to improve carotenoid content of tubers. In particular, to overcome the zeaxanthin deficiency of human diet, Römer et al. (2002) down-regulated the synthesis of zeaxanthin epoxidase specifically in tubers through antisense technology and co-suppression approaches. It should be noted that genetic engineering complements conventional breeding strategies to improve quality traits. Classical breeding is still important when progeny testing is necessary to identify the integration events that result in stable expression, or when backcrosses are required to introduce the transgene from the transformed plant into the desired genetic background. In addition, the genotype x environment interaction is as true with transgenic as with conventional breeding lines. Dunwell (2000) reported that it takes about 10-12 years (from gene identification and cloning to multi-site field testing) before final release of a new cultivar. For cold induced sweetening tolerant genotypes through transgenic approachwas attempted in two different ways: (1) reducing vacuolar acid invertase (INV) by over expression of tobacco invertase inhibitor gene (Nt-Inhh) in varieties Kufri Chipsona-1, Kufri Jyoti and Kufri Badshah. The transgenics produced showed acceptable chip colour even after cold storage (Pandey et al., 2009). (2) Silencing of INV gene expression at the post-transcriptional level through artificial micro-RNA (ami RNA), RNA interference (RNAi), or ribonuclease P (RNase P)-mediated silencing. In addition to this, reduction of UDP-glucose pyrophosphorylase (UGPase) was also attempted. This was done by silencing the gene either via introduction of intron containing inverted repeatgene construct against UGPase promoter (silencing at the transcriptional level) or through the introduction of ami RNAs against UGPase gene (silencing at the post-transcriptional level) (Pandey et al., 2009). RNAi, amiRNA and RNase P-mediated suppression of invertase activity has generated transgenic lines in variety Kufri Chipsona-1. These lines exhibited good cold chipping attributes even after 3 months of cold storage at 4°C.

Generally, any plant breeding program is based on two fundamental steps, (1) production of genetic variability through either conventional or innovative methods, and (2) efficient selection within the variability created. There are two methods to create variability, sexual hybridization and genetic engineering. Similarly, breeding varieties for value addition comprises of following major steps:

- Identification and evaluation of parental lines with desirable attributes

- Combining the desired attributes with high yielding base variety through hybridization or genetic manipulations

- Evaluation of segregating population/clones and selection for stability of yield and desired processing traits

Inheritance of processing especially related to quality traits is poorly defined and genetically least researched. The available information on various quality traits of interest is presented below.

Morphological attributes

Tuber shape: Tuber shape is a varietal character but it is also influenced by environmental conditions and cultural practices. The shape of the tuber is scored by determining the length/breadth ratio (I). Tuber shape ranges from compressed to very long, the most common types being; round (I < 1.4), oval (1.5 < I <1.9) and long (I > 2.0). Inheritance of the tuber shape is yet not finally concluded due to contradictory results of the researchers. Some researchers considered long tubers as being dominant to round tubers, but others suggested that round shape was dominant to either oval or long oval (Dale and Mackay, 2007)

Growth cracks: The causes of growth cracks are not well understood. It is attributed to fluctuating water stress or changes in tuber growth rates. In addition to the abiotic stresses, there is also a heritable component, which does not result in clearly delineated groups, suggesting that a number of minor genes may be involved. Therefore, for selection of this character one has to rely on observations in various environments over a number of years, as is presently practiced.

Hollow heart: Hollow heart is a physiological defect resulting from an internal cavity of varying dimensions. It is found more frequently in the larger tubers. Development of hollow heart is often associated with periods of rapid tuber growth, which may have been preceded by a period of moisture or nutritional stress. The expression of this trait is affected by various environmental and genetical factors. Jansky and Thompson (1990), in a study of tetraploid segregating population found positive correlation between hollow heart and both mean yield and tuber size and a negative correlation with tuber number. The genotypes with hollow heart defect are known to transgress it in the progenies and hence, the use of susceptible parents for this trait may be avoided in a breeding programme. Multi-location testing over the years helps to select the progenies devoid of hollow heart.

Brown center: It is often a precursor to hollow heart and sometimes referred to as incipient hollow heart. Brown center is characterized by browning due to cell death in the pith area of the tuber. It is documented in the literature that cool temperatures during tuber initiation induce brown center. As on date very little is known about this disorder and control seems to be bit difficult.

Internal brown spots: Internal brown spot (IBS) is a physiological disorder characterized by brown/rust coloured spots or blotches distributed irregularly throughout the tubers. Its development is reported to be induced by a number of environmental factors, including restricted water supply and high temperatures during the growing season and by uneven growth rates. It is also associated with a low concentration of calcium in the affected tubers. There is a variation for this character between the cultivars. Without a reliable test that can assess a large number of genotypes, the segregating material may be tested over the years on many locations.

Greening: Potato tubers exposed to light in the field, during or after the harvest develop a green pigmentation, initially at the surface and subsequently throughout the whole tuber. This condition is caused by the formation of chlorophyll. Greening of potatoes is often associated with an increased level of glycoalkaloids, which impart a bitter taste to the tubers and are poisonous. Excessive greening can reduce the marketability of the tubers. Tuber greening is inherited in a quantitative manner with most of the variation being additive. However, broad sense heritability was sufficiently large to permit effective selection within the potato breeding programmes.

Biochemical attributes

Dry matter: Starch constitutes about 70% of the dry matter (DM) content of potato tubers. There is considerable variation among cultivars for DM content and this trait is also influenced by the environment during the growing season as well as storage. Dry matter has been associated with many other quality parameters such as texture, suitability for processing and susceptibility to mechanical damage. It is polygenically controlled and responds to selection, but the environmental effects are so large that much effort is required to achieve significantly higher dry matter content that are currently available in the known varieties. Too high DM content also badly affects certain culinary properties, such as disintegration on boiling (sloughing), so that selection for high DM content *per se* is seldom a breeding objective. Moreover, high DM has a negative correlation with total tuber yield.

Reducing sugars: Starch is the major carbohydrate source in potato tubers, yet small but varying amounts of sugars, namely sucrose, glucose and fructose are also present. In terms of quality for processing, these sugar levels are critical, particularly the hexose sugars, glucose and fructose. A vast quantity of potatoes is processed into fried products, *viz.* chips and French fries. On frying at high temperatures, sugars react with α-amino acid groups of nitrogenous compounds giving rise to dark coloured compound and the reaction is known as Maillard reaction. This results in a dark coloured, bitter tasting product. Reducing sugar content of 0.1% per 100g fresh weight is ideal for processing into chips and higher than 0.33% is unacceptable. Sugar content of tubers varies considerably among seasons, locations and also varieties. The chip colour has high heritability thus allowing fairly reliable prediction of parental values based on progeny means. The storage temperature has an important influence on sugar content, with storage at 4°C resulting in cold-induced sweetening. However, high reducing sugar levels can be avoided by storage at 8-10°C. But the storage at such temperatures for prolonged period results in an excessive growth of sprouts hence requires the use of sprout suppressing chemicals. There is little evidence for genetic variation for resistance to cold-induced sweetening in the cultivated potato. The glucose content of the tubers seems to be the most important factor in determining fry colour, but this can be more accurately predicted when fructose content is also taken into account. Sucrose levels do not correlate well with either of the hexose sugar levels or fry colour.

Enzymic browning: It is an important problem resulting in increased costs through losses, additional labour for sorting and preventive measures during processing. The enzymic browning occurs because of the oxidation of tyrosine and other orthodihydric phenols by polyphenoloxidase (PPO). Tyrosine oxidation initiates the subsequent formation of a dark or black melanin pigment. To control this, the processors use various chemicals like chelating agents, reducing agents, bisulphites or sulphydryl compounds. Various environmental factors influence the observed degree of enzymic browning, though there is limited information about the factors responsible for it. A good relationship was observed in some cases between the rainfall and amount of tyrosine in the tubers. Tyrosine content in the tuber has a high predictive value with regard to degree of enzymic discolouration and the content of chlorogenic acid influences the speed of reaction. Variation in the degree of observed enzymic browning is exhibited between varieties and appears to be continuous in nature, suggesting that there are no major genetic factors operating.

Anthocyanins and carotenoids: Cultivated forms of potatoes contain varying amounts of anthocyanins and caretenoids in tuber skin and flesh (Gross 1991, Mazza and Miniati 1993). Outside of center of the origin of cultivated potato, in the Andes of South America, it is rare to find cultivars with anthocyanin pigments conferring red or purple flesh. Genetic control of presence and absence of antocyanins is monogenic, although the completeness of anthocyanin distribution in pigmented flesh may be under complex genetic control (Brown *et al.*, 2003). In general consumer interest in potatoes with red or purple flesh has been increasing over the past decade, in part because of novel appearance, and in part because of the perceived benefits of higher antioxidant content (Scalbert *et al.*, 2005). Thus development of new cultivars substantially higher concentrations offers a new rationale to promote consumption.

Glycoalkaloids: Potato tubers contain small quantities of naturally occurring steroidal glycoalkaloids, a class of potentially toxic compounds, found throughout the family Solanaceae. Approximately 95% of the total glycoalkaloids present in potatoes are -á-solanine and á-chaconine, both of which are structurally similar, but differently glycosylated forms of the aglycone and solanidine. The distribution of the glycoalkaloids in the tubers is not uniform. Periderm and cortex have higher concentrations but lower towards the pith. Cultivars vary with respect to their inherent glycoalkaloid content; at lower levels it is suggested that they may enhance potato flavour, but at higher concentration (above 15 mg 100 g $^{-1}$ fresh wt.) they impart bitterness and levels above 20 mg 100 g^{-1} fresh wt. are considered unsuitable for human consumption resulting in symptoms typically associated with food poisoning. However, the glycoalkaloids are thought to confer a degree of protection to the plant against various pathogens. A number of factors can influence the level of glycoalkaloids in potato tubers, including cultivar, climate, storage environment, maturity, damage, temperature and exposure to light. The varieties developed under Indian potato breeding programme have lower glycoalkaloid content. There are conflicting reports available in the literature about the inheritance of this trait. Some reports suggest that it is inherited in a dominant manner with a couple of genes controlling this trait while others indicate its polygenic inheritance. Nevertheless, it is important that initial content of parental material may be assessed before starting a breeding programme particularly with wild species.

INDIAN BREEDING PROGRAMME FOR PROCESSING POTATOES

In Europe and USA, where the potato processing industry is well established and highly developed, potato varieties have been bred specifically for the purpose. Such efforts in our country, however, had not been made till 1989 as all the varieties have been bred for consumption as fresh potato. Therefore, despite large quantities of potato available in the market, the availability of processing quality potato was poor. These varieties generally have a low dry matter (17-19%) and high reducing sugar content (above 250 mg/100 g fresh tuber wt.) and were, therefore, not suitable for processing. Of these, three varieties, *viz.* Kufri Chandramukhi, Kufri Jyoti and Kufri Lauvkar were identified by the chipping industry to be better than the others. The produce of these varieties from various locations, however, was not found suitable for processing. Only a few regions, *viz.* Indore and Ujjain districts in Madhya Pradesh, Nagrota in foothills of Himachal Pradesh, parts of central Uttar Pradesh *(rabi*crop), Hassan, Belgaum and Dharwad districts in Karnataka *(kharif)* produce acceptable quality potatoes having higher dry matter and low reducing sugars suitable for processing. The supply of fresh potatoes for processing from these regions starts from the end of December and continues till April-May, while lesser quantities of potatoes during August to October are transported from Nagrota (HP) and Karnataka. Despite large quantities of fresh potatoes available in the market, the supply of potatoes suitable for processing (chipping) was limited.

Efforts to introduce European and north American processing and table varieties has not been successful in the country mainly because these varieties were bred for long day conditions whereas, the crop in India is grown under short winter days. Keeping in view the demand of quality raw material for processing, ICAR-Central Potato Research Institute launched a breeding programme in 1990 for developing indigenous processing varieties.

Under this programme, crosses were made at ICAR-Central Potato Research Station, Kufri (HP) in hills (31° N 77° E; 2402 m above msl) and raising and evaluating of segregating progenies and selection of suitable genotypes was done at ICAR-Central Potato Research Institute, Campus, Modipuram (U.P.) in the plains (29° N 76° E; 222 m above msl). Assessment of segregating populations for as many characters as possible in the initial stages was done to eliminate undesirable genotypes facilitating better assessment of the selected genotypes in subsequent generations (easily identifiable characters were considered for this purpose). Extensive evaluation of genotypes was done in subsequent generations for tuber dry matter, reducing sugars, chip colour and yield. Finally the selected hybrids are tested under industrial processing conditions. The parameters being tested under industrial conditions include specific gravity, UC (undesirable colour), ED (external defects), ID (internal defects), TPOD (total potato defects) and HC (Hunter Colour). The variety having specific gravity =1.080, UC=<5, TPOD=<15, and HC=>58 is considered acceptable for chipping. These were assessed for adaptability and processing quality at different locations in the major potato growing regions and also compared for quality against exotic processing varieties. The general procedure for evaluating the hybrids in different stages of selection is given in Table 3.

The breeding programme for development of indigenous varieties culminated in the release of two processing varieties, Kufri Chipsona-1, Kufri Chipsona-2 in 1998 in India. Availability of these varieties and the standardization of elevated temperature (10-12°C) storage technology for storing potatoes for processing resulted in rapid development of potato processing industry in the country (Gaur et al., 1999). In the year 2006, Kufri Chipsona-3 was developed with lesser defects and higher processing grade tubers.

Table 3: Outline of breeding programme for processing purposes

Year	Stage	No of cultures†		Tuber material*	Selection criteria
		Planted	Selected		

0 Germplasm evaluation/selection of parents & crossing to obtain true potato seeds of desired cross

Year	Stage	Planted	Selected	Tuber material*	Selection criteria
1	F_1	30,000 Seedlings	2,000	Single hill 5	The seedlings are raised in the field under aphid free conditions. At harvest genotypes showing long stolons, irregular tuber shape, deep eyes, cracking and undesirable tuber colour are rejected. Yield is not considered for selection at this stage.
2	F_1C_1	2,000	500-600	5 30	The cultures are planted in short rows with control varieties after every 30 row. At full growth weaklings and diseased cultures are rogued out. At harvest rejections are done on the basis of undesirable shape and colour. Four tubers of each culture are dipped in Brine solution for specific gravity estimation and cultures with < 1.080 specific gravity are rejected. Within family comparisons of cultures are made for tuber bulking and uniformity in size.
3	F_1C_2	500-600	50-100	30 120	Plant the cultures in double observational row along with control varieties. Inferior cultures are rejected as in F_1C_1 and cultures producing uniform size tuber of desired shape are retained. Selected cultures are tested for chips/French fries colour and cultures having acceptable colour (<3) are selected.
4	F_1C_3	50-100	20-30	120 500	Clones with poor storability after cold store are rejected. The remaining hybrids are planted in RBD (3 rows of 15 tubers/replication, 3 replication) along with control Kufri Chipsona-1, Kufri Chipsona-3, Kufri Frysona and other local control. The trial is harvested at 90 days after planting and culture with high yield, reducing sugars (<0.1%), chip/fry colour (<3) and dry matter (>20 %) are retained.

(Contd.)

5	F_1C_4	20-30	10-15	500 1000	Clones with poor storability after cold store are rejected. as in F_1C_3. The remaining cultures along with controls are planted in preliminary yield trial in RBD (4 rows of 60 tubers/ replication, 9 replications). The hybrids are harvested at 75, 90 days and 105 days after planting and compared with control for yield (at least 10% more), chip colour better than controls, dry matter (>20%) and reducing sugars (<0.1%). Promising hybrids are retained. Those showing short dormancy (sprouting within 4-5 weeks) are not retained for further trials.
6	F_1C_5	10-15	5-6	1000 4000	Clones with poor storability after cold store are rejected. The selected hybrids are tested in confirmatory yield trial (RBD, 9 replications, 120 tubers/8rows/per replication) at 75, 90 and 105 days after planting. Best genotypes are selected on the basis of yield, chip colour, dry matter and reducing sugar. The selected hybrids are further tested for bulk frying at industrial level.
7	F_1C_6	5-6	2-3	1000 4000	The confirmatory yield and quality trial is repeated as in previous generation. Additional tubers are used for multiplication. The produce of best performing hybrids is kept for further trial. At this stage processing quality are also determined during storage at elevated temperature. Keeping quality also assessed at room temperature. The superior hybrids are introduced in AICRP for multi-location testing.
8	F_1C_7	2-3	2-3	2000 10,000	The hybrids are multiplied at Seed Preparatory Unit Modipuram, Meerut and disease free seed material of uniform size and same physiological age is supplied AICRP Centres for conducting evaluation trials. The hybrids are also tested for different seasons for their adaptability at Modipuram.

(Contd.)

9	F_1C_8	2-3	2-3	4q 20q	Preliminary yield trials are conducted at AICRP Centres and processing quality is tested in laboratory at Modipuram, Meerut.
10	F_1C_9	2-3	2-3	10q 50q	Preliminary yield trials are conducted at AICRP Centres and processing quality is tested in laboratory at Modipuram, Meerut.
11	F_1C_{11}	2-3	1-2	20q	Promising hybrids are tested in On Farm trials for two years at all AICRP Centres and processing quality is tested in laboratory at Modipuram, Meerut.
12	F_1C_{10}			100q	

The results of AICRP and on-farm trials are discussed in AICRP-Potato workshop. The best hybrid is recommended for release.

† No is approximate

*Number of tuber of each clone evaluated/multiplied

After the development of Kufri Chipsona-3, the efforts were directed to develop a processing variety suitable for hills which could fulfil the demand of raw material for processing industry after July/ August. The produce of the popular variety of hills, Kufri Jyoti could not meet the processing standards due to poor and inconsistent quality of tubers and breakdown of late blight resistance. This has led to the development of variety Kufri Himsona suitable for the hills in 2008 which ensures off season supply of raw material to the processing industry (Pandey *et al.*, 2008).

Non availability of specialized processing variety for French fries was another impediment for the processing industry. The demand of French fry was being met either by importing frozen fries or preparing French fries from indigenous variety Kufri Chiposna-1 which produced low French fry grade tuber yield. Keeping in view the long felt demand of French fry variety, the institute developed and released Kufri Frysona for north Indian plains in 2009.

All the indigenous processing varieties in the country were of medium to long duration, therefore, the target was focused on development of early bulking late blight resistant processing varieties for Karnataka, West Bengal and Madhya Pradesh. In Karnataka, out of total major potato area is under Kharif crop (mainly Hassan and Chickmangalore).The prevalent variety Kufri Jyoti is not suitable because of poor processing quality. The exotic cultivar Atlantic is highly susceptible to late blight and has hollow heart and cracking in tubers. West Bengal also needs a variety with early maturity because of short potato season due to harvest of paddy in November. Malwa region in Madhya Pradesh also needs an early maturing processing variety. To feed the industry during the lean period in September & October, *Kharif* produce from the plateau region is being used. Field resistance to late blight in such a variety will be the boon for potato cultivation. The region specific processing variety Kufri Chipsona-4 was developed and released in 2010 for plateau region of Karnataka, West Bengal and Madhya Pradesh.

Another French fries variety Kufri FryoM with early maturity and suitable for cultivation in North-western and Central plains of India has also been developed and released in 2018.

The salient features of the processing varieties (Figure 1) are given below.

Kufri Chipsona-1 (MP/90-83): The variety is a selection from the progeny of the cross CP2416 x MS/78-79. The female parent CP2416 was the Mexican genotype MEX 750826, while the male parent - 'was from Indian potato breeding programme which involved KufriJyoti and EM/H-1601 in its parentage. The variety produces white-cream ovoid tuber with shallow eyes and white-cream flesh. The variety produces nearly 30-35 t/ha yield under north Indian plains of the country. Its dry matter content is 20-23%and reducing sugar content is 10-75 mg/100 g fresh weight. The variety yields processing grade tubers to the tune of 60-75%. The variety has good late blight resistance and is well adapted to Indo-Gangetic plains. The tubers possess excellent keeping quality. The variety is suitable for making chips, French fries and flakes.

Kufri Chipsona-2 (MP/91-G): The variety is a selection from the progeny of the cross CP2346 (F-6 from Peru) x QB/B 92-4. The female parent CP2346 is an accession from the germplasm collection received from International Potato Centre, Lima, Peru, while the male parent is from an earlier breeding programme involving in its parentage an Indian variety Kufri Red and the variety Navajo from USA. The variety produces white-cream round tuber with medium eyes and cream flesh. The variety produces nearly 30-32 t/ha yield under north Indian plains of the country. The tubers of this variety have reducing sugars in the tune of 30-100 mg/ 100 g fresh weight and dry matter content is 22-24%.The variety has high degree of late blight resistance. The processing grade percentage is 65-75and is highly suitable for chip making.

Kufri Chipsona-3 (MP/97-583): The variety is a selection from the progeny of the cross MP/91-86 x Kufri Chipsona-2. The female parent MP/91-86 is a promising selection from the processing breeding programme. The variety produces white-cream ovoid tuber with shallow eyes and white flesh. The variety produces nearly 30-35 t/ha yield under north Indian plains of the country. The tubers of this variety have reducing sugar content of 10-100mg/100 g fresh weight and dry matter content 20-23%. The variety has moderate resistance to late blight and is suitable for chips and flakes preparation. The variety yields reasonably good process grade tubers to the tune of 70-80%.

Kufri Himsona (MP/97-644): The variety is a selection from the progeny of the cross MP/92-35 x Kufri Chipsona-2. The female parent MP/92-35 is a selection from processing breeding programme. The variety produces white-cream ovoid tuber with shallow eyes and white flesh. The variety produces nearly 15-20 t/ha yield under hilly regions of the country. The tubers have dry matter content of 21-24% and reducing sugars 10-80 mg/ 100 g fresh weight. The variety has a high degree of resistance to late blight. The variety is suitable for chips and flakes making

Kufri Chipsona-4 (MP/01-916): The variety is selection from the progeny of cross Atlantic x MP/92-35. It produces high yield with higher proportion of chip grade tubers. It has early maturity with field resistance to late blight, thus helping farmers in saving on costly fungicides. The variety produces white-cream round-ovoid tuber with shallow eyes and white flesh. The variety produces nearly 18-22 t /ha yield during kharif crop in Karnataka and 30-35 t/ha tuber yield in rabi crop in plains of the country. It is suitable for preparation of Chips owing to its round-ovoid shape, high dry matter (>20%) and low reducing sugars (40-80 mg/ 100 g fresh weight.). It will fill the void of a suitable chipping variety from Karnataka, West Bengal and Madhya Pradesh where processors are in need of variety combining high tuber yield and high level of late blight resistance. Long dormancy and good keeping quality will help storage of this variety for longer period thus ensuring round the year availability of raw material to chipping industry.

Kufri Frysona (MP/98-71): The variety is selection from the progeny of MP/ 90-30 x MP/90-94. It produces attractive white-cream long oblong tubers with shallow eyes and white flesh. The variety produces nearly 30-35 t/ha tuber yield in rabi crop in plains of the country. It possesses very good field resistance against late blight disease and has reasonably good frost tolerance. It is a good keeper under country store conditions and possess longer tuber dormancy period of more than 8 weeks. It possesses high tuber dry matter (22%), low reducing sugars (30-80 mg/100 g fresh weight) and very good quality French fries can be prepared. The industrial testing has shown the superiority of this hybrid for French fry making in terms of taste, texture and colour.

Kufri FryoM (MP/04-578): The variety is selection from the progeny of Kufri Chipsona-1 x MP/92-35. It produces attractive white long oblong tubers with shallow eyes and white flesh. The variety produces nearly 30-35 t/ha tuber yield in plains of the country Tubers do not show deformities like cracking or hollow heart. It has field resistance against late blight disease. It is a good keeper under country store conditions and possess longer tuber dormancy period of more than 10 weeks. It possesses 20% tuber dry matter, low reducing sugars (50-90 mg/100 g fresh weight) and very good quality French fries can be prepared.

Fig. 1: Indigenous potato processing varieties developed by CPRI: Kufri Chipsona-1 **(A)**, Kufri Chipsona-3 **(B)**, Kufri Chipsona-4 **(C)**, Kufri Himsona **(D)**, Kufri Frysona **(E)** and Kufri FryoM (F)

In the absence of availability of good quality raw material in the vicinity of industries, several processing units (Fritolay, ITC, Haldiram etc.) located in North- western plains (Punjab and Uttrakhand) and South west region (Andhra, Karnataka, Maharashtra and Tamilnadu) of country had to transport large quantities of potatoes from faraway places like MP and Gujarat involving very high transportation cost. The development of indigenous varieties has largely helped the industry to get their quality processing potato grown locally from nearby areas. Growing potatoes in south and western zones of our country have many challenges like shrinking potato production window, high night temperature, bacterial wilt / soft rot and late blight infestation. Although potato produced in south west region would accumulate less reducing sugars in tubers and can be source of raw material during off season (produce of *Kharif* Crop) but due to shorter growing window and diseases tuber may not be suitable for long term storage. Besides, tubers produced in the region may also accumulate more glycoalkaloides due to higher mean growing temperatures. Addressing all these challenges through new widely adapted genotype is little difficult but these challenges can be properly addressed through integrated approach of genotypes and best management strategy.

Successful cultivation of processing varieties and its storage at the intermediate storage temperature (10-12°C with CIPC) has provided suitable raw material to the processing industries to run them round the year. This would not only reduce the cost of production of fried products but would also provide quality products to the consumers at reasonable price.

WAY FORWARD

The potato value addition or processing sectors is well developed in USA, Canada and Europe where 30-67% of the total produce is processed into various products. In contrast, India is in developing phase and on its way to have a crispy revolution due to emerging growth and rapid progress in potato processing sector. Due to availability of suitable raw material and adoption of improved storage technologies, the potato processing activity both in the organized and unorganized sector has increased from <1% in late nineties to about 7.5% in 2009-10. Processing has helped to reduce the post-harvest losses. The reduction in postharvest losses will not only result in economic gains to the farmers but also provide better food and nutritional security to the country. Besides, in India, there has long been a demand and liking for specific variety due to their skin colour (red skinned varieties *viz.*, Kufri Lalima, Kufri Sindhuri are preferred in Eastern plains), flesh colour (white flesh varieties Kufri Bahar, Kufri Badshah in North-West Plains) and also due to better taste and texture (Kufri Chandramukhi, Kufri Chipsona-1, Kufri Jyoti *etc.*) in various regions of the country. Although, these varieties fetch premium price but there has not been concerted efforts to exploit these qualities into sustainable commercial benefits by the farmers/producers and its entire business chain. Keeping in view the future demand of raw material for processing in India and utilization of more potato produce towards value addition, varieties with specific attributes need to be developed with good yield potential (30-35t/ ha), high dry matter (21-23%), low reducing sugars (150mg/ 100g fresh weight), high concentration of anthocyanins, rich in minerals suitable for growing under early to medium crop conditions (75-90 days).

To keep pace with the future needs of growing processing industry in India, suitable cultivars need to be developed with the following desirable traits:

- Varieties with the processing quality demanded by the manufacturers of chips, French fries and flakes. They should be resistant to low temperature sweetening so that tubers can be stored at 2 to 4°C to control the development of diseases, reduce weight loss and sprouting in store, with reduced reliance on sprout inhibiting chemicals.

- Varieties for the table/ specialty potatoes segment for supermarkets. The tubers must be resistant to after-cooking-blackening should have attractive skin, good flavour and desirable, low levels of glycoalkaloids and varieties for baking, canning, scooping *etc.*

REFRENCES

Brown CR, Wrolstad R, Durst R, Yang CP and Clevidence B (2003) Breeding studies in potatoes containing high concentrations of anthocyanins. *American Journal of Potato Research* **80**:241-250

Chakaborty S, Chakaborty N and Datta A (2000) Increased nutritive value of transgenic potato by expressing a nonallergenic seed albumin gene from *Amaranthus hypochondriacus*. *Proceedings of the National Academy of Sciences of the United States of America* **97**:3724-3729

Dale MFB and Mackay GR (2007) Inheritance of table and processing quality. In, Potato Genetics. Bradshaw JE and Mackey GR (eds), *CAB International*: 285-315

Dunwell JM (2000) Transgenic approaches to crop improvement. *Journal of Experimental Botany* **51**:487- 496

Gaur PC, Pandey SK, Singh SV and Kumar Dinesh. 1999. Indian Potato Varieties for Processing. Central Potato Research Institute, Shimla, India. Technical Bulletin- 50: 25p

Gross J (1991) Pigments in vegetables: chlorophyll and carotenoids. Van Nostrand Reinhold, New York

Jansky SH and Thompson DM (1990) Expression of hollow heart in segregating tetraploid families. *American Potato Journal* **67**:695-703

Luthra SK, Pande PC and Singh BP (2004) Perspective planning for developing potatoes for export. In, Processing and Export Potentials of Indian Potatoes.. Khurana SM Paul, Singh BP, Luthra SK, Kumar NR, Kumar Devendra and Kumar Dinesh (eds), International Potato Conference and Fest-'04, CPRI Campus, Modipurarm, Meerut. Indian Potato Association, Shimla:18-27

Mazza G and Miniati E (1993) Anthocyanins in fruits, vegetables and grains. CRC Press, Boca Raton, FL.

Mosley AR and Chase RW (1993) Selecting varieties and obtaining healthy seed lots. In, Potato Health Management, APS Press.

Pandey SK, Singh SV, Gaur PC, Marwaha RS, Kumar D, Kumar P and Singh BP (2008) Chipsona varieties: A success story. Central Potato Research Institute, Shimla, India. Technical Bulletin- 89: 40p

Pandey SK, Marwaha RS, Kumar Dinesh and Singh SV (2009) Indian potato processing story: industrial limitations, challenges ahead and vision for the future. *Potato Journal* **36**(1- 2): 1-13

Römer S, Lübeck J, Kauder F, Steiger S, Adomat C and Sandman G (2002) Genetic engineering of a zeaxanthin-rich potato by antisense inactivation and co suppression of carotenoid epoxidation. *Metabolic Engineering* **4**:263-272

Scalbert A, Johnson IT and Saltmarsh M (2005) Polyphenols: antioxidants and beyond. *American Journal of Clinical Nutrition* **81**(1):215S–217S

6

Potato: Genome Sequencing and Applications

Virupaksh U Patil[1], Ayyanagouda Patil[2], Jagesh K Tiwari[1]
Vanishree G[1] and SK Chakrabarti[1]

[1]*ICAR-Central Potato Research Institute, Shimla - 171 001*
Himachal Pradesh, India
[2]*University of Agriculture Sciences, Raichur- 584 104, Karnataka, India*

INTRODUCTION

Potato (*Solanum tuberosum* L.) is the most important non-grain food crop of the world and is central to global food security, offering higher yields in calories per acre than any grain. More than 2 billion people worldwide are estimated to be dependent on potato for food, feed or income. Global potato breeding concentrates on the objectives of increasing the yields through resistance to various biotic and abiotic stresses along with improving quality traits but the cultivated potato is an auto-tetraploid (2n=4x=48), clonally propagated, highly heterozygous, and suffers acute inbreeding depression. All these characters complicate genetics/genomic studies as well as breeding efforts to improve important traits such as disease/pest resistance, processing quality and nutritive value. This has lead to the need of high quality, well annotated genome sequence of potato along with high resolution maps for aiding the identification of allelic variants for important agronomic traits in potato. Multiple linkage maps have been constructed for potato using various molecular markers like RFLPs, AFLPs and SSRs in an effort to develop markers for marker assisted breeding and facilitate map based cloning. These earlier efforts got unequivocal boost when an international collaboration of 13 countries *viz.* Argentina, Brazil, China, Chile, India, Ireland, The Netherlands, New Zealand, Peru, Poland, Russia, the United Kingdom and the United States came together to form the Potato Genome Sequencing Consortium (PGSC) with aim to decipher the potato genome (Xu *et al.*, 2011). The potato genome sequence has provided a major boost to gaining a better understanding of potato trait biology, underpinning future breeding efforts. This would greatly help in enhancing the breeding techniques by implementing the genome assisted methods for imparting the greater tolerance to various biotic and abiotic stresses and ultimately increasing the yields. The knowledge of the complete genome sequence is an invaluable resource for the identification of genes and variant/novel alleles of genes for every trait of interest to potato breeders. This is very much required for feeding the ever burgeoning global population.

SEQUENCING TECHNIQUES

The development of DNA sequencing strategies began with the works by Sanger *et al.* (1977) that was responsible for the introduction of first automated DNA sequencers led by Caltech and subsequently commercialized by many firms. These are now categorized as first-generation sequencing technologies. Despite their popularity as a 'gold standard' among the research community, they suffer certain limitations. With the advances made in the field of micro-fluidics, imaging and detection power, computational tools, unconventional sequencing technologies with increased throughput and lower sequencing cost are continuously emerging. The completion of the first human genome drafts in the year 2000 was just the start of the modern DNA sequencing era, which resulted in further invention, improved development towards new advanced strategies of high-throughput DNA sequencing, which were collectively called the 'High-Throughput Next Generation Sequencing' (HT-NGS) technologies. The first of these NGS technologies was pyrosequencing, which was followed by Illumina and SOLiD (Sequencing by Oligo Ligation and Detection). The horizons and expectations have broadened due to the technological advances in the field of genomics, especially the HT-NGS and its wide range of applications such as: chromatin immune-precipitation coupled to DNA microarray (ChIP-chip) or sequencing (ChIP-seq), RNA sequencing (RNA-seq), whole genome genotyping, *de novo* assembling and re-assembling of genome, genome-wide structural variation, mutation detection, single cell transcriptome sequencing and carrier screening, DNA library preparation, paired ends and genomic captures, and the sequencing of the mitochondrial and chloroplast genome. Besides the advances in sequencing techniques, the past decade will be remembered as the decade of the genome research when many NGS technologies have emerged and are still emerging with the aim of reduced sequencing cost and higher throughput. Since the publications of the first composite genomes of humans, many draft genomes from plants and animal species have been published (www.ensembl.org/info/about/species.html), including the potato genome. For the genome sequencing of potato, three main sequencing technologies were used, namely, Illumina, Pyrosequencing and the Sanger sequencing, details of the potato genome sequencing and its applications for crop improvement are discussed in the present chapter.

SEQUENCING THE POTATO GENOME

The knowledge of the genome sequence in potato facilitates advanced breeding targeting many important agronomic traits. To overcome the key issue of heterozygosity to generate a high-quality draft potato genome sequence, an unique homozygous form of potato called a doubled monoploid was derived using classical tissue culture techniques. The draft genome sequence from this genotype, *S. tuberosum* group Phureja DM1-3 516 R44 (DM), was used to integrate sequence data from a heterozygous diploid breeding line, *S. tuberosum* group Tuberosum RH89-039-16 (RH) (Xu *et al.*, 2011). Whole genome sequence of an important wild potato species *S. commersonii* has also been sequenced (Aversano *et al.*, 2015). The potato genome consists of 12 chromosomes and has a (haploid) length of approximately 840 million base pairs, making it a medium-sized plant genome. Initially 78,000 bacterial artificial chomosomes (BACs) clone libraries were built and fingerprinted to align them into 7000 physical map contigs. Approximately 30,000 BACs were anchored to the Ultra High Density genetic map of potato, composed of 10,000 unique AFLP markers. Fluorescent *in situ* hybridization experiments on selected BAC clones confirmed these anchor points. The seed clones provide the starting point for a BAC-by-BAC sequencing strategy and later it was complemented by whole genome shotgun approaches using both 454 GS FLX and Illumina GA2 instruments. The heterozygosity of RH limited the progress of physical mapping and complicated the assembly of the genome sequence; therefore, sequencing of DM was started simultaneously. In 2011, the findings of genome and transcriptome sequencing of DM and RH was published in Nature in 'Genome sequence and analysis of the tuber crop potato' (Xu *et al.*, 2011) (Figure 1). In the later sections we will discuss the strategies and major conclusions of this project and also applications of the genome sequence knowledge for potato improvement.

S. phureja (DM) Genome Sequencing and Assembly

The nuclear and organellar genomes of DM were sequenced using a whole-genome shotgun sequencing (WGS) approach. In total, 96.6 Gb of raw sequence data was generated from two next-generation sequencing (NGS) platforms, Illumina Genome Analyser and Roche Pyrosequencing, as well as the conventional Sanger sequencing technologies. The genome was assembled using SOAPdenovo, resulting in a final assembly of 727 Mb, of which 93.9% is non-gapped sequence. Analysis of the DM scaffolds indicated 62.2% and much of the unassembled genome is composed of repetitive sequences. Libraries from DM genomic DNA were constructed for Illumina Genome Analyser II (paired end 200 and 500 bp; mate pair 2, 5 and 10 kb insert size) and Roche 454 platforms (8 and 20 kb) for sequencing using standard protocols. A high-quality

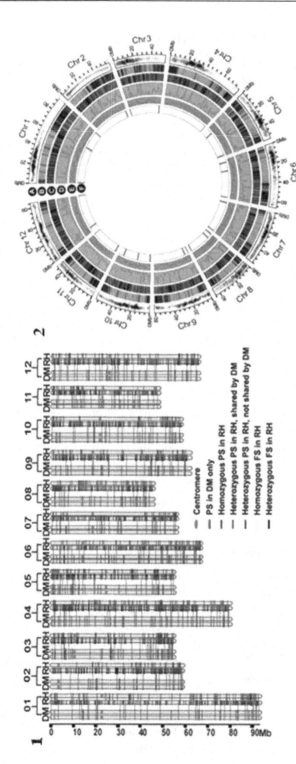

Fig. 1: The relative chromosome size and composition of DM (*S. phureja*) and RH (*S. tuberosum*). 2. A) Chromosome karyotype, B) Gene density (genes/Mb), C) Repeats coverage (%), D) Transcription state, E) GC content (%) and F) Subtelomeric repeats distribution in potato genome (Source: Xu *et al.*, 2011)

potato genome was constructed using the short read assembly software SOAPdenovo (Version 1.014). The 69.4 Gb data of GAII paired-end short reads were first assembled into contigs, which were sequence assemblies without gaps composed of overlapping reads. For the DM v3.0 assembly, 95.45% of 880 million usable reads were mapped back to the assembled genome by SOAP 2.20 using optimal parameters. Approximately 96% of the assembled sequences had more than 20-fold coverage. The overall GC content of the potato genome is about 34.8% with a positive correlation between GC content and sequencing depth. The DM potato possesses few heterozygous sites and 93.04% of the sites can be supported by at least 90% reads, suggesting high base quality and accuracy (http://solanaceae.plantbiology.msu.edu/).

S. tuberosum (RH) Genome Sequencing and Assembly

Whole-genome sequencing of genotype RH was performed on the Illumina GAII platform using a variety of fragment sizes and reads lengths, resulting in a total of 144 Gb of raw data. These data were filtered using a custom C program and assembled using SOAPdenovo 1.03. Additionally, four 20-kb mate-pair libraries were sequenced on a Roche/454 Titanium sequencer, amounting to 581 Mb of raw data. The resulting sequences were filtered for duplicates using custom Python scripts. The RH BACs were sequenced using a combination of Sanger and 454 sequencing at various levels of coverage. Consensus base calling errors in the BAC sequences were corrected using custom Python and C scripts. Sequence overlaps between BACs within the same physical tiling path were identified using megablast from BLAST 2.2.21 and merged with megamerger from the EMBOSS 6.1.0 package. Using the same pipeline, several kilobase-sized gaps were closed through alignment of a preliminary RH whole-genome assembly. The resulting non-redundant contigs were scaffolded by mapping the RH whole-genome Illumina and 454 mated sequences against these contigs using SOAPalign 2.20 and subsequently processing these mapping results with a custom Python script. The scaffolds were then ordered into superscaffolds based on the BAC order in the tiling paths of the FPC map.

The *ab intio* predicted a total of 39,031 protein coding genes and RNA Seq data revealed alternative slicing in at least 9,875 (25.3%) genes. As observed in other plants genomes, there was an inverse relationship between gene density and repeatative sequences. The orthologous clustering revealed at least 3,372 potato specific genes and 2,642 *asterid* specific genes. A total of 15,235 genes are predicted to be involved in tuberization process. Moreover, genes involved in starch biosynthesis are 3-8 fold highly expressed during the transition stage from stolon to tuber. Potato is susceptible to a wide range of pests and pathogens and identification of disease resistance genes very important to impart resistance in the crop. The DM potato contains 408 *R* genes of NBS-LRR domain, 57 Toll/interleukin 1 receptor/plant R (TIR) domain, and 351 non TIR types. The R genes of late blight like *R1, RB, R2, R3a, Rpi-blb2* and *Rpi-vnt1.1* were all present in the assembly (Xu *et al.*, 2011).

S. commersonii Genome Sequencing and Assembly

Potato has the largest gene pool in any crop with more than 200 tuber bearing wild species of *Solanum* native to south, central and north America. Although, these genetic resources provide great potential to breed improved cultivars, linkage drag typically limits the use of wild potato species since many exotic genes imparting undesirable traits can be co transferred with desirable genes. Very recently *S. commersonii*, a tuber bearing wild species possessing several resistance traits not found in cultivated potato including resistance to root knot nemetode, soft rot and black leg, bacterial and verticillium wilt, potato virus X, tobacco etch virus, common scab, late blight and freezing tolerance. Despite bearing such characters, this has not been exploited in breeding programs as both are sexually incomputable with *S. commersonii* having 1 emdosperm balance number (EBN) and *S. tuberosum* having 4 EBN. Using various strategies breeders have successfully overcome the sexual isolation barriers to introgress *S. commersonii* genetic material into cultivated potato. Despite such efforts very little progress has been made. To exploit the species to a greater extent for important traits, whole genome sequence of *S. commersonii* (accession PI243503) from Inter Regional Potato Introduction Station, Sturgeon Bay, Wisconsin was sequenced using shot gun sequencing method. Illumina HiSeq1000 sequencing platform was used and assembled using SOAPdenovo and gaps were closed using GapCloser v 1.12 (a SOAP suit tool). The gene space of the assembled genome was assessed by aligning CEGs to assebly using blast with a 65% identity threshold. Reads were aligned to the assembled genome using SOAP-aligner v 2.21 with standard parameters. Estimation of genome size of *S. commersonii* was done using flow cytometry analysis of propidium iodide stained nuclei of *S. commersonii* and analyzed. The cytological study for karyotype mapping of *S. commersonii* was done using flouroscence *in situ* hybridization using a telomeric DNA probe. The mitotic metaphase chromosomes were stained in blue by DAPI (4', 6-Diamidino 2 phenylindole). The telomeric probe, a $(TTTAGGG)_4$ oligonucleotide labelled at 5' end with carboxytetra methyl rhodamine (TAMRA) generated signals at the ends of each chromosome (Aversano *et al.*, 2015).

The cold tolerant *cmm1t* clone was used producing 145.93 Gb of sequence reads which yielded 88 Gb quality data after filtering covering roughly around 105X of the *S. commersonii* genome size of 830 Mb. All contigs were assembled into 64,665 scaffolds (>1Kb) of which 4,833 containing 50% of assembly were 44.3 Kb or longer (N50 = 44.3 Mb). Using interactive mapping and *S. tuberosum* as reference 12 pseudomolecules were created. The distribution of gap length varied from 1 to 8,369 bp. The GC content of the coding sequence was 34.5% and in total 243 (98%) core eukaryotic genes were found in *S. commersonii* genome. About 383 Mb of repetitive sequence were identified accounting for 44.5% of the assembly. The repetitive fraction is dominated by long terminal repeat - retrotransposon (LTR-RTs) with 34%. A total of 37,662 genes were predicted and nearly 20,500 *S. commersonii* genes were assigned to Gene Ontology (GO) terms and more than 4,900 proteins were annotated with a four digit EC number indicating more than 24% of the predicted proteome of *S. commersonii* has enzymatic function. A total of 22 tRNAs, 40 Rrnas, 18,882 long non coding RNAs (lncRNAs) and 1,703 miRNA precursors were predicted. Total 942 pathogen recognition genes (*R*) of which 286 coiled-coil nucleotide binding site (NBS)-leucine-rich repeats (LRR), 71 NBS, 143 Toll/interleukin1 receptor (TIR)-NBS and 37 TIR genes are found to be present in the *S. commersonii* (Aversano *et al.*, 2015).

S. tuberosum dihaploid 'C-13' Genome Sequencing and Assembly

An androgenic dihaploid potato 'C-13' (*S. tuberosum*) was sequenced using Illumina technology at the ICAR-CPRI, Shimla. Dihaploid C-13 was developed at the institute level by anther culture from the potato cv. Kufri Chipsona-2. Total of 344.794 million reads of very high quality were generated, which resulted in 80.68% reference assembly with the reference potato genome. Overall, C-13 genome assembly size with consensus sequence 809.59 Mb and 30,241 genes were identified, of which 15,538 genes were characterized by the gene ontology (GO) terms. Further, a total of 1,122,388 SNPs and 48,145 indels (insertion/deletions) were detected in C-13 genome in comparison to the reference potato. In both SNPs and indels, higher homozygous (65.78% and 90.16%) were observed than heterozygous (34.22% and 9.84%), respectively. Orthologs analysis with protein sequences revealed 33 core clusters in C-13, potato genome and wild species (*S. commersonii*). As expected, phylogeny analysis using protein sequences revealed very close association of C-13 with *S. tuberosum*. This preliminary investigation provides genomic resources like genes, SNPs, indels, 77,068 SSRs markers and candidate genes for application in potato breeding and genomics using dihaploid lines.

COMPARATIVE GENOMICS

Comparative genomics is a branch of genomics that compares the genomic features of different organisms. The genomic features may include the DNA sequence, genes, gene order and regulatory sequences. Sequence alignment of

71 Mb of euchromatic regions from the tomato reference genome to their counterparts in potato revealed 8.7 % nucleotide divergence with an average of one indel per 110 bp. Comparison with potato enabled the identification of over 160 orthologous gene pairs, each single copy tomato gene being represented in potato by up to four genes, reflecting the polyploid origins of *S. tuberosum* (Fray *et al.*, 2016). Comparative studies among the *Solanum* sps. have revealed that *S. commorsonii* genome shows lower level of heterozygosity than *S. tuberosum*. The wild potato contains less number of genes with 37,662 compared to the 39,031 predicted protein coding genes of cultivated. But the wild species has more predicted genes than tomato (34,727). A catalog of 942 non redundant pathogen recognition proteins were predicted from wild species whereas 1,406 gene were reported from tuberosum, though the genome size of both the species are almost similar with tuberosum being 844 Mb and commersonii being 830 Mb. A total of 9,894,571 reliable single nucleotide polymorphism (SNPs) were identified of which 12,412 SNP containing genes displayed expressional variation between DM and *S. commersonii*, whereas 3.67 million SNPs were identified between the DM and RH of cultivated potato. Comparative studies also concluded that *S. commersonii* has lower amount of repetitive DNA (44.5%) compared to cultivated with 55%. This may predict different genome dynamics in these two species since their separation from common ancestor. Roughly 17,300 (44%) and 16,821 (42%) *S. commersonii* genes showed one to one orthology with genes from *S. tuberosum* and *S. lycopersicum*, respectively but only 7058 (18%) with genes from more distinctly related astrid *Mimilus gullatus*. The average exon size of *S. commersonii* (239.4bp) is less than *S. tuberosum* (353.5bp) and *S. lycopersicum* (261.4bp). Contrastingly average intron size of *S. commersonii* (603bp) is greater than both of them. It has 17,297 genes orthologs in *S. tuberosum* and 16,821 in *S. lycopersicum*. The divergence of cultivated potato from tomato happened around 7.3 million years ago (PGSC, 2011) whereas, cultivated and wild potato found to have diverged/separated around 2.3 million years ago. Based on the studies between cultivated potato (DM and RH) indicate that the complement of homologous alleles in DM may be responsible for its reduced level of vigour. Even the expression of genes varies greatly between the species for example genes coding for enzymes involved hydrolytic and phosphorolytic starch degradation α Amylase (10 – 25 fold) and β Amylase (5 – 10 fold) show higher expression in DM tubers compared to the tubers from the RH. This indicates the more primitive nature of DM and higher yields of cultivated RH may have attained by selection for decreased activity of the hydrolytic starch degradation pathway.

APPLICATIONS OF POTATO GENOME SEQUENCE

Potatoes are an excellent, low-fat source of carbohydrates, with one-fourth the calories of bread. An average serving of potatoes with the skin on provides about 10 percent of the recommended daily intake of fiber (https://cipotato.org/crops/potato/). Some of the important constraints that affect potato production include biotic stresses like late blight, bacterial wilt, common scab, potato leafroll virus, potato virus Y, potato virus A, cyst nematode, colarado potato beetle and potato tuber moth; abiotic stresses like heat and drought stress (Patil *et al.*, 2012). The important processing quality traits that affect the market value include high starch, shallow eye, shape of tuber (round shape for chips and long for fries), reduced cold sweetening. Conventional breeding methods are of primary importance but are too slow (10–15 years) because they are essentially based on crossing and several generations of selection in field evaluation and phenotypic selection. Using marker assisted selection it is possible to screen traits that are extremely difficult expensive or time consuming to score or measure such as tolerance to drought, salt, mineral deficiency, root traits and resistance to specific races of pathogen. Conventional phenotypic selection involves destructive sampling of tubers for analysis of quality traits like Zn and Fe content which can be avoided by DNA markers.

The potato genome sequencing has been completed by potato genome consortium in the year 2011. The DM Whole Genome Shotgun project has been deposited at DDBJ/EMBL/GenBank under the accession AEWC00000000 and the *S. commersonii* under the accession number GCHT00000000. Genome sequence and annotation can be obtained and viewed at http://potatogenome.net. The availability of potato genome sequence assists in development of genome wide molecular markers, construction of high density genetic and physical maps, and comparative genome analysis for identification of homologous genes and evolutionary relations, gene discovery and reference sequence for sequencing, alignment and gene annotation of related crop species. The application of the genome editing like *CRISPR-CAS9* tools demands the complete genome sequence and its functional characterization.

Genomics and genotyping

The recent and tremendous reduction in costs associated with HT sequencing have enabled the development of genetic markers with a single nucleotide resolution that can be rapidly assayed on hundreds to thousands of individuals. These molecular markers can be used in applications such as marker-assisted breeding, quantitative trait loci (QTL) determination, genome-wide association analyses (GWAS), as well as, evolutionary and diversity studies. Genotyping arrays have been the most common tool for high-throughput SNP genotyping in the last decade. Arrays have been developed for multiple platforms (including Infinium and Axiom) and offer many advantages over low-throughput gel-based genotyping platforms: a relatively low cost per sample, automation and standardization that makes it easy to analyze and compare the results of many individual samples. For example, one of the most popular is the Infinium 8,303 Potato Array. As its name suggests, this array contains 8,303 SNP markers chosen to provide roughly even distribution across all 12 potato chromosomes. This platform has proven useful in a number of studies, including genetic mapping of important agricultural traits, retrospective analysis of potato breeding and taxonomic studies. A second recently developed SNP platform is the SolSTW array. It includes a total of 14,530 SNP markers, the majority of which were selected from a previous sequence based genotyping experiment. As opposed to the Infinium array that used the transcriptome of only six elite cultivars as the main source for markers, the majority of the markers in the SolSTW array are derived from a broad sequencing study that included 84 unique individuals and include chloroplastic DNA (Galvez et al., 2016).

As an alternative to genotyping arrays, several new sequencing based genotyping methods have emerged, leveraging high-throughput, short read sequencing to genotype hundreds of individuals simultaneously at thousands of genetic loci. The two most common methods are: genotyping-by-sequencing (GBS) and RAD-seq. Both techniques have become popular in the agricultural genomics and ecological genetics communities respectively. In each case, a small subset of the genome is sequenced at low coverage, providing a relatively cheap tool to identify molecular markers. This reduced representation of the genome is constructed by digesting the genome with restriction enzymes (GBS) or digestion in combination with physical shearing (RAD-seq). The reduced representation libraries from many individuals can be DNA barcoded, pooled, and then sequenced in the same experiment, greatly reducing the cost per sample. Post sequencing analyses can be performed using available software packages and tools, including TASSEL-GBS, UNEAK, Stacks, Haplotag and GBS-SNP-CROP. While there are many benefits to using GBS or RAD over genotyping arrays, including no requirement of a complete reference genome, no array ascertainment bias, and the ability to identify multiple types of genetic markers. The main obstacle is the sparse genotype matrix that is missing genotype calls, produced during the computational step that calls and filters SNPs and indels. This is due to the tradeoff of sequencing coverage and depth among multiplexed DNA samples. Despite this hurdle, GBS has been successfully implemented in genetic mapping studies of diploid crops such as maize, wheat and barley and polyploid crops such as alfalfa. In potato, there has been limited application of GBS for molecular marker development perhaps due to the highly heterozygous, tetraploid genome. A modified form of GBS has been used in potato and these markers were then successfully used in analyses to determine population structure, sequence diversity, chloroplast type and genetic association (Uitdewilligen *et al.*, 2013). The successful use of GBS in tetraploid potato cultivars opens the door to future studies exploring the wider diversity of commercial and non-commercial potato varieties. Similar studies in other *Solanaceae*, such as tomato, show the potential benefits of using this technique to explore wild species diversity. Additionally, it has been recently reported that GBS can be used to aid in the analysis of diploid potato mapping populations. Marker assisted selection (MAS) increases the efficiency of breeding. To date most MAS studies in potato have relied on low-throughput molecular markers, including amplified fragment length polymorphism (AFLP) and simple sequence repeats (SSRs) that have been associated with traits with relatively simple genetic basis such as disease resistance. Regardless of trait, many of these low-throughput, gel-based, markers in their current form are not suitable for large scale screening of progenies, which would be required for application in a breeding program. The successful use of high-throughput genotyping platforms, in potato, opens the door to exploring the wider diversity of potato genetic variation, and the practical application of MAS in breeding programs. Ultimately, the genome-wide marker information could be used to go beyond MAS at a few loci, to being able to predict the phenotype solely from marker genotypes at all marker loci using whole genome selection methods.

Genome re-sequencing and genetic diversity

Whole genome re-sequencing can reveal important differences between cultivated potato varieties and related wild species, especially at a large scale. Traditionally, the cost of resequencing entire populations of samples has been prohibitive, and thus, there has been a need for novel solutions to genotype large collections of potato germplasm. However, most cultivated potatoes are polyploid and highly heterozygous which could mean that the genomes of potato landraces and native cultivars might differ significantly from the potato reference genome. The complexity of the potato genome has made the genetic differences between these populations difficult to discern. Although different approaches have been employed to classify potato germplasm (morphological, molecular, cytometric), taxonomy remains challenging due to varying ploidy levels, sexual and asexual reproduction, the ease of interspecific hybridization, and introgressions from various wild species. Current research programs on genetic resources are working on sorting potato taxonomy and making modifications as needed.

Since the release of the potato reference genome, significant amounts of data have been collected using high-throughput sequencing and SNP arrays.

Identification of novel alleles and their potential utilization is a key factor to assist breeding programs in developing improved varieties in order to advance this important crop. High-throughput sequencing data was recently used to identify copy number variations (CNVs) within a panel of 12 potato monoploids containing diverse genetic backgrounds (Hardigan *et al.*, 2016). Using CNVnator, a program developed to detect CNVs by comparing sequencing read depth to a reference genome, the prevalence of CNVs in potato varieties was confirmed. A recent *de novo* assembly of a diploid wild potato species (*S. commersonii*) revealed significant differences in the distribution of SNPs, a lower degree of heterozygosity, fewer zones of repetitive DNA, and novel genes, when compared to the potato reference genome. This study, along with additional experiments performed in other *Solanaceae* crops such as tomato, highlight the potential benefits to further sequence, assemble and analyze close potato varieties and close relatives. An alternative way to overcome challenges associated with assembling new plant genomes is to take advantage of the reference genomes that are currently available as a way to reduce the need for more data. The results show that using a reference-guided approach effectively increases the coverage of the resulting assemblies enabling the discovery of previously unknown variants. If the purpose of a genome re-sequencing study is to identify all the non-redundant DNA sequences in a particular population, a novel approach has been developed that utilizes a metagenome-like assembly strategy. Briefly, the procedure consists of sequencing all the individuals of the population at a very low coverage, and then using this data in addition to a well annotated reference to identify unique sequences that are present in at least two of the individuals. The effectiveness of this technique has been shown in rice (*Oryza sativa*) where 1,483 accessions were sequenced enabling the assembly and mapping of most of the known agronomically important genes that were previously absent from the Nipponbare rice reference genome. In future studies where the detection of large-scale structural variants is not important, this approach can reduce the amount of sequencing data required while still enabling the discovery of novel sequences found in a sub-set of individuals in a population.

Regulatory motif and *R* gene utilization

The availability of a high-quality potato reference genome and transcriptome have, in turn, enabled the development of techniques that allow an accurate quantification of gene transcripts that will aid in the understanding of the complexities potato genetics. This includes the analysis of the *cis*-regulatory elements that are flanking genes, which are important because many polymorphisms associated with crop domestication are found in these regions. Studies performed in Arabidopsis and maize have shown that flanking regions can contain potential binding sites for elements regulating important phenotypic characteristics such as nitrogen (N) response and assimilation. Therefore, a greater understanding of gene regulatory mechanisms in potato will provide important information for breeding and genetic modification. The identification and characterization of regulatory elements has remained a challenge. Techniques such as ChIP-sequencing can reveal the binding sites of regulatory elements, such as transcription factors, by taking advantage of chromatin immuno-precipitation (ChIP) along with DNA sequencing. Regulation at the translational level also involves sequence motifs and proteins binding to them (RNA binding proteins). Gene regulatory regions also contribute to phenotypic variation and regulatory motif discovery will be important in understanding the impact of genetic variation in regulatory regions. There are a limited number of studies on potato gene regulation. Recent approaches have leveraged increasing amounts of sequencing data to characterize not just a promoter region, but also individual motifs and their binding regulatory elements to provide a better understanding of how gene regulation is carried out at a molecular scale. Several conserved occurrences of previously validated motifs were identified and they had associations with plant functions such as light and sucrose responsive transcriptional regulation, transcription enhancers, and response to abiotic stress. The promoters for the pathogenesis-related PR-*10a*, *chitinase C*, stolon-specific *Stgan, snakin-1, granule-bound starch synthase (GBSS1)*, and *chalcone isomerase* have also been characterized in a similar fashion. Three software packages that have been used successfully in plants are: Weeder, MEME and Seeder. To increase the probability of discovering and predicting regulatory elements, it is common to analyze a single dataset using different software, which compensates for the strengths and weaknesses of each algorithm. The aggregated results obtained from these tools can then be used to search curated motif databases. A recent study conducted using a *de novo* motif discovery approach was able to identify nine putative *cis*-regulatory motifs in the upstream flanking region of nitrogen responsive genes in three potato cultivars. However, future research on motifs and regulatory elements must also take into account the diversity of potato cultivars and varieties, which requires a deeper knowledge of the genetic differences between them.

Potato and its related species furnish a diversity of resistances to diseases and pests. Classical genetics studies have been conducted on inheritance and *R* genes have specific gene name designations. These *R* genes have been mapped to different chromosomal locations. For example, *R1, R3, R6* and *R7* from *S. demissum* providing resistance against late blight pathogen *Phytophthora infestans* have been mapped on to chromosome 11 and *R2* on the chromosome 4. Other *R* genes like R_{ber} from *S. berthaultii* (chromosome 10) and *Rblc/RB* from *S. bulbocastanum* (chromosome 8) have been widely used for introgression into the cultivated. Similarly, *Gro1.4* (chromosome 3), *Gro1.2* (chromosome 10) against *Globodera rostochiensis* and *Gpa4* (chromosome 4), *Gpa5* (chromosome 5) providing resistance against another root nematode *Globodera pallid* are also used in breeding programs. Many viral resistant genes like *Rx1* from *S. andigena, Rx2* from *S. acaule* and Nx_{phu} from *S. phureja* against Potato Virus X (PVX), Ry_{adg} and Na_{adg} from *S. andigena* against Potato Virus Y (PVY) and PVA respectively, Ry_{sto} from *S. stoloniferum* against PVY, *Ns* against PVS and N_0 against PLRV have been very well mapped and widely used in breeding. But till date no QTL mapping for quantitative virus resistance is reported in potato. The key class of genes that comprise the vast majority of plant *R* genes contain a nucleotide-binding site (NBS) and leucine-rich repeat (LLR) domain, and they could be used as a genetic marker set. Many of the *R* genes for different pests and diseases are independent. The systematics of the *R* gene structure was postulated using 438 NB-LRR-encoding sequences. Linking the existing 82 disease and pest resistance loci using 738 NB-LRR disease resistance gene homologues and further physical localization of 428 *R* genes has been achieved (Jupe *et al.* 2014). These studies have supported the implementation of marker-assisted selection (MAS) for potato breeding.

GENOMIC ASSISTED SELECTION

QTL mapping

The advent of NGS technologies enable researchers to generate high density SNP genotype data for mapping studies. Hackett *et al.* (2013) proposed use of SNP dosage data for QTL mapping in polyploid species. In polyploid species, SNP data usually contain a new type of information, the allele dosage, which is not used by current methodologies for linkage analysis and QTL mapping. They further extended existing methodology to use dosage data on SNPs in an auto-tetraploid mapping population. The SNP dosages are inferred from allele intensity ratios using normal mixture models. The steps of the linkage analysis (testing for distorted segregation, clustering SNPs, calculation of recombination fractions and LOD scores, ordering of SNPs and inference of parental phase) are extended to use the dosage information. For QTL analysis, the probability of each possible offspring genotype is inferred at a grid of locations along the chromosome from the ordered parental genotypes and phases and the offspring dosages. A normal mixture model is then used to relate trait values to the offspring genotypes and to identify the most likely locations for QTLs. Significant QTLs for tuber glucose concentration and tuber fry colour were detected on chromosomes 4, 5, 6, 10, and 11. Collectively, these QTLs explained between 24 and 46% of the total phenotypic variation for tuber glucose and fry colour, respectively. A tetraploid potato population was mapped for internal heat necrosis

(IHN) using the Infinium® 8,303 potato SNP array and QTL for IHN were identified on chromosomes 1, 5, 9 and 12 that explained 28.21% of the variation for incidence and 25.3% of the variation for severity.

Marker assisted selection (MAS)

MAS refers to the use of DNA markers that are tightly-linked to target loci as a substitute for or to assist phenotypic screening. An example of the successful development of marker-assisted breeding in tetraploid potato is described by Moloney *et al.* (2010), who selected for resistance to the potato cyst nematode *G. pallida*. Gebhardt *et al.* (2006) developed potato clones with multiple pathogen resistance traits by applying PCR-based markers to combine Ry_{adg} (resistance to PVY), *Gro1* (resistance to the nematode *G. rostochiensis*) and *Rx1* (resistance to potato virus X), or *Sen1* (resistance to potato wart, *Synchytrium endobioticum*). Amylose contents were significantly increased (28–59%) potato by introducing a recessive allele (gene marker: *IAm*) from wild potato (*S. sandemanii*) into cultivated potato through marker-assisted crossing. Identification of more markers using NGS technologies and their linkage to targeted traits will enlarge the process of MAS and marker assisted breeding.

Bulk segregant analysis by next generation sequencing technologies

Michelmore *et al.* (1991) developed a bulk segregant analysis a rapid method for identification of markers linked to specific gene or genomic region. The basis of this method is that all alleles must be present when DNA is bulked from a group of plants sharing the same phenotype. Consequently, two bulked pools of segregating individuals differing for a trait will differ only at the locus that harbours that trait. The biggest advantage of BSA over "regular" QTL analysis is that there is no necessity for genotyping and phenotyping each of hundreds of individuals in a segregating population. Instead, by grouping plants according to extreme levels of a particular trait and extracting DNA from these two bulks, the process of genotyping is reduced to only two DNA samples to be analysed. BSA is emerging as a method for genetic mapping that has a particularly good compatibility with genome re-sequencing. BSA is an approach for gene mapping in a single bulked sample from a single biparental cross, but it can also be used for three-way, four-way and multiparental crosses, including those developed with special designs such as diallele design. Genome re-sequencing of the two pools plus two parents is a cost-effective way of getting high density genotyping data. Sequence data are mapped to a reference sequence and base distribution across the genome is analyzed. Detection of trait-associated variants in pooled sequence data involves use of statistical analysis to compare observed base distributions in the pools with that predicted by parental base distributions. BSA was successfully used in potato to map steroidal glycoalkaloid content in tetraploids. In another study BSA was employed using NGS data on wholegenome sequencing of these bulks and the resistant parent to identify and validate the haplotype-specific SNPs associated to a dominant *Rpi-cap1* gene from *S. capsibaccatum*. The BSA combined with *de novo* assembly and haplotype-based variant calling pipelines can be applied for the identification of causal genomic locus and haplotype-specific allelic variations associated with trait specific bulk.

Genomic selection

The important objective of any crop improvement program is yield which is affected by a large number of genes with small individual effects. With the availability of potato genome sequence significant efforts have been applied to the development of genomic resources to improve potato breeding through identification of a large number of SNPs. The availability of these genome-wide SNPs is a prerequisite for implementing genomic selection for improvement of polygenic traits such as yield. The auto-tetraploid and heterozygous genetic nature of potato, the rate of decay of linkage disequilibrium, the number of required markers, the design of a reference population, and trait heritability affect this process. Slater *et al.* (2016) compared the expected genetic gain from genomic selection with the expected gain from phenotypic and pedigree selection and found that genetic gain can be substantially improved by using genomic selection.

Genomic prediction is routinely used to estimate breeding value in animal breeding programs. Recently the concept has been exploited for genome estimated breeding value in many plant species. Its usefulness for breeding of polypoid crop like potato has to be evaluated in detail. Stich and Inghelandt (2018) evaluated prospects of genomic prediction of key performance traits in a diversity panel of tetraploid potato modelling additive, dominance, and epistatic effects, the effects of size and make up of training set, number of test environments and molecular markers on prediction accuracy, and the effect of including markers from candidate genes on the prediction accuracy. With genomic best linear unbiased prediction (GBLUP), BayesA, BayesCδ, and Bayesian LASSO, four different prediction methods were used for genomic prediction of relative area under disease progress curve after a *Phytophthora infestans* infection, plant maturity, maturity corrected resistance, tuber starch content, tuber starch yield (TSY), and tuber yield (TY) of 184 tetraploid potato clones or subsets thereof genotyped with the SolCAP 8.3k SNP array. The cross-validated prediction accuracies with GBLUP and the three Bayesian approaches for the six evaluated traits ranged from about 0.5 to about 0.8. The evaluation of the relative performance of genomic prediction vs. phenotypic selection indicated that the former is superior, assuming cycle lengths and selection intensities that are possible to realize in commercial potato breeding programs. Elshire *et al.* (2011) evaluated the germplasm for late blight (n=1,763) and common scab (n=3,885) collected in seven and nine years, respectively with 4,110 SNPs and reported moderately high genomic heritability estimates for late blight and common scab. For late blight, small but statistically significant gains in prediction accuracy were achieved using a model that accounted for both additive and dominance effects. Using whole-genome regression models SNPs were located in previously reported hotspots regions for late blight, on genes associated with systemic disease resistance responses, and a new locus located in a WRKY transcription factor for common scab. In a study, Sverrisdottir *et al.* (2016) generated genomic prediction models for starch content and chipping quality in tetraploid potato to facilitate varietal development. Genotyping-by sequencing was used to genotype 762 offspring, derived from a population of biparental crosses of 18 tetraploid parents and 74 breeding clones representing a test panel for model validation. Genomic prediction models were generated from 1,71,859 SNPs to calculate genomic estimated breeding values. Cross-validated prediction correlations of 0.56 and 0.73 were obtained within the training population for starch content and chipping quality, respectively. Results suggest that genomic prediction is feasible; however, the extremely high allelic diversity of tetraploid potato necessitates large training populations to efficiently capture the genetic diversity of elite potato germplasm and enable accurate prediction across the entire spectrum of elite potatoes.

Genome Editing and Genetic Engineering

The application of genome editing tools like CRISPR/Cas9 requires prior functionally well characterized genome sequence. The availability of potato genome sequences allows for the application of genome editing to induce site specific mutations. The genome editing tools utilizes the specialized sequence specific nucleases that can insert, replace or remove nucleotides from genome with high degree of specificity. The relatively simplicity, ease and high specificity of CRISPR/Cas9 provide innovative tools in plant breeding to induce mutations at predetermined genomic loci. CRISPR/Cas9 systems targeting the potato granule bound starch synthase I (*GBSSI*) gene examined the frequency of mutant alleles in transgenic potato plants. The mutants that exhibited targeted mutagenesis in the *GBSSI* gene showed characteristics of low amylose starch in their tubers (Kusano *et al.*, 2018). Many of the potato species are self-incompatible. Wang *et al.* (2018) applied CRISPR/Cas9 system to knockout self-incompatible gene *S-RNase* in diploid potato species. Cytidine base editors (CBEs) are CRISPR/Cas9 derived tools recently developed to direct a C-to-T base conversion. Veillet *et al.* (2019) targeted the acetolactate synthase (ALS) gene in tomato and potato by a CBE and successfully and efficiently edited the targeted cytidine bases, leading to chlorosulfuron-resistant plant. Later they used transcription activator-like effector nucleases (TALENs) to knockout *VInv* within the commercial potato variety, Ranger Russet. Vacuolar invertase gene (*VInv*), which encodes a protein that breaks down sucrose to glucose and fructose under cold storage. CRISPR/Cas systems provide bacteria and archaea with molecular immunity against invading phages and foreign plasmids. The class 2 type VI CRISPR/Cas effector Cas13a is an RNA targeting CRISPR effector that provides protection against RNA phages. Transgenic potato lines expressing Cas13a/sgRNA (small guide RNA) constructs showed suppressed PVY accumulation and disease symptoms.

Important traits are easy to find in wild relatives of potato, but their introduction using traditional breeding can take 15-20 years. This is due to sexual incompatibility between some wild and cultivated species, a desire to remove undesirable wild species traits from adapted germplasm, and difficulty in identifying broadly applicable molecular markers. Fortunately, potato is amenable to propagation via tissue culture and it is relatively easy to introduce new traits using currently available biotech transformation techniques. For these reasons, potato is arguably the crop that can benefit most by modern biotechnology. Transgenic potato plants over expressing Arabidopsis cystathionine ã-synthase (AtCGS) in up-regulates a rate-limiting step of methionine biosynthesis and increases tuber methionine levels. Kamrani *et al.* (2016) transformed multiple gene-silencing vector pBIPhLSAR1-IR possessing a part of starch phosphorylase L and starch-associated R1 genes into potato cultivars Agria and Marfona to reduce the accumulation of reducing and non-reducing sugars in potato tubers stored at low temperature. Transgenic potato plants (*Désirée*) expressing a 300 bp hairpin loop nucleotide sequence targeting the potato vacuolar invertase gene (*VInv*), under the constitutive promoter were generated. Tubers collected from transgenic lines showed a significant reduction in reducing sugar content after 180 days of cold storage, without showing any measurable off-target effects on plant morphology and tuberization compared to non-transformed control plants. Transgenic plants encoding the *AC1* gene in three different orientations, viz. sense, antisense and hairpin loop conferred extreme resistance to the geminivirus causing apical leaf curl disease in potato. The better response of the potato for agrobacterium transformation and tissue culture has been exploited widely for developing transgenic potatoes with improved traits.

Computational tools for potato improvement

The advent of NGS technologies and endeavours such as 1000 sol genome sequencing has tremendously increased the volume and complexity of sequence data available. Computer databases are necessary tools for organizing the vast amounts of biological data currently available and making it easier for researchers to locate relevant information. Three database repositories, the European Molecular Biology Laboratory (EMBL), the DNA Databank of Japan (DDBJ) and GenBank (National Centre for Biological Information) jointly created the International Nucleotide Sequence Database Collaboration (INSDC) to collect and disseminate the burgeoning amount of nucleotide and amino acid sequence data that was becoming available. The sheer volume of the raw sequence data in these repositories has led to attempts to reorganize this information into various kinds of smaller, specialized databases. Such databases include various genome browsers, model organism databases, molecule- or process-specific databases, and others. With a uniform interface, users can navigate the whole genome using the same genomics coordinate system, and make comparative analysis across different lineages such as primates, mammalians, vertebrates and plants. There are several web-based genome browsers viz. Ensembl, NCBI Map viewer, Phytozome and Gramene platforms including SPUD database for potato (http://solanaceae.plantbiology.msu.edu/). These genome browsers integrate sequence and annotations for dozens of organisms and further promote cross-species comparative analysis. Most of them contain abundant annotations, covering gene model, transcript evidence, expression profiles, regulatory data, genomic conversation, etc. (Wang *et al.,* 2018). The Sol Genomics Network (SGN, http://solgenomics.net) is a web portal with genomic and phenotypic data, and analysis tools for the *Solanaceae* family and close relatives. SGN hosts whole genome data for an increasing number of *Solanaceae* family members including tomato, potato, pepper, eggplant and tobacco. The database also stores loci and phenotype data, which researchers can upload and edit with user-friendly web interfaces. Tools such as BLAST, GBrowse and JBrowse for browsing genomes, expression and map data viewers, a locus community annotation system and a QTL analysis tools are available. A new tool was recently implemented to improve Virus-Induced Gene Silencing (VIGS) constructs called the SGN VIGS tool. With the growing genomic and phenotypic data in the database, SGN is now advancing to develop new web-based breeding tools and implement the code and database structure for other species or clade-specific databases.

CONCLUSION

The elucidation of the reference potato genome, including the annotation of around 39,000 protein-coding genes, has opened up opportunities to rapidly identify candidate genes in regions associated with a trait of interest. The genome sequence also provides a catalogue of candidate resistance genes in the potato genome, radically enhancing our ability for rapid discovery and introgression of

R-genes in potato. Given the pivotal role of potato in world food production and security, the potato genome provides a new resource for use in breeding. Many traits of interest to plant breeders are quantitative in nature and the genome sequence will simplify both their characterization and deployment in cultivars. Whereas much genetic research is conducted at the diploid level in potato, almost all potato cultivars are tetraploid and most breeding is conducted in tetraploid material. Hence, the development of experimental and computational methods for routine and informative high-resolution genetic characterization of polyploids remains an important goal for the realization of many of the potential benefits of the potato genome sequence.

REFERENCES

Aversano R, Contaldi F, Ercolano MR, Iorizzo M, Grosso V, Tatino F, Xumerle L, Molin AD, Avanzato C, Ferrarini A, Delledonne M, Walter S, Cigliano RA, Gutierrez SC, Gabaldón T, Frusciante L, Bradeen JM and Carputo D (2015) The *Solanum commersonii* genome sequence provides insights into adaptation to stress conditions and genome evolution of wild potato relatives. *The Plant Cell* **27**: 954-968

Elshire RJ, Glaubitz JC, Sun Q, Poland JA, Kawamoto K, Buckler ES and Mitchell SE (2011) A robust, simple genotyping-by-sequencing (GBS) approach for high diversity species. *PLoS One* **6**: e19379

Frary A, Doganlar S and Frary A (2016) Synteny among *solanaceae* genomes: The Tomato Genome, Compendium of Plant Genomes, DOI 10.1007/978-3-662-53389-5_12

Galvez JH, Tai HH, Barkley NA, Gardner K, Ellis D and Stromvik MV (2016) Understanding potato with the help of genomics. *AIMS Agriculture and Food* **2**(1): 16-39.

Gebhardt CD, Bellin H, Henselewski W, Lehmann J, Schwarzfischer JPT and Valkonen (2006) Marker-assisted combination of major genes for pathogen resistance in potato. *Theoretical and Applied Genetics* **112**(8): 1458-1464

Hackett AC, McLean K and Bryan GJ (2013) Linkage analysis and QTL mapping using SNP dosage data in a tetraploid potato mapping population. *Plos One* **8**(5): e63939

Hardigan MA, Crisovan E, Hamilton JP, Kim J, Laimbeer P, Leisner CP, Manrique-Carpintero NC, Newton L, Pham GM, Vaillancourt B, Yang X, Zeng Z, Douches DS, Jiang J, Veilleux RE, Buell CR (2016) Genome reduction uncovers a large dispensable genome and adaptive role for copy number variation in asexually propagated Solanum tuberosum. *Plant Cell* **28**: 388-405.

Jupe AG, Witek FK, Andolfo G, Jupe F, Etherington GJ, Ercolano MR and Jones JD (2014) Defining the full tomato NB-LRR resistance gene repertoire using genomic and cDNA RenSeq. *BMC Plant Biology* **14**: 120

Kamrani M, Kohnehrouz BB and Gholizadeh A (2016) Effect of RNAi-mediated gene silencing of starch phosphorylase L and starch-associated R1 on cold induced sweetening in potato. *The Journal of Horticultural Science and Biotechnology* **91**(6): 625-633 DOI: 10.1080/14620316.2016.1208544

Kusano H, Ohnuma M, Mutsuro-Aoki H Asahi Takahiro, Ichinosawa Dai, Onodera Hitomi, Asano Kenji, Noda Takahiro, Horie Takaaki, Fukumoto Kou, Kihira Miho, Teramura Hiroshi, Yazaki Kazufumi, Umemoto Naoyuki, Muranaka Toshiya and Shimada Hiroaki (2018) Establishment of a modified CRISPR/Cas9 system with increased mutagenesis frequency using the translational enhancer dMac3 and multiple guide RNAs in potato. *Scientific Reports* **8**, 13753 (2018) doi:10.1038/s41598-018-32049-2

Michelmore RW, Paran and Kesseli RV (1991) Identification of markers linked to disease-resistance genes by bulked segregant analysis: A rapid method to detect markers in specific genomic regions by using segregating populations. *Proceedings of the National Academy of Sciences USA* **88**: 9828-9832.

Moloney C, Griffin D, Jones PW, Bryan GJ, McLean K, Bradshaw JE and Milbourne D (2010) Development of diagnostic markers for use in breeding potatoes resistant to *G. pallida* pathotype Pa2/3 using germplasm derived from *S. tuberosum ssp. andigena* CPC 2802. *Theoretical and Applied Genetics* **120**:679-689

Patil VU, Gopal J and Singh BP (2012) Improvement for bacterial wilt resistance in potato by Conventional and Biotechnological Approaches. *Agriculture Research* **1**(4): 299-316

Sanger F, Nicklen S and Coulson AR (1977) DNA sequencing with chain terminating inhibitors. *Proceedings of the National Academy of Sciences USA* **74**: 5463-5467

Slater AT, Cogan NOI, Forster JW, Hayes BJ and Daetwyler HD (2016) Improving genetic gain with genomic selection in autotetraploid potato. *The Plant Genome* **9**(3):1-15

Stich B and Inghelandt VD (2018) Prospects and potential uses of genomic prediction of key performance traits in tetraploid potato. *Frontiers in Plant Science* **9**:159

Sverrisdottir E, Byrne S, Sundmark EHR, Johnsen HO, Kirk HG, Asp T, Janss L and Nielsen KR (2016) Genomic prediction of starch content and chipping quality in tetraploid potato using genotyping by sequencing. *Theoretical and Applied Genetics* **130**: 2091-2108.

Uitdewilligen JG, Wolters AMA, D'hoop BB, Borm TJ, Visser RG and Van Eck HJ (2013) A next-generation sequencing method for genotyping-by-sequencing of highly heterozygous autotetraploid potato. *PLoS One* **8**: 10-14

Veillet F, Perrot L, Chauvin L, Kermarrec MP, Debast AG, Chauvin JE, Nogue F and Mazier M (2019) Transgene-Free genome editing in tomato and potato plants using Agrobacterium mediated delivery of a CRISPR/Cas9 Cytidine base Editor. *International Journal of Molecular Sciences* **20**(2): 402

Wang H, Park H, Liu J and Sternberg PW (2018) An efficient genome editing strategy to generate putative null mutants in *Caenorhabditis elegans* using CRISPR/Cas9. *G3: Genes, Genomes, Genetics***8**(11), 3607–3616 doi:10.1534/g3.118.200662

Xu X, Pan S, Cheng S, Zhang B, Mu D, Ni P, Zhang G, Yang S, Li R, Wang J, Orjeda G, Guzman F, Torres M, Lozano R, Ponce O, Martinez D, Cruz GD, Chakrabarti SK, Patil VU, Skryabin KG,Kuznetsov BB, Ravin NV, Kolganova TV, Beletsky AV, Mardanov AV, Di Genova A, Bolser DM, Martin DM, Li G, Yang Y, Kuang H, Hu Q, Xiong X, Bishop GJ, Sagredo B, Mejía N, Zagorski W, Gromadka R, Gawor J, Szczesny P, Huang S, Zhang Z, Liang C, He J, Li Y, He Y, Xu J, Zhang Y, Xie B, Du Y, Qu D, Bonierbale M, Ghislain M, Herrera Mdel R, Giuliano G, Pietrella M, Perrotta G, Facella P, O'Brien K, Feingold SE, Barreiro LE, Massa GA, Diambra L, Whitty BR, Vaillancourt B, Lin H, Massa AN, Geoffroy M, Lundback S, DellaPenna D, Buell CR, Sharma SK, Marshall DF, Waugh R, Bryan GJ, Destefanis M, Nagy I, Milbourne D, Thomson SJ, Fiers M, Jacobs JM, Nielsen KL, Sønderkær M, Iovene M, Torres GA, Jiang J, Veilleux RE, Bachem CW, de Boer J, Borm T, Kloosterman B, van Eck H, Datema E, Hekkert Bt, Goverse A, van Ham RC and Visser RG (2011) Genome sequence and analysis of the tuber crop potato. *Nature* **475**(7355): 180-195

7

True Potato Seed: Achievements and Opportunities

Jagesh Kumar Tiwari[1], Salej Sood[1], Vinay Bhardwaj[1], SK Luthra[2], Dalamu[3], Sundaresha S[1], Hemant B Kardile[1], VU Patil[1], Vanishree G[1], Vinod Kumar[1], Shambhu Kumar[4] and SK Chakrabarti[1]

[1]*ICAR-Central Potato Research Institute, Shimla-171 001*
Himachal Pradesh, India
[2]*ICAR-Central Potato Research Institute Regional Station*
Modipuram-250 110, Uttar Pradesh, India
[3]*ICAR-Central Potato Research Institute Regional Station Kufri*
Shimla-171 012, Himachal Pradesh, India
[4]*ICAR-Central Potato Research Institute Regional Station, Patna-801506*
Bihar, India

INTRODUCTION

Potato is a vegetatively propagated crop mainly through tubers. Seed is the costliest input, which constitutes nearly half of the production cost. Due to the high cost of seed, seed replacement rate is low, which results in build up of various tuber-borne diseases. To maintain high tuber yield, multiplication is done through "Seed Plot Technique" (SPT), which involves seed production in low aphid pressure period to minimize transmission of viruses and also adopt a set of cultural practices aimed at minimizing disease spread. The production of healthy seed tubers through SPT is expensive and the low rate of tuber multiplication (normally 6-8 times) provides only a limited quantity of quality tubers. Further, the low aphid areas suitable for producing healthy seed in the country lie in the northern plains and high hills (> 1800 m above mean sea level). Changing climate scenario is likely to shrink the window for potato cultivation and increase pathogen/vector attacks across the country thereby seriously limiting quantity and quality of seed in the country. An alternative technology of true potato seed (TPS) or use of botanical seed has shown great potential for producing both disease-free healthy (quality) and cheaper planting materials for commercial potato production. This chapter describes the classical TPS technology and two novel potential TPS-based technologies in potato propagation.

CLASSICAL TPS TECHNOLOGY

The concept of exploring the potential of TPS to raise commercial potato crop from TPS in India was first realized in the early 1950s by Dr. S. Ramanujam, the first Director of the ICAR-Central Potato Research Institute, Shimla. The advantages conceived by him were: i) Requirement of TPS in small quantity and diversion of tubers meant for seed towards human consumption, ii) Freedom of TPS crop from viral diseases common in seed tuber crops, and iii) Elimination of storage losses in seed tubers. However, due to high heterogeneity of crop morphology and late maturity in the potato variety *Phulwa*, used for evaluation of this technology did not give encouraging results (Gaur 1999; Upadhyay *et al.*, 1996). Moreover, selfing in this variety resulted in pollen sterility and low productivity. At present, TPS is being used by farmers in at least a dozen locations, including China, India, Nepal, Bangladesh, Vietnam, Peru, Nicaragua and Venezuela. However, TPS-related research activity has progressively declined (Almekinders *et al.*, 2009). Since the early period of TPS research at CIP, Dr. Mahesh Upadhyay led an active TPS breeding programme in India. Over time, a strong TPS programme has been established in the north-eastern hill (NEH) region especially in Tripura state.

The TPS is a botanical seed as a result of sexual reproduction, which develop as a small seed inside the fruit known as a berry. Depending upon genotype and environment, a single potato plant may have 50-100 berries. A berry contains 200-250 seeds and 1 g TPS may contain 1500-2000 seeds. In India, TPS is produced through bi-parental crosses. Hybrid TPS population perform better than open-pollinated families as regards yield. There are three hybrid TPS populations, *viz.*, TPS C3 (JT/C-107 × EX/A-680-16), HPS-1/13 (MF-1 × TPS-13), and 92-PT-27 (83-P-47 × D-150) which have been recommended for commercial potato cultivation in the country and their salient features are described below (Gaur 1999):

- TPS C3 (JT/C-107 × EX/A-680-16): The population is recommended for cultivation at the AICRP (All India Coordinated Research Project on Potato) workshop held in 1991. It is a medium maturing population with a yield range of 35-40 t/ha in 90-100 days crop raised from seedling tubers. It produces medium sized white-yellow, round-ovoid tubers with fleet to medium eyes. This population is resistant to both early and late blight and phoma leaf spot.

- HPS-1/13 (MF-1 × TPS-13): The population is also recommended for cultivation in 1991. It is a medium to late maturing population with a yield range of 35-40 t/ha in 90-100 days crop raised from seedling tubers. It produces medium sized yellow, ovoid-round tubers with fleet to medium eyes. This population is generally resistant to both early and late blight and phoma leaf spot.

- 92-PT-27 (83-P-47 × D-150): This population was recommended in 2001 and has shown 10 % higher yield potential than TPS C-13 and HPS-1/13. It is a medium maturing population with early and fast bulking habit. Tuberlets are used for growing normal crop in 90-100 days, however seedling crop matured in 110-120 days. It produces large sized white-cream, round-ovoid tubers with shallow to medium deep eyes. This population is vigorous, resistant to phoma leaf spot, and both early and late blight.

POTATO PRODUCTION THROUGH TPS

The cultivated potato is a complex tetraploid that exhibits tetrasomic inheritance. Potato cultivars are highly heterozygous genetically but show great uniformity owing to vegetative propagation. The segregating population developed from crossing between cultivars is a heterogenous collection of genotypes for various traits. Thus, for TPS production, the parents producing relatively homogenous off-springs (progenies) are desirable. The main objective of TPS breeding is to involve the parents with desirable traits in crossing, which would result in segregating population with an acceptable level of variation in plant morphology and tuber characters (Figure 1 and 2).

The hybrid TPS population produced through hybridization and is used for potato production. There are two methods of utilization of TPS for commercial potato production: a) Raising crops from transplanted seedlings, and b) Raising crops from seedling tubers. In the first method, a commercial crop can be raised in the first year itself while in the second method, the seedling tubers produced in the first year are used as planting material for raising a commercial crop in the second year. The yield is higher when the crop is raised from seedling tubers. It may be adopted in the northern Indo-Gangetic plains where winter is harsh, crop duration is short and healthy seed tubers can be produced through "Seed Plot Technique". However, in areas, where production and storage of healthy tubers is costly/difficult, temperatures are mild during crop season (15-25°C±5°C) and crop duration is long (more than 120 days) seedling transplant is successful.

Seedling transplanting

Preparation of nursery bed

Normally 150 g TPS and a nursery bed area of 75 sq. m. is required for raising seedlings for a transplanted crop in one hectare. Seedling beds may be temporary or permanent with brick lined boundary. The width of the bed is kept at one metre to facilitate manual cultural operations while the length may be kept as per requirement. Substrate is prepared by mixing soil free from pathogens and well decomposed FYM in 1:1 proportion and fertilizers are added @ 4-5g N, 6-8g P_2O_5 and 10g K_2O per square meter of nursery area.

Seedling raising

One-year-old TPS is better for seedling raising because fresh TPS has a dormancy period of about 4-6 months. The TPS should be treated with 0.01% Benlate/Bavistin or any systemic fungicide to avoid damping off disease at seedling stage. The nursery bed is watered a day before TPS sowing and prepared to fine tilth for better germination. Seeds are sown in rows 5 cm apart at a depth of 0.5 cm and covered with a thin layer of FYM before watering. Beds are kept moist by watering twice a day till germination. Shading of the nursery area may be done up to 15 days after sowing if day temperature is > 30°C. Foliar application of 0.1% urea (1 g urea in 1 lit water) at alternate days after two leaf stage is beneficial for quick growth and vigour of seedlings until transplanting at 4-5 leaf stage (achieved after 25-30 days of sowing). The best time for raising TPS nursery in the North-eastern hill regions is in the month of March (summer crop) and July-August (autumn crop).

Seedling transplanting

Fields are prepared as per local practices. Half dose of nitrogen and full doses of P and K are applied in rows. Remaining half of the nitrogen is given in two split doses at the time of earthing up of seedlings. Seedlings are transplanted on the 20 cm high ridges at 45-50 cm apart. Furrows are irrigated immediately after transplanting and thereafter at required intervals.

Harvesting and storage

Haulms are removed after 90 days of transplanting and tubers are harvested in another 15 days after dehaulming. Tuber size of 10-40 g are selected and used as seed. Tubers > 40g size may be disposed off for table purpose use. The seed size tubers are soaked in 3% boric acid solution for 30 min., dried in the shade and stored in the cold storage for planting in the next crop season.

Seedling tuberlets

The agronomic and cultural practices needed for commercial potato production from different sizes of seedling tubers are similar to the traditional method of propagation through seed tubers. Potato growers can raise the crop from smaller size seedling tubers with little modifications of the existing cultural practices.

Brick-bed method for production of seedling tuberlets

In the nursery bed method the seeds have to be planted at a very shallow depth. This necessitates continuous watering to ensure that the soil surface does not dry out. Therefore, to avoid such tedious manual watering, a brick-bed method has been designed. In this method the bricks absorb water continuously though capillary pores in it. A bed of medium quality bricks is prepared on polythene sheets and its sides are raised to enable ponding water around the brick bed. The bricks must be able to absorb water very quickly from the bottom and move through capillary action from the bottom to the top. On the top of brick bed a thin layer of fine soil and FYM mixture is spread out. The soil and FYM mixture must be fine and should have been moistened about ten days early so as to allow for weed germination which are then removed and then the soil and FYM mixture is dried and made into a fine tilth. This pre-germination of weeds would greatly reduce the problem of weeds on the TPS nursery beds. TPS may be broadcasted on the thin soil layer on top of the brick bed and then covered by another thin layer of pre-weeded soil and FYM mixture. Water is filled around the brick beds to allow the water to move to the top of the bed through capillaries. Pre-weeded fine soil and FYM mixture may be broadcast frequently i.e. almost at twice weekly intervals after seedling emergence.

Field preparation and planting

Field preparation, recommended doses of fertilizers (N, P and K) are applied as above and pre-planting irrigation of field is necessary for uniform and early growth. Well sprouted tubers are planted as per planting season with appropriate inter and intra row spacing. Small size tubers are planted at closer spacing while large ones at wide spacing to maintain almost similar stem density per unit area. Studies have shown that spacing of within row 20, 15, 10 and 5 cm for 30-35, 25-30, 20-25 and 15-20 mm size of seedling tubers, respectively is optimum while the between row spacing is maintained at 60 cm. The rests of the cultural practices are similar as adopted by the potato growers.

Seed rate and tuber production

For planting one hectare crop, 1.6 tonne of seedling tubers (15-35 mm size) are required as compared to 2.5-3.0 tonne of seed tubers (25-50 mm size). The reduction in planting material for a unit area reduces the storage and transportation costs significantly. Moreover, the produce obtained from TPS crop can be used for three successive clonal generations without any significant reduction in yield.

Fig. 1: Hybrid TPS production technology

Fig. 2: TPS utilization technology

ADOPTION OF TPS TECHNOLOGY

TPS technology is highly useful for areas where quality seed tubers are costly or not available or cannot be produced due to high vector pressure (especially aphids) and shortage of cold storages. Such areas include Karnataka, Maharashtra, Orissa, Madhya Pradesh, Bihar and the states of NEH region. This technology could further be expanded in the areas where good quality tubers can be produced but shortage of breeder/ foundation/ certified seed necessitates continued use of degenerated seed. It can provide a cheap source of planting material for the small and marginal farmers and can play a significant role in bringing down the cost of commercial potato cultivation. Though, congenial climatic condition and assured irrigation are necessary for growing potato crop from transplanted TPS seedlings, seedling tubers on the other hand can be used as seed material in all traditional potato growing areas in the country. If this technology picks up, the yields in low productivity regions like NEH will certainly increase sharply giving a boost to average national productivity.

In view of the problems and constraints like inadequate and high price of quality seed tubers and thereby high production cost faced by the small and marginal potato growers in NEH region, TPS is a viable solution to mitigate these problems. States like Meghalaya, Manipur, Nagaland and Arunachal Pradesh grow potatoes in two seasons. Hence, seedling tubers produced in one season could be stored easily under ambient conditions in diffused light and utilized for potato production in the following crop season. Tripura has taken a lead in the NEH region in potato production through TPS. State Department of Horticulture (Tripura) has already started production of hybrid TPS to the tune of about 300 kg /crop season. Over time, a strong TPS programme has been established by Tripura state with a remarkable diffusion and availability of TPS in seed stores, and currently, an estimated 1,500 ha is planted with TPS in Tripura State alone. Utilization of TPS as planting material in the region has lead to realization of average yield of about 17 tonnes/ha.

Advantages of TPS technology

TPS can be used as an alternative technology to reduce the pressure on seed tubers and also cost of potato production. TPS in place of traditional seed tubers offers several advantages, some of which are listed below:

- Quality planting material: TPS is absolutely diseases free planting material except potato spindle tuber viroid (PSTVd) and potato virus T (PVT), which are transmitted through TPS.

- Low cost of production: Cost of planting material from the TPS is merely 1/10th of the cost of seed tuber and hence is affordable to small and subsistence farmers.

- Easy storage and transportation: TPS is convenient and inexpensive to store for planting. It can be stored even for 8-10 years in a refrigerator (4-10°C) by maintaining seed moisture content between 3-5%.

- High profit: The cost of production of potato using TPS is approximately 55% less compared to the seed tuber but yield levels are generally at par.

- Fast multiplication: Multiplication of quality planting material through TPS is much faster than the traditional methods, since the different stages of multiplication as in that through seed tubers $viz.$, I^{st}, IInd, IIIrd etc. are not required.

- Flexibility of planting system: Potato grower has many options of planting systems such as direct seedling, nursery beds and brick-bed tuberlet methods.

Disadvantages of TPS technology

- *Long dormancy:* TPS has long dormancy periods varying from 6-12 months. Therefore, either 1 year old TPS or hormone (GA_3) treated fresh TPS should be used for crop raising.

- *Long crop maturity:* TPS crop takes 15-20 days more to mature compared to the crop raised from seed tubers.

- *Non-uniformity:* Crops grown from TPS are less uniform in emergence, plant type, tuber shape, size and colour and crop maturity compared to the tubers produced from a clone or variety.

- *Labour intensive:* TPS crop is labour intensive especially during the initial phases of seedling growth and establishment of transplanted crop.

DIPLOID HYBRID POTATO BREEDING

In conventional tetraploid potato breeding program, the four sets of 39,000 genes known to occur in potato genome are randomly reassorted in each generation. The deleterious alleles cannot be identified and removed nor all the desirable alleles could be fixed. Therefore, maximum heterozygosity was thought to be critical for high yield in autotetraploid potato (Jansky and Spooner, 2018). It is believed that increased heterozygosity can improve vigour and yield but genomics data contradicts this belief. The heterozygosity in modern day cultivars is no different to that of old cultivars, created 10 to 15 generations ago. Studies of potato landraces which range in ploidy also revealed that tuber yield is unrelated to ploidy and gene action and heterozygosity have a greater effect on plant vigour and productivity. Some diploids crosses and landraces show high yield potential. Moreover, the tetraploid cultivated potato has originated from diploid species, *S. stenotomum* and *S. sparsipilum*. Selfing in autoteraploid potato leads to high inbreeding depression due to expression of deleterious alleles. The attempts to produce inbred lines in autotetraploid potato were terminated in mid twentieth century due to low yields and poor fertility (Jansky et al., 2016). The breeders therefore thought of potato breeding at the diploid level rather than at tetraploid level. Most tuber-bearing *Solanum* species are diploid. These species are valuable to potato breeding as sources of resistance genes and of allelic diversity to improve quantitative characters. Most of these species can be crossed only with diploid forms of *S. tuberosum*. Hougas *et al.* (1958) emphasized two major advantages of this approach: (1) direct gene transfer from the wild and cultivated diploid *Solanum* species to *S. tuberosum*, and (2) disomic instead of tetrasomic inheritance of characters, a potential improvement of breeding efficiency.

The genomic data of potato diploid species showed that most of the deleterious mutations were line specific and there will be a high degree of heterosis when two inbred lines from different lineages are crossed. Although some large-effect deleterious recessive mutations exist in potato, they can be effectively purged by recombination (Zhang *et al.*, 2019). Many aspects need to be considered for transition of potato from tetraploid to a diploid crop, effect of the ploidy reduction on agronomic traits, cell size reduction and its affect on starch granule size, tuber sugar content, potato texture, and other quality characteristics. Comparison of diploids with their tetraploid counterparts showed that diploids grow faster and mature earlier which makes them better choice in comparison to tetraploids. Diploid breeding through TPS provides an exciting opportunity to revolutionize potato research and the potato industry. New genes can be stacked into well-established inbred lines and seed increase can occur in months rather than years. The TPS based breeding system allows the farmers to save their own seed if it is not hybrid seed. The main benefit of homozygosity and diploid breeding is fixing traits under recessive control, which otherwise is almost impossible in heterozygous tetraploid potato.

Most of the wild species of potato are diploids and are sexually compatible with cultivated potato at the diploid level (Spooner *et al.*, 2014). Hybrids between diploid wild and cultivated potato are often surprisingly well-adapted and agronomically acceptable. In diploid breeding efforts, it will be important to produce wild × wild, wild × cultivated, and cultivated × cultivated hybrids. It is impossible to predict the interactions of such genomes together that may never have been together before (Jansky and Spooner, 2018). There are two ways in which the potato can be made a diploid crop. First one is to reduce the ploidy of cultivated tetraploids through anther culture or wide hybridization method and second is to use the diploid cultivated and wild genepool for development of inbred lines. It has been observed long back that dihaploids in potato show severe inbreeding depression, poor fertility and poor agronomic traits. However, crosses between dihaploids and diploid wild species could result in trait improvement at diploid level.

The major hinderance of using diploid potato species is gametophytic self-incompatibility. However, the identification of a dominant self-incompatibility inhibitor (*Sli*) in the sexually compatible wild species *S. chacoense* opened new doors to explore inbred line breeding in potato (Hosaka and Hanneman, 1998). Later, self-compatibility was also found in some other *Solanum* species. Like allogamous crops, inbreeding depression is observed in diploid species upon selfing (Phumichai *et al.*, 2005; Lindhout *et al.*, 2011) but the challenge is to maintain the male fertility. Jansky *et al.* (2016) have suggested that male sterility may be a barrier in advanced generations of cultivated x wild species, therefore backcross breeding to introgress small chromosomal regions from wild species into a cultivated background may be a more effective strategy. Androgeneisis could be a better option for the development of inbred lines but its application is limited by its responsiveness in some genotypes only and also instant homozygosity may reduce the survival of plants.

The efforts are needed to transfer self-compatibility gene in diploid species to develop inbred lines. The next step is evaluation of inbred lines for tuber traits and quality traits and combining ability studies to exploit heterosis in potato. This approach has been demonstrated by Lindhout *et al.* (2011), where they developed diploid inbred lines with acceptable tuber traits and intercrossed F_3 uniform plants to develop hybrids (Figure 3). They have selected good inbred lines and developed hybrids with high tuber yield potential. Inbred improvement is a continuous process through accumulation of desirable alleles in them particularly for disease resistance and quality. This pioneer work set the stage for potato breeding through TPS at diploid level. Later in USA, work on development of inbred lines at diploid level resulted in the development of M6 line, which is highly vigorous, fertile and homozygous for a dominant self-incompatibility inhibitor. In addition, efforts have been made to develop recombinant inbred lines using six diploid populations at University of Wisconsin, Madison, USA (Jansky *et al.*, 2016). One of the inter-specific population developed from DM 1-3 and M6 have been used for mapping studies. Recombinant inbred lines are important resource for gene discovery and mapping. Once the inbred lines are available with different backgrounds, it will be easy to introgress desired genes. The hybrids made from inbred lines will result in heterozygous homogenous hybrid TPS for cultivation like single cross hybrids in maize.

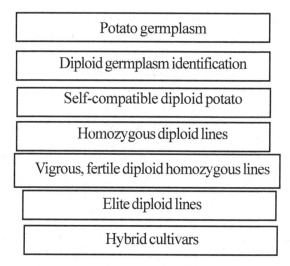

Fig. 3: Diploid hybrid potato breeding scheme

VARIETY SPECIFIC TPS THROUGH GENOME EDITING

An alternate way to use TPS as propagating material in potato is through knocking down the function of meiotic genes through genome editing approach. In this approach, the heterozygosity can be fixed at the tetraploid level, instead of developing homozygous diploid inbred lines. This is a genome engineering approach, where the recombination and independent assortment is blocked to get parental type gametes during meiosis. The meiotic cell division in the reproductive tissues is converted into mitotic cell division. The gametes thus produced are unreduced gametes i.e. they have the same ploidy (2n) as the somatic mother cells instead of gametic chromosome number (n). This is clonal propagation of heterozygous genotypes through botanical seeds.

Apomixis is an asexual reproductive process through seeds that bypasses meiosis and fertilization, to produce offspring genetically identical to the mother (Spillane *et al.,* 2004). Apomixis is common in the *Poacae* family, but is absent from the most important crops like potato. In crops like potato, it is possible to create apomixis *de novo* by targeted modification of the sexual reproduction mode, shown in *Arabidopsis* (Marimuthu *et al.,* 2011). In apomixis, the first component is apomeiosis, which is the conversion of the meiotic division into a mitotic-like division, leading to the formation of functional, diploid clonal gametes. Several meiotic mutants have been identified in *Arabidopsis*, maize or rice that produce apomeiosis-derived gametes, but most of these mutations lead to almost complete male and female sterility. However, *Arabidopsis MiMe (Mitosis instead of*

Meiosis) genotype, in which meiosis is turned into a mitotic-like division is associated with both high fertility and production of clonal diploid gametes at a very high frequency (D'Erfurth *et al.,* 2009). *MiMe* is the combination of mutations in three genes (*SPO11-1*, *REC8* and *OSD1*), each mutation impairing one of the three main processes that distinguish meiosis from mitosis. These three mutations affect, respectively, the three pillars of meiosis, homologous recombination, monopolar orientation of sister chromatids and the occurrence of a second division. Initially, the mutation in *spo11-1* abolishes meiotic recombination, mutated *REC8* causes the separation of sister chromatids at first meiotic division, instead of the distribution of homologous chromosomes and *osd1* causes the skipping of the second meiotic division (D'Erfurth *et al.,* 2009). The *MiMe* meiosis thus occurs without recombination and distributes sister chromatids in a single round of division, mimicking a mitotic division (D'Erfurth *et al.,* 2009). The process lead to production of clonal male and female gametes with doubled ploidy and in each generation of selfing the ploidy gets doubled. Crossing a *MiMe* plant as male or female with a line whose genome is eliminated following fertilization (lines expressing modified CENH3) leads to the production of clonal offspring (Marimuthu *et al.,* 2011, Ravi and Chan, 2010).

Although *MiMe* technology has been proven to be very efficient in the model species *Arabidopsis*, a major question has been about its transferability to other crop species. *SPO11-1* and *REC8* are widely conserved among eukaryotes, both in sequence and in function, making the identification of their homologs in other plant species easily feasible. Other genes known to be essential for recombination initiation are also well conserved among plants and all these meiosis genes are typically represented by a single copy. However, *OSD1* appears to be a plant-specific gene and exhibits a more complex phylogeny.

MiMe plants produce 2n gametes that are genetically identical to their parents. However, fertilization still occurs normally, leading to the doubling of ploidy in the next generation. The next challenge is to trigger the development of a *MiMe* gamete into an embryo and then a plant without the contribution of another gamete. One approach is to use genome elimination, where the chromosome set from one parent is removed after fertilization in the zygote. *MiMe* combined with genome elimination indeed leads to normal ploidy instead of ploidy doubling in each sexual cycle (Marimuthu *et al.,* 2011). However, the CENH3 manipulation used in *Arabidopsis* to induce genome elimination has not been transferred in other species to date, except in maize but with a lower frequency (Kelliher *et al.,* 2016).

The creation of *MiMe* triple mutant is time-consuming and the approach was earlier restricted to genotypes/species with large associated mutant resources.

However, recent breakthrough in genome editing by CRISPR/Cas9 may remove many obstacles, making apomeiotic conversion a feasible reality in crop plants. Programmable sequence-specific endo-nucleases that facilitate precise editing of endogenous genomic loci are now enabling systematic interrogation of genetic elements and causal genetic variations in a broad range of species, including those that have not previously been genetically tracta-ble. A number of genome editing technologies have emerged in recent years, including zinc-finger nucleases (ZFNs), transcription activator–like effector nucleases (TALENs) and the RNA-guided clustered regularly interspaced short palindromic repeats (CRISPR) - Cas nuclease system (Cho *et al.,* 2013).

Cas9 is a nuclease guided by small RNAs through Watson-Crick base pairing with target DNA, representing a system that is markedly easier to design, highly specific, efficient and well-suited for high-throughput and multiplexed gene editing for a variety of cell types and organisms. Similarly, to ZFNs and TALENs, Cas9 promotes genome editing by stimulating a double stranded breaks (DSB) at a target genomic locus (Urnov *et al.,* 2010). Upon cleavage by Cas9, the target locus typically undergoes one of two major pathways for DNA damage repair the error-prone non-homologous end joining (NHEJ) or the homology directed repair (HDR) pathway, both of which can be used to achieve a desired editing outcome. In the absence of a repair template, DSBs are re-ligated through the NHEJ process, which leaves scars in the form of insertion/deletion (indel) mutations. NHEJ can be harnessed to mediate gene knockouts, as indels occurring within a coding exon can lead to frameshift mutations and prema-ture stop codons.

Targeted nucleases are powerful tools for mediating genome alteration with high precision. The RNA-guided Cas9 nuclease from the microbial CRISPR adaptive immune system can be used to facilitate efficient genome engineering targeting meiotic cell cycle genes in potato to produce apo-meiotic genotypes and CENH3 variant for haploid induction. Once both the mutants are ready the hybridization can lead to sexual production of clonal seeds (botanical seeds) in potato.

CONCLUSIONS

TPS technology is scientifically sound, technically feasible, economically viable and eco-friendly. TPS provides an opportunity even to small farmers to generate high quality planting material and assures high yields with low inputs as compared to the tuber seed of unknown health status. Although there are some bottlenecks that hinder adoption of TPS among the farmers, those could be overcome by further improvement in parental lines by the introgression of desired traits from diverse genetic resources and adjusting the planting dates as well as agronomic practices according to the local conditions. The TPS systems have encountered major problems such as poor germination rates, non-uniform tubers, long dormancy and increased irrigation needs. Finally, TPS technology remains to be seen as a ready-to-use technology with the economic comparisons of TPS vs. seed tubers in the dynamics of cropping system for potato production. In the era of modern genomics tools, production of clonal seeds through diploid breeding and genome editing tools holds a great promise for growing potato from botanical seeds instead of the tubers. The clonal seed production technology could make TPS a more attractive option for potato breeders and growers and such an approach would assist the expansion of potato acreage in the tropics.

REFERENCES

Almekinders CJM, Chujoy E and Thiele G (2009) The use of true potato seed as pro-poor technology: the efforts of an international agricultural research institute to innovating potato production. *Potato Research* **52**: 275-293

Cho SW, Kim S, Kim JM and Kim JS (2013) Targeted genome engineering in human cells with the Cas9 RNA-guided endonuclease. *Nature Biotechnology* **31**: 230-232

D'Erfurth I, Jolivet S, Froger N, Catrice O, Novatchkova M and Mercier R (2009) Turning meiosis into mitosis. *PLoS Biology* **7**: e1000124

Gaur PC (1999) Manual for true potato seed (TPS) production and utilization. Central Potato Research Institute, Shimla, India. Extension Bulletin- 30

Hosaka K and Hanneman RE (1998) Genetics of self-compatibility in a self- incompatible wild diploid potato species *Solanum chacoense*. I. detection of an S locus inhibitor (Sli) gene. *Euphytica* **99**: 191-197

Hougas RW, Peloquin SJ and Ross RW (1958) Haploids of the common potato. *Journal of Heredity* **49**: 103-106

Jansky SH and Spooner DM (2018) The evolution of potato breeding. *Plant Breeding Reviews* **41**: 169-214

Jansky SH, Charkowski AO, Douches DS, Gusmini G, Richael C, Bethke PC, Spooner DM, Novy RG, Jong HD, Jong WSD, Bamberg JB, Thompson AL, Bizimungu B, Holm DG, Brown CR, Haynes KG, Sathuvalli VR, Veilleux RE, Miller JC, Bradeen JM and Jiang JM (2016) Reinventing Potato as a Diploid Inbred Line–Based Crop. *Crop Science* **56**: 1-11

Kelliher T, Starr D, Wang W, Jamie McCuiston, Heng Zhong, Michael L. Nuccio, and Barry Martin (2016) Maternal haploids are preferentially induced by CENH3-tailswap transgenic complementation in maize. *Frontiers in Plant Science* **7**:1-11

Lindhout P, Meijer D, Schotte T, Hutten RCB, Visser RGF and VanEck HJ (2011) Towards F1 hybrid seed potato breeding. *Potato Research* **54**: 301-312

Marimuthu MP, Jolivet S, Ravi M, Pereira L, Davda JN, Cromer L, Wang L, Nogué F, Chan SWL, Siddiqi I and Mercier R (2011) Synthetic clonal reproduction through seeds. *Science* **331**(6019): 876

Phumichai C, Mori M, Kobayashi A, Kamijima O and Hosaka K (2005) Toward the development of highly homozygous diploid potato lines using the self- compatibility controlling Sli gene. *Genome* **48**: 977-984

Ravi M and Chan SWL (2010) Haploid plants produced by centromere-mediated genome elimination. *Nature* **464**: 615-618

Spillane C, Curtis MD and Grossniklaus U (2004) Apomixis technology development-virgin births in farmers' fields? *Nature Biotechnology* **22**: 687-691

Spooner DM, Ghislain M, Simon R, Jansky SH and Gavrilenko T (2014) Systematics, diversity, genetics and evolution of wild and cultivated potatoes. *The Botanical Review* **80**: 283-383

Upadhyay MD, Hardy B, Gaur PC and Ilangantilenke S (1996) Production and utilization of true potato seed in Asia. CIP, Lima, Peru and ICAR, New Delhi, India.231p

Urnov FD, Rebar EJ, Holmes MC, Zhang HS and Gregory PD (2010) Genome editing with engineered zinc-finger nucleases. *Nature Reviews Genetics* **11**: 636-646

Zhang C, Wang P, Tang D, Yang Z, Lu F, Qi J, Tawari NR, Shang Y, Li C, and Huang S (2019) The genetic basis of inbreeding depression in potato. *Nature Genetics* **51**(3): 374-378

8

Potato Physiology for Crop Improvement

Bandana[1], Brajesh Singh[2], Devendra Kumar[1], Som Dutt[2]
Milan K Lal[2], Sushil S Changan[2] and N Sailo[3]

[1]*ICAR- Central Potato Research Institute Regional Station*
Modipuram-250 110, U.P., India
[2]*ICAR- Central Potato Research Institute, Shimla-171 001*
Himachal Pradesh, India
[3]*ICAR- Central Potato Research Institute Regional Station, Shillong*
Meghalaya, India

INTRODUCTION

Potato is a highly versatile crop grown in varying environmental conditions *viz.* plains, hills and plateau in India. It is grown under short days (SD) in the plains in autumn season and under long days (LD) during summers in hills. Therefore, the pattern of growth, duration of crop as well as the yield varies according to the environment. Besides, the potato cultivars differ from each other in relation to their morphological characters, response to the environmental factors and finally in yield and the dry matter accumulation. The biochemical reasons for such variations are not clear, but it seems that genetically determined differences in the synthesis of growth regulators such as auxins, cytokinins, gibberellic acid (GA) and abscissic acid (ABA) play an important role and their proportion at different growth stages affect the growth and development and thus, the yield of the cultivar (Minhas and Singh, 2003). The present chapter deals with the physiological factors, which determine the growth, development and yield of potato crop.

GROWTH OF POTATO PLANT

Potato is normally cultivated as a vegetatively propagated crop and the basic material used for multiplication, is its tuber which is a modified meristem. It can also be grown through true seeds (TPS) obtained from its fruits. The potato tubers have several eyes (growing points) and it can be planted as whole tuber or as cut-seed pieces, with every piece having one or more eyes. At optimum physiological age, these eyes develop into hairy structures called sprouts, which have apical meristem and are capable of growing into potato plant. The plant with its root and shoot (stem and leaves) system grows during the crop season and produces stolons which grow underground and finally swell at their terminal ends in the form of tuber. A major portion of the photosynthate is translocated to these tubers at the time of bulking which is stored in tubers in the form of starch and other biochemical constituents. Under appropriate conditions (mainly long days), potato plants produce flowers, fruits and seeds (true potato seed or TPS) as well. After full maturity the tubers are harvested and they may either be used as food or stored as seed. If used as seed, these are stored till the next crop season, the stored food material in tuber helps to maintain the vital life processes (respiration and germination). Through these phases the life cycle of a potato plant is completed. The detailed physiology of the different phenological events is described below.

Sprout

Freshly harvested potato tubers do not sprout even under suitable environments because they are dormant or in rest period. Every potato cultivar has specific dormancy duration and in Indian cultivars this duration varies from 6 to 8 weeks. The physiological age and proportion of endogenous growth regulators, which is further influenced by the environmental interactions, determine the dormancy duration. As soon as the tubers attain proper physiological age, the dormancy is terminated and the buds on the tubers develop into sprouts (Figure 1a). The temperature and humidity at which the tubers are stored play a key role in sprouting behaviour. Temperatures ranging from 18-20°C, high humidity and darkness are appropriate conditions for sprout growth. Each sprout consists of three parts *viz.* base, middle portion and apical tip. The roots arise from the base of the sprouts and at the time of tuber initiation the stolons are also formed from the base. The middle portion of the sprout gives rise to stem and the apical part contains the meristematic tissue, which causes growth and differentiation in potato plant. Morphologically, the pigmentation of sprout may vary from white to green, pink, red, violet or purple and is a characteristic of the cultivar. This pigmentation is caused by anthocyanin and if the sprouts are exposed to light they also develop chlorophyll pigmentation. The sprouting behaviour of Indian cultivars has been studied in great detail. The temperature, humidity and gaseous environment under which the sprout growth takes place and the climatic conditions in which the crop is planted have a bearing on the sprouting behaviour. Very low (5°C or below) and very high (30°C and above) temperatures are known to reduce sprout growth. Similarly, low levels of humidity (30% or below) reduce the sprout growth. However, in crease in the CO_2 concentration up to

15% has been found to reduce the dormancy duration and enhance the sprout growth particularly at 20-25°C temperature (Singh and Ezekiel, 2001). A potato cultivar when grown under short days enhances sprout growth early as compared to that when grown under long days. Thus, the sprout growth pattern besides being a cultivar characteristic is dependent on environmental factors as well, which determine the potato plant stand at later stages.

Root

The potato root system is adventitious and has been classified into basal roots (arising from stem), stolon roots (from stolons), junction roots (from the joint of stem and stolon) and tuber roots (from the tuber bud base). As far as water and nutrient absorption is concerned the basal and junction roots are more important. Addition of nitrogen up to 60 ppm has been known to stimulate root growth whereas, higher doses retards it. Root growth is stimulated at 75% soil moisture, whereas, the levels below this retard it (Burton, 1989). It has generally been found that the early maturing cultivars have shallow root system whereas the late maturing cultivars have deeper roots. For the better productivity of the potato crop it is desirable that the plants have good root system, higher root biomass and volume, *etc.* for proper absorption of water and nutrients from the soil.

Shoot

Above ground green haulms act as the source of photosynthate production and stem and leaf are its components. Large size tubers produce more stems than small size tubers, though the cultivar differences exist and environmental interactions also affect it. The potato crop when grown in hills produces taller stems in comparison to when grown in plains. The number of stems appearing above the ground is increased by increasing the concentration of mineral ions in the soil. These stems arise from the first and higher order buds differentiated during the growth of the mother tuber. The optimum temperatures for stem elongation and branch production ranges between 25-30°C and haulm growth is hastened by high amounts of radiation. On an undeveloped bud, about 12-15 leaves are present which later on develop into fully grown leaves, however, a stem could have up to about 40 leaves on one axis. The axillary branches also arise from the stem nodes like leaves. The number of leaves per plant is more in hills under long days than under short days. The optimum temperatures for leaf expansion are about 25°C. Nitrogen fertilization increases the leaf area and leaf area duration (LAD) and delay the leaf senescence. The leaves are sites for photosynthetic activity and more number of photosynthetically active leaves are desirable for better productivity of the crop (Harris, 1992). The leaf area index (LAI) of 2.5-3.0 is optimum for high bulking rate of tubers. It is the early stage of plant growth when LAI plays an important role in increasing the photosynthetic efficiency and if plant achieves higher LAI at initial stage and also maintains it for the longer duration, the final yield is influenced positively (Khurana and McLaren, 1982). However, increasing LAI beyond 4.0 is not beneficial since it results in increasing the area of unproductive shaded leaves (Singh *et al.*, 2001). The leaf angle is also important for light interception; fully grown potato plants intercept as much as 95% of the incident radiation mostly by the top half of the plant (Singh *et al.*, 2000).

Stolon

The stolons are modified lateral shoots that arise from the nodes at the base of the shoots below the soil. They have elongated internodes, small-scale leaves, hooked tips and are devoid of chlorophyll (Figure 1b). Tubers develop from the sub-apical region of stolons by accumulation of starch. Hence there are two distinct processes, stolon formation and tuberization. The stolon formation is adversely affected by low levels of nutrients, whereas, moist, dark conditions and GA favour stolonization. The temperature on the other hand has great influence on the stolon formation and the night temperatures of less than 17°C are desirable for stolonization.

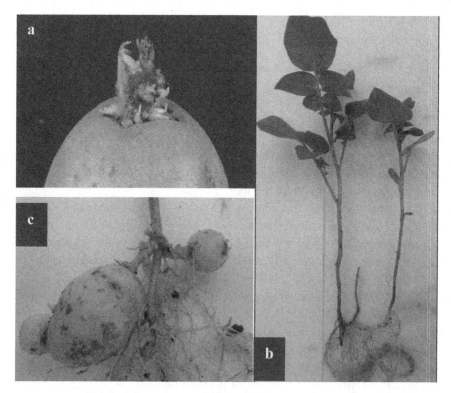

Fig. 1: a) Potato sprout; b) Potato plant; c) Tuber formation

Tuber

Potato tuber is a modified stem possessing leaves and axillary buds. The end of tuber where it joins stolon is called as the heel end or stem end while the other end is called as the rose or bud end. The eyes present on tubers are the nodes of modified stem and contain axillary buds and scale leaves. Potato tubers originate from stolons and its initiation is referred as tuberization when the stolon starts expanding radially due to cell division and cell enlargement (Figure 1c). The initiation of young tubers at the sub-apical region of the stolons usually occurs 5-7 week after planting under short days when the plants are 15-20cm high, whereas, under long days, tuberization is delayed by about 15-25 days. The potato tubers contain all the characteristics of the stems including dormant buds (eyes) with detectable scale leaves (eye brows). During the bulking process, about 75-80% of the total dry matter produced in leaves is translocated to the growing tubers. The number of tubers per hill is a varietal characteristic. Tuberization is influenced by several environmental factors like moisture, temperature, day length and nutrient availability. The moisture stress adversely affects the growth of tubers and decreases the yield. The night temperature of less than 17°C has been found optimum for balanced growth of the tubers and temperatures beyond it adversely affect the process of tuberization (Ewing, 1981). Short days favour tuberization whereas, long days delay the process. High levels of nitrogen encourage extensive haulm growth and delay tuber initiation.

FACTORS ASSOCIATED WITH GROWTH AND YIELD

It is indeed difficult to provide optimum conditions for the growth of potato crop in the field, though these may be available for some period of the crop growth. Therefore, full potential of potato cultivars is not exploited in the field. The factors associated with potato crop growth and yield are discussed below.

Temperature

All the growth processes of a plant are influenced by the prevailing temperatures and affect plant development and yield.

The potato plant growth starts with the sprouting in seed potatoes and ends with the production of tubers in the next crop. The dormancy and sprout growth of seed tubers are temperature dependent and dormancy duration is shortened and sprout growth is enhanced between 18-20°C storage temperatures. The temperatures above or below this level are not favourable for the dormancy breaking and sprout growth.

Once the sprouting begins, the temperature of soil becomes important factor for further growth of sprouts and root system. It has been found that soil temperature of about 20°C is favourable for the plant emergence and development of root system in potato. The shoot growth is retarded at temperatures of 10°C and below and is optimum between 25-30°C. The number of stems per hill increases with increase of temperature from 20°C day/ 10°C night to 30°C day/ 20°C night temperature (Burton, 1989). The leaf area is also affected by temperature like number of stems. The life span of the potato crop increases with temperature under long days but under short days it is adversely affected. The LAI is positively correlated with temperature up to 30°C, but for the individual leaf size and longevity of leaves the optimum temperature is around 20°C. As regards induction of flowering, the optimum temperatures range is 16-28°C.

The stolon initiation is strongly influenced by temperature and is delayed by high temperature. The tuber initiation is delayed, impeded or even inhibited at higher temperatures, particularly at high night temperatures (above 22°C). Higher temperatures also reduce the tuber bulking and thus the final yield of the potato plant (Kumar *et al.*, 2003).

The yield of the potato crop is a result of photo assimilates production in the plant through the process of photosynthesis. Different optima have been reported for gross and net photosynthetic productivity. The optimum temperature for gross photosynthesis in potato plant ranges between 24 to 30°C whereas, for net photosynthesis it is less than 25°C. Though the plant response to photosynthesis is a genetic character and plants have adaptive capacity, but higher temperatures reduce the rate of photosynthesis as a result of accelerated senescence, chlorophyll loss, reduced stomatal conductance and inhibition of dark reaction (Rykaczewska, 2013). The partitioning of dry matter is also influenced by the temperature and it has been found that in heat sensitive cultivars high temperatures result in more starch accumulation in the leaves and low rate of its translocation to the tubers than in tolerant genotypes thus decreasing the harvest index.

Light

Light is one of the most crucial factors influencing the potato plant growth and photosynthesis. The light intensity though affects only the process of photosynthesis and thereby final yield, the photoperiod (light availability period) greatly influences almost all the growth and developmental phases of potato plant. In India potato crop is grown primarily in the North Indo-Gangetic plains under short days (less than 12h photoperiod) in autumn/winter season, whereas, in the hills and plateau it is grown during long days (more than 12h photoperiod) in summer/ *kharif* season. The growth and development of potato plant is therefore, different in these two environments and are being discussed below.

Like temperature, the photoperiod also influences most of the growth and development processes in potato plant. The effect of photoperiod during storage of seed tuber is small, but after the dormancy is broken, photoperiod may have a significant effect. The length of the stem and number of leaves is generally more under long days compared to short days. The life span of entire shoot is increased by long days in comparison to short days and therefore, the crop duration of a particular cultivar is less under short days (SD) and more under the long days (LD). The leaf area increases rapidly under SD and the specific leaf area is generally higher in comparison to that under LD. The overall influence of photoperiod on growth shows that long photoperiods stimulate the development of haulm by increasing branching and number of leaves per stem, but reduce leaf size and specific leaf area. During the early phases of plant development, stolon initiation is stimulated by short days, while it is delayed by long days. Tuber initiation is also favoured by short photoperiods. The rate of dry matter production per unit of intercepted light and the rate of photosynthesis per unit leaf dry weight is known to increase under short photoperiods. The short day conditions increase the daytime assimilate transport resulting in high harvest index (about 80%) whereas, the harvest index, is reduced under long days (about 60-65%).

In India, more than 90% of the potato crop is grown in the sub-tropics during a mild and short winter and most of the Indian potato varieties are SD adapted having lower critical day length values, below which they do not produce good yields. The experiments with potato cultivars planted under both SD and LD conditions have shown better tuberization under short days compared to LD. The leaves are the sites for photoperiodic reception and after meeting the required short day cycles, the tuberization stimulus is produced and transported to the lower parts (Figure 2). Once this requirement is fulfilled the plant exposed to long days produces higher yields. The number of tubers is usually higher under long days, though the growth rate of individual tubers is slow in comparison to SD.

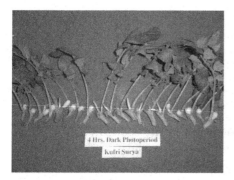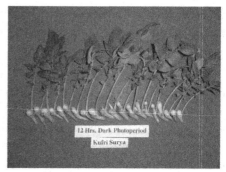

Fig. 2: Tuberization in response to photoperiod showing formation of more tubers under 12 hrs dark period compared to 4hrs.

Photosynthesis

The unit rate of photosynthesis may be described as the amount of carbon dioxide fixed per unit leaf area, per unit time under saturated light intensity by the potato leaf. It is often termed as the light-saturated net photosynthetic rate (P_{max}).

More than 90% of the dry weight of potato plant is derived from the photosynthetically fixed carbon dioxide. The potato crop has C3 photosynthetic metabolism and the P_{max} is greatly influenced by the genetic and climatic factors. The peak photosynthesis occurs during the forenoon (10AM to 1PM) whereas; during the mid afternoon a depression in P_{max} is observed due to water stress by the sudden increase in stomatal resistance. This phenomenon is known as the diurnal variation in the net photosynthetic rate. The rate of photosynthesis is generally higher in tuberized potato plants, which is due to the presence of an active sink. However, it is generally observed that the P_{max} is more in young plants and it decreases with the age of the crop.

Potato genotypes differ with respect to their photosynthetic efficiency and this information can be utilized during potato breeding for a higher photosynthetic and productivity potential in potato crop. The P_{max} though varies among the genotypes, it does not correlate well with the potato yield since it is not the sole determinant of yield. The rate of photosynthesis per unit ground area by the whole canopy for the full season as well as the partitioning of assimilates towards tubers is more important determinant of yield. However, higher P_{max} is an essential feature for high yields and therefore, the parental lines with higher P_{max} can be used as selection criteria to introgress it in varieties for better productivity.

Potato, like other C3 plants, is influenced by the availability of CO_2 in the atmosphere. With the increased CO_2 concentration in the atmosphere, its entry into the sub-stomatal cavity is enhanced and it results in increased biomass productivity and yield. At present the CO_2 concentration in atmosphere has reached nearly 400 ppm from the pre-industrial level of about 280 ppm. It has been established that, production of C3 plants would increase by about 10-40% by doubling the CO_2 concentration. Root zone CO_2 enrichment in potatoes has been found to increase the dry matter production up to the maximum of 25%. The CO_2 thus can be termed as a limiting factor for photosynthesis in potato plant and it is likely that the increase in atmospheric CO_2 will positively affect the potato yield, provided the other factors do not become limiting.

Most of the Indian potato cultivars achieve maximum photosynthetic rate between 500-800 ¼Em⁻²s⁻¹ PARi (photosynthetically active radiation incidental). The interception of light by the canopy plays major role in achieving P_{max}, and penetration of light to the whole canopy if increased, may result in higher P_{max}. The stomatal conductance is also influenced by the light intensity and is directly correlated. The leaves on the upper canopy, which receive more light, have greater stomatal conductance and higher P_{max} at a given light intensity than the lower leaves under shade.

Prolonged drought results in a decline in minimum and maximum fluorescence, resulting in premature senescence and a decline in the chlorophyll light-harvesting complex. Therefore, regular supply of irrigation is desirable for effective photosynthetic rate of the potato crop. In an experiment comparing net photosynthetic rates in irrigated and riverbed cultivated potato crop, it has been observed that the mid-day depression in the net photosynthetic rate in irrigated crop occurs due to increased stomatal resistance and low water potential (less than -0.6 MPa) (Minhas and Singh, 2003). The temperature also influence the process of photosynthesis and the optimum range varies between 15 to 30°C for P_{max}. Beyond 30°C temperature the P_{max} declines substantially. The stomatal conductance has been found to reach its maximum by 24°C and remains at this level up to 30-35°C temperature. The temperatures lower than 15°C result in low P_{max}, which may be due to the lower activity of the biochemical processes involved in photosynthesis.

Respiration

Respiration is a physiological process that sustains cellular life in which the organic substrates are broken down to produce energy, which is used in the biosynthetic events needed for growth and development. It is a gaseous exchange process where the tissue takes up the oxygen and CO_2 is released. The respiration in potato takes place both in tubers and plants. Just after harvest the respiratory rate of potato tuber is generally high and then declines rapidly. The rate of respiration in potato tuber increases after wounding or slicing. It is high in immature tubers than the mature tubers and in dormant tuber it is extremely low. The thick periderm of the tuber acts as a barrier for gaseous exchange. Following the initial decrease, respiratory activity during storage at constant temperature changes little with time until sprouting begins, when there is an increase in respiration rate. At the whole plant level the rate of respiration in potato is like any other plant and the plant organs utilize the accumulated carbohydrate as substrate for generating energy to meet their requirement of growth and metabolism.

Dry matter partitioning

The dry matter produced as a result of photosynthetic process is utilized by the plant for its growth and development. For better tuber yield, it is important that a substantial part of the assimilated carbon is translocated to the tubers. It has been found that under short days more dry matter is partitioned to the tubers whereas under long days more dry matter is partitioned to the foliage. The dry matter distribution in plant parts is dependent on the source-sink interaction. The sink may either be haulms or tubers and the capacity to attract the photoassimilate by the sink is known as sink strength. The sink strength also influences the rate of photoassimilation and it has been found that after tuberization the sink strength of the tuber increases and as a result the rate of photosynthesis also increases. Generally, before tuber initiation the major part of the photoassimilate is translocated to the haulms for growth and development processes and after the tuber initiation a substantial amount of assimilate is partitioned to the tubers. For example, at 52 days the proportion of dry matter accumulated in leaves and stems of potato was 54% that decreased to about 15% by 80 days of crop growth, thereby showing greater partitioning of dry matter to the tubers rather than the haulms at maturity.

Yield components

The yield in potato plant is dependent on genotypic, phenotypic and environmental factors. The morphological factors are important as most of the physiological processes responsible for tuber yield depend on them. Therefore, it is pertinent to mention the desirable traits of the plant parts in the yield formation.

The number of stems produced per tuber is influenced by the genetic factors, seed tuber size and environment. It is desirable that a potato seed tuber gives rise to more number of stem so that more foliage is supported by it, thereby producing more photoassimilate for growth, development and yield. It is a well-known fact that large size tubers bearing more eyes give rise to more number of stems. For better productivity of the potato plant it is desirable that the plant should have higher leaf area per unit land area, so that higher photosynthetic activity can be taken up in the given area and more dry matter is produced (Li,

1985). The increase in leaf area at initial growth stages and its maintenance over a longer period of crop growth is favourable for higher yields in potato. The cultivars attaining higher LAI at early stages have shown higher harvest index and yield. The leaf area duration is another parameter for estimation of plant growth and expected yield. It has been established that effective participation of leaves for longer duration in the process of photosynthesis results in higher productivity. The leaves having low efficiency act as sink rather than source for the dry matter accumulation and are detrimental for the productivity. Hence, the more number of photosynthetically active leaves for the overall duration of the crop are desirable for increased productivity. Similarly, for better productivity of the potato plant, more number of tubers with bigger size is important. When the potato plant is healthy and active, a large amount of photosynthate produced in foliage is translocated to the tubers leading to growth in size of the tuber. The strong tuber sink influences the photosynthetic accumulation positively and thus more number of tubers cause an increase in the accumulation process. The ultimate result is higher yields with more number of tubers and bigger size.

Plant growth regulators

The plant growth regulators influence the growth and developmental processes of potato like any other plant. These plant growth regulators (PGR's) either acts as hormones or they affect the biosynthesis or the action of plant hormones. PGR's are of two types *viz.* primary and secondary. The primary growth regulators include auxins, GA, kinins and ABA, whereas, secondary growth regulators are chlorocholine chloride (CCC), maleic hydrazide (MH), dormins, *etc.* These regulators do not operate in isolation but their relative concentrations at particular stage influence the growth stimulus and accordingly the plants respond to these stimuli. Thus the interactions between these primary and secondary growth regulators (be it antagonistic, additive or synergistic) results in growth and development of the plant parts individually or collectively as a system. The internal hormonal balance can be influenced by external factors and is the path through which environment affects the expression of genetic information into metabolic activity, ultimately expressing in the form and function of the plant. In potato plant various growth and developmental processes depend on such hormonal balance and are influenced by the surrounding environment. Here a general profile of balance of these regulators during some important stages of potato plant growth and tuberization is discussed.

Plant growth

The potato tubers remain dormant for few weeks after harvesting depending on their genetic constitution and the environment under which they are stored.

The dormancy break is induced by the hormonal regulation. The content of ABA decreases in the peels of potato tubers at the termination of dormancy whereas, concentration of GA increases. This results in increasing GA/ABA ratio and finally in termination of dormancy (Harris, 1992). It has therefore, been recommended that if the dormancy has to be artificially broken for subsequent planting of seed in desired areas, a double treatment consisting of 1 ppm GA and 1% thiourea is given to the tubers. Auxins per se, however, have not been shown to influence dormancy but influences only sprout growth after dormancy break. Cytokinins like GA are generally considered to be growth promoters, and are capable of breaking dormancy when used as external treatment. After the dormancy period is over the sprouts emerge and elongate and this process is affected positively by GA. Auxins play an important role in the development of root system in potato, which are formed from the base of sprouts. Thus GA, auxins and cytokinins together influence the growth of potato plant and each process requires a particular combination of these growth regulators.

Tuberization

It is generally believed that the specific tuber forming stimulus responsible for tuberization may be related to a cytokinin since application of cytokinins have shown tuber induction and starch accumulation in *in-vitro* grown potato stolons. This is further strengthened by the presence of high levels of cytokinins in the stolon tips and newly initiated tubers. GA on the other hand inhibit tuber formation and cause elongation of stolons. Natural tuberization is inhibited by ethylene as well, and it does not even allow stolons to elongate. ABA stimulates the tuberization process. Studies in India on application of exogenous ABA could not induce tuberization in potato. However, it is widely believed that higher ABA/GA ratio is important for tuberization stimulus and once this stimulus is obtained, higher cytokinin activity helps in rapid cell division for tuber growth in potato. The overall sequence for the role of PGR's may be that GA promotes stolon formation, IAA helps in cell enlargement in the subapical region, ABA causes cessation of apical growth and cytokinin helps in cell division at later stages of development of the young tubers (Burton, 1989). Jasmonic acid also plays a role during tuber initiation in potatoes. Application of various plant growth substances has been found to increase the yield components in potato. The number of tubers per plant has been found to increase with the application of auxins (IBA at 10 ppm concentration). Plant growth regulators (such as auxins and cytokinins) are also known to increase the proportion of medium and large sized tubers in potatoes and consequently the number of small sized tubers is less. However, such results are not uniform and in general the attempts to increase the potato yield through application of growth regulators have not yielded consistent results.

ABIOTIC STRESS AND POTATO PRODUCTION

The higher plants including potato are subjected to a large number of environmental and biological stresses. These adverse conditions may interfere with normal growth and development and finally result in poor quality and yield of the crop. Among the environmental factors affecting growth and development of potato plant, high temperature and low moisture availability are the major stresses affecting the crop growth and yield adversely. The effect of these two major stresses on potato plant growth and yield are discussed further.

Water stress

Water stress may affect potato plants through its effect on radiant energy interception, conversion of light energy into dry matter, partitioning of assimilates and tuber dry matter concentration. Water stress reduces both the rate and the duration of leaf growth along with reduction in number and size of leaves, therefore, the radiant energy interception is decreased. Drought has adverse effects on the functioning of photo system II, and thus the rate of photosynthesis. Prolonged drought reduces the activity of light harvesting complex and down regulation of carboxylating enzymes. The partitioning of dry matter is also affected by drought, though, genotypic differences are there. The potato root system is shallow and under water stress conditions it is not able to extract water from soil effectively as compared to cereal crops. Thus the potato plants respond by increasing root: shoot ratio under drought and the balance of growth is favoured towards root growth.

Potato is considered as a sensitive crop to water stress requiring about 400mm of water for optimum yield. The extent however varies with the time of stress, its duration and its severity. Transient water stress, induced by high evaporative demand may have accentuated, but only temporary effect even in well watered plants during hot, sunny conditions may be observed. However, prolonged drought and gradual depletion of soil moisture may have chronic effects on plant growth and may lead to premature senescence of the plant. Stomata of potato leaves close at relatively lower water deficits (leaf water potential or LWP of -3 to -5 MPa) resulting in reduced transpiration, whereas, in crops like sorghum, soybean *etc.* stomatas close at -11 to -13 MPa LWP. Adaptation to water stress may involve several physiological and morphological characters and their importance may vary according to the type of water stress experienced by the plants.

Plant growth

Drought after planting may delay or even inhibit germination. The leaf expansion rate is one of the first processes affected badly by water stress and negligible expansion occurs below -5 MPa LWP. If the potato plant is facing water scarcity from the time of plant emergence, radiation interception is reduced. The aspects related to adverse affects of water deficit have been described in previous sections also. The genotypic differences in response to drought may be attributed to their sensitivity to increasing soil moisture deficit. The tolerant genotypes generally have large rooting system that enables them to exploit available soil moisture and thus the leaf expansion is sustained in such conditions. Negative correlation between reduction in stomatal conductance and reduction in leaf growth has been observed, suggesting that genotypic differences in the ability to maintain leaf growth under drought may be linked to stomatal control of leaf water stress. Apart from reduction in leaf growth, water stress leads to the early senescence resulting in decreased LAI. At first, lower leaves start to wilt and drop off, followed by middle and top leaves and concomitantly development of new leaves is inhibited.

The quantity of roots produced by potato is only about one third of that produced by cereal crops and most of it remains in the surface layers (up to a depth of 50-60cm). These rooting characteristics may limit the ability of potatoes to extract available soil moisture effectively and may contribute to the relative sensitivity of the crop to water stress thus potato plants require frequent irrigation. As the soil profile dries, the soil water potential experienced by the crop tends to decrease more rapidly in shallow- rooting crop species than in deeper rooting crops. This may be the reason for potato to be a sensitive crop to water stress. Under stress conditions potato plant has been found to increase the root: shoot ratio to sustain the water requirement of the plant.

Photosynthesis

The plant responds to water stress through stomatal closure and shuts off the supply of CO_2 for photosynthesis. Indian cultivars have shown a decrease of 61-86% in stomatal conductance of stressed plants. The sugar concentration increases in the leaf tissue so as to increase the osmotic potential of the plant which leads to feed back inhibition of photosynthesis since leaf tissues give indication of saturation with photoassimilates. However, prolonged drought results in decline in minimum and maximum fluorescence indicating premature senescence and a decline in the chlorophyll light-harvesting complex. It may also result in a reduction in photosynthetic capacity as a result of down regulation of carboxylating enzymes.

Yield and quality

The yield of potato crop depends on the dry matter production, its partitioning to tubers and dry matter accumulation in the tubers. Both of these processes have been found to be affected by drought. The partitioning of dry matter or harvest index is found to generally decrease in the water stressed plants in comparison to the irrigated plants. However, time of drought relative to tuber initiation may be critical in determining patterns of assimilate partitioning. Water stress early in the season reduces the numbers of stolons and tubers initiated on them. Severe and prolonged stress from early in the season may result in reduced tuber initiation causing assimilate to be partitioned to other organs. However, if water stress occurs after tubers have been initiated, then partitioning to tubers is promoted and maturity is advanced. The work on Indian potato cultivars has also shown that the most critical stage in crop production under water stress is the stolon initiation stage, which represses the yield by 30-65%.

Tuber dry matter content, which is an important component of tuber yield and quality, is dependent on the dry matter accumulation into tubers and tuber water content. The dry matter generally increases from the time of tuber initiation until maturity, except when stress induces a remobilization of starch. Drought conditions generally cause an increase in the tuber dry matter content. The overall response of water stress on potato yield depends on its effect on dry matter production per unit area, dry matter partitioning to tubers and the dry matter concentration in the tubers. The yield of potatoes is affected adversely by the imposition of water stress condition; however, the intensity differs with the stage of stress and its severity.

Besides dry matter, other quality characteristic such as shape of the tuber and reducing sugars content are also influenced by water stress during vegetative growth. Shape defects such as dumb-bell shaped, knobby to pointed end tubers and chain tubers result by short-term stress during bulking phase as a result of second growth of the tubers. Knobby tubers are formed when secondary lateral growth occurs in one or more eyes of the tuber. Bottleneck, dumbbell and pointed end tubers result from secondary longitudinal growth and are characterized by a constriction along the longer tuber axis. These kinds of malformations have been associated with periods of moisture stress followed by return to normal soil water availability and rapid growth. Like dry matter content, reducing sugars also increases in the tubers in response to stress as a result of less water potential in the tuber and therefore, they become unfit for processing industry.

Drought tolerance

Adaptation to water stress conditions may be either through drought avoidance or drought tolerance and it may involve several physiological and morphological characters, the relative importance of which may vary according to the type of water stress experienced by the crop. Potato cultivars considered to be drought tolerant have been found either to have greater threshold of soil moisture deficit or are less sensitive to soil moisture deficit and sometimes both. Such cultivars sustain leaf expansion with increasing soil moisture deficit and also develop relatively large root system for better exploitation of available soil moisture. The work on recovery of leaf growth after a period of water stress in Indian potato cultivars has shown that plant response varies with the cultivars, in some cultivars minimum reduction in leaf growth occurred and the recovery from stress on re-watering enhanced the leaf growth. Some cultivars showed moderate reduction in growth and re-watering caused sufficient recovery of leaf growth, whereas, other cultivars had great reduction in growth and re-watering could not result in sufficient recovery of leaf growth. This shows that leaf growth response varies with the genotype and their tolerance limit.

Heat stress

Potato is well adapted to mean temperatures of approximately 17°C, and is considered as a crop of cool and temperate climate. Higher temperatures like those encountered at lower elevations in the tropics and subtropics cause severe yield reduction and high temperatures are considered as a major physiological constraint for potato production in those areas. In Mediterranean regions, as well as in sub-tropical parts of Asia and Africa, potatoes are exposed to high day and night temperatures and a comparatively dry atmosphere, where yield of potatoes is limited by high temperatures. Low night temperatures of less than 17°C are desirable for tuberization and night temperatures above 20°C are detrimental with no tuberization occurring at 25°C and above (Ewing, 1981). Due to this reason vast areas of southern India are considered unsuitable for potato cultivation. The effect of heat stress on plant growth, photo-assimilation and yield are discussed here.

Plant growth

Soil temperatures above 25°C have been found to delay emergence and reduce plant survival. The rate of sprout growth is slowed by high temperatures leading to delayed emergence. The plants produce more number of thin stems and the number of leaves is increased but the leaf expansion rate is slow, resulting into smaller leaves. The highest number of leaves is obtained at about 30°C, but for leaf growth the optimum temperature is lower (about 25°C). Thus the higher temperatures adversely affect the total leaf area of the plant. Therefore, at higher temperatures the leaf/stem ratio declines. To sum up, high temperatures have positive effect on number of stems, number of leaves and specific leaf area, but the optimum temperatures for size of individual leaves and their longevity are relatively low, hence the overall effect is adverse on plant growth and development.

Photosynthesis

Temperature has a pronounced effect on photosynthesis in potato as described in previous section on photosynthesis. The photosynthetic system in potato has a high adaptive capacity to high temperatures, but low photosynthesis at very high temperature results due to accelerated senescence, chlorophyll loss, reduced stomatal conductance and inhibition of dark reaction. Above 38°C temperature a drastic reduction in photosynthetic efficiency has been found in potatoes. Daily crop growth, which depends upon photosynthesis and respiration, is optimum between temperatures of 15 and 23°C. At higher temperatures the crop growth is reduced and the light use efficiency is lowered. Thus the overall effect of higher temperature is reduction in photosynthesis of potato plant (Ewing, 1981).

Yield and quality

The yield of potato plant depends on the dry matter production, its partitioning to tubers, tuberization and tuber bulking. All of these phases are affected adversely by high temperature. High temperatures adversely affect dry matter production, which is dependent on haulm size, rate of net photosynthesis and leaf area duration.

To obtain harvestable tubers, the plants must be induced to tuberize, and must also be stimulated to partition major part of the dry matter to the below ground plant parts. Temperature has an immediate effect on assimilate partitioning (Figure 3). It has been established that heat sensitive cultivars accumulate more starch in leaves and the rate of transport of sucrose to tubers is low in comparison to tolerant genotypes grown under heat stress. This may be either due to lower photoassimilate synthesis or also due to utilization of substantial amount of photoassimilate by shoots itself. High temperatures strongly reduce the harvest index in potato. Stolon initiation is significantly influenced by temperature; it has been found that under high temperature the stolon development is delayed. Branching and growth of stolons is stimulated but tuberization is delayed. High temperatures delay, impede or even inhibit tuber initiation. High temperature thus leads to slowing down or inhibition of tuber enlargement and tubers tend to revert to stolons, finally resulting in low yield of poorly shaped tubers. Potato crop can give good yields even at day temperatures of 30-35°C provided night temperatures are below 17°C, but if night temperatures go beyond 22°C, there is very little tuberization even when the day temperatures are low (25-27°C).

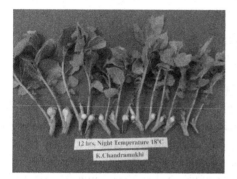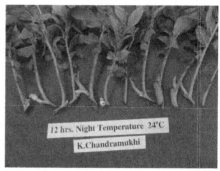

Fig. 3: Tuberization in response to night temperature showing formation of more tubers at 18°C than 24°C.

High temperatures stimulate synthesis of GA, thus reducing the activity of ADP-glucosepyrophosphorylase and starch production. Thus the tubers grown under high temperatures will have low starch content and increased sugar content. This also affects the dry matter concentration in tubers; the specific gravity of tubers grown under high temperatures is low. Second growth and other physiological disorders such as heat sprouts, growth cracks, translucent ends and heat necrosis are enhanced by high temperatures (Li, 1985).

Heat tolerance

Potato tuber formation is badly affected due to high night temperatures of above 20°C and therefore, the most important characteristic of a heat tolerant potato genotype should be its ability to tuberize under high night temperatures of above 20°C. Better foliage growth and higher LAI are also characteristics of heat tolerance. In addition, higher tuberization in the leaf bud cuttings, low shoot/root ratio, early maturity, higher rate of photo-assimilate partitioning from leaves to tubers, higher cell membrane thermo-stability, higher stability of chlorophyll, slower rate of elongation of internodes and low GA/ABA ratio during tuberization may also be important features of a probable heat tolerant genotype. One of these attributes *viz.* tuberization at high night temperatures of 22°C is being well exploited for selection of seedlings for development of heat tolerant genotype, one such genotype named as Kufri Surya has already been developed, which has potential to tuberize up to 22°C temperature and few more are in advanced stages of trial for release as a variety.

REFERENCES

Burton WG (1989) The Potato. Longman Scientific and Technical, New York, USA: 742p

Ewing EE (1981) Heat stress and tuberisation stimulus. *American Potato Journal* **58**: 31-49

Harris PM (1992) The potato crop. Chapman and Hall, London, UK: 909p

Khurana SC and McLaren JS (1982) The influence of leaf area, light interception and season on potato growth and yield. *Potato Research* **25**: 329-342

Kumar D, Minhas JS and Singh B (2003) Abiotic stress and potato production. In, The Potato-Production and Utilization in Sub-tropics. Khurana SMP, Minhas JS and Pandey SK (eds), Mehta Publishers, New Delhi: 314-322

Li PH (1985) Potato Physiology. Academic Press, London, UK: 586p

Minhas JS and Singh B (2003) Physiology of crop growth, development and yield. In, The Potato-Production and Utilization in Sub-tropics. Khurana SMP, Minhas JS and Pandey SK (eds.), Mehta Publishers, New Delhi: 292-300

Rykaczewska K (2013). The impact of high temperature during growing season on potato cultivation with different response to environment stresses. *American Journal of Plant Sciences*, **04**(12): 2386-2393.

Singh B and Ezekiel R (2001) Effect of carbon di-oxide concentration on sprout growth and sugar content of potato tubers stored at 20°C. *Journal of Indian Potato Association* **28**(1): 145-146

Singh B, Ezekiel R and Sukumaran NP (2000) Influence of leaf angle alterations on photosynthetic rate and tuber yield in potato. *Journal of Indian Potato Association* **27**: 91-96

Singh B, Ezekiel R and Sukumaran NP (2001) Dry matter accumulation and tuber yield in potato as affected by defoliation. *Journal of Indian Potato Association* **28**: 251-255

9

Precision Agriculture in Potato Production

Manoj Kumar¹, Preeti Singh², Brajesh Nare³ and Santosh Kumar⁴

¹ICAR- Central Potato Research Institute Regional Station
Modipuram-250110, UP, India
²ICAR- Central Potato Research Institute, Shimla-171001 Himachal Pradesh
India
³ICAR- Central Potato Research Regional Station, Jalandhar-144003
Punjab, India
⁴ICAR-RMRSPC (ICAR-IIMR), Begusarai-851129, Bihar, India

INTRODUCTION

Potato (*Solanum tuberosum* L.) is one of the major food crops which recorded total global production of 388.1 million metric tonnes (mt) in 2017-18. Out of the major staple food crops, potato production (388.1 mt) is exceeded only by maize (1077.98 mt), wheat (761.88 mt) and rice (494.88 mt) (FAOSTAT, 2017). According to the three years (2012-2014) averages, globally, India ranks 3rd in area and 2nd in terms of production next to China. The nutritional quality, high productivity and acquiescence for inclusion in intensive cropping systems of this short duration crop reflects its great potential in modern agriculture to feed the exponentially rising population in the developing countries. But, the intensive cultivation of potato crops urges increased use of fertilizers, pesticides and other chemicals leading to high input costs with plateauing yields. Blanket dose of fertilizers as well as imbalance use of nutrients not only increases the cost of farm inputs but also degrades soil condition and causes severe environment pollution. Indiscriminate use of insecticide as well as fungicide is very common in potato crop which contaminate the environment and deteriorate product quality. Therefore, the chemical inputs need to be optimised based on actual requirement of the crops for sustainable crop production. Furthermore, potato production is associated with a high tillage practices, number of tillage operation would depend on soil type, previous crop etc.

Precision Agriculture (PA) illustrate the strategic resource management to increase productivity and economic returns with reduced input cost. It utilizes advanced information technology tools to decipher spatial and temporal variability in soils, plants and surrounding environmental conditions within a field, which further integrate agricultural practices to meet site-specific requirements. Aim of precision farming is to uphold precision and accuracy while applying site specific farm inputs and as per requirement of the plants. Potato, being a high-value crop, holds a great opportunity for the implementation of PA due to the high cost of farm inputs used for various field operations such as seed, fertilizers, and agrochemicals as well as cost incurred in various field operations soil preparation, planting, harvesting, grading and handling. Furthermore, potato yield and quality are highly susceptible to crop management and environmental conditions with highly uncertain rainfall under the changing climate scenario. Hence, appropriate and on time application of inputs and resources is imperative to move up productivity at decreasing input cost and increasing profit.

In such situation, precision farming (PF) can help in in-situ specific and automated application of crop inputs to manage in-field variability resulting in increased yield with higher revenue. Précised application of input in potato results in not only lowering the input requirement and crop damage (e.g. reduced herbicide application causes less damage to the crop) but also increases the quality of produce leading to more uniform tuber size distribution with high specific gravity of tubers which may result in higher revenue in return. The United States Department of Agriculture (USDA) call this kind of agriculture 'as needed' farming and define it as 'a management system that is information and technology based, is site specific and uses one or more of the following sources of data: soils, crops, nutrients, pests, moisture or yield, for optimum profitability, sustainability and protection of the environment'. It is a new management technology which uses geo-referenced information for the control of agricultural systems. It is based on the detailing of geo-referenced information through the application of monitoring processes and integration of characteristics of soil, plant, and climate.

TOOLS AND PROCESSES IN PRECISION AGRICULTURE

First step to proceed on path of PA agriculture is to assess the intra field variability in soil and plant with respect to desired parameter precisely and accurately (tagged with GPS location) followed by precise application of inputs based on measured variability. Various types of soil and crop nutrient assessment techniques have been developed. The conventional assessment of plant nutrient requirement can be determined by visual diagnosis of plants, plant and soil tests, nutrient omission plot technique etc. Water requirement of crop is assessed by soil-water balance model, metrological data, by estimating crop coefficient, soil data etc. Disease and pest assessment in crop fields are done by monitoring the crop, diagnostic kits, counting the spores of the pathogen, ELISA, PCR etc. However, now in the precision agricultural technologies, proximal or remote sensors can be used to determine the spatial variability of soils and crops with improved efficiency. During the period crop is in field, the data are collected through sensing instruments such as soil probes, reflectance sensors like the Chlorophyll meter, the Green Seaker, the Crop Circle, the FieldScan, the Dualex etc.

Precision agriculture technologies involve various tools and techniques like GIS, GPS, variable rate applicator, UAV *etc*. First step to proceed on path of PA agriculture to assess the variability followed by precise application of inputs based on measured variability. Global positioning system (GPS) is being used for providing precise location coordinates. GPS is satellite navigation system gives continuous ground position information in real time. Having precise location information at real time allows soil and crop parameter measurements to be mapped, user can return to specific location for sampling or for application of inputs. GPS receiver with electronic yield monitors generally used to collect yield data across the land in precise way. Some of the uses in agriculture are - variable rate planting, variable rate fertilizer application, field mapping for records and insurance purposes and mapping yields. The data collection technologies are grid soil sampling, yield monitoring, RS and crop scouting. During crop production, the data are collected through sensing instruments and then recorded and stored in a computer system for future action and generated maps (using GIS) used for acquisition of information and for making strategic decisions (using DSS) to control variability.

Geographic information systems (GIS) which are used for generating maps are basically computer hardware and software that use feature attributes and location data to produce maps. An important function of an agricultural GIS is to store layers of information related to soil and crop. Geographically position can be added and mapped in GIS, this will help in easily viewing and analysis of data. Decision support system (DSS) are used to smoothen the decision-making process for management, operations, planning, or optimal solution path recommendation. These are software-based systems that gather and analyse data from a variety of sources.

Variable rate technology is considered core and heart of the PA (Figure 1). There are two types of variable rate technology for site specific application of inputs: map-based and sensor-based or real time based. However, both methods have their own advantages and limitations. The first site-specific management method *i.e.* map based method is the use of geo-referenced digital maps to represent crop yields, soil properties, pest infestations, and variable-rate application plans. These geo-referenced digital maps have digital information about the particular soil grid in each pixel. Further these maps are integrated with the variable rate applicator. Whereas, in variable rate with real time sensing input is applied on real time basis.

Several studies have shown that use of site-specific variable input management strategies are capable to save input to a considerable level when compared with uniform management strategy and it has been concluded that, variable rate input application utilizing site-specific management are more cost effective and viable than conventional practices. Hence, it is evident that using precision agriculture technology leads to reduced use of input leading to reduced cost, improved output quality and quantity and reduced environmental pollution. This in turn helps in improving total factor productivity of agriculture farm.

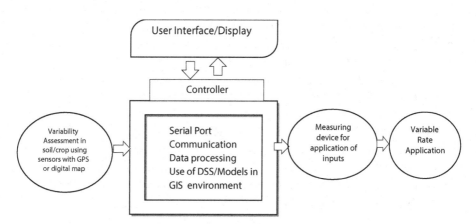

Fig. 1: Diagrammatic depiction of VRT approach.

Using Unmanned Arial Vehicle in PA

Airborne remote sensing technology is very useful to acquire high resolution images for spatial analyses of plant stress due to nutrients, diseases and pest infestation in the field. A high-resolution multispectral imaging system can be used with unmanned aerial vehicle (UAV) for variability assessment and site-specific application of the agro-chemicals. Main limitation of UAVs are their inability to carry heavy weight. However, an ultra-low volume variable rate aerial applicator could be used with UAV for low volume precise application of the chemicals. It could minimize the waste of inputs, increase input use efficiency, reduce health problem, reduce input cost and finally improves total factor productivity. Current uses of UAV in agriculture are in precision fertilizer programme planning, weed and disease control programmes, tree and land mapping and crop spraying.

UAVs are being used in two ways-

Sensors and Visions

UAV is used for capturing images of the particular field or plant. Observation of individual plants, patches, gaps and patterns over the landscapes can be done which was not previously possible with traditional remote sensing techniques.

An UAV based image analysis software was developed for potato crop emergence assessment (Li *et al.,* 2019). High correlation was found between the image analysis and manual counting for plant emergence. Compared with traditional manual assessment, this new approach can save the time required to estimate crop emergence, average canopy cover and crop emergence uniformity.

Sprayer system

This system is used for site-specific pest management and vector control. Multi spectral camera monitors and scans the whole crop field and then generates a spatial map. This map illustrates the condition of the crop through normalized difference vegetation index (NDVI) and then the type of pesticides/fertilizers to be applied on the crop can be evaluated (Mogili and Deepak, 2018). The UAV system provides good spatiotemporal capabilities and when combined with appropriate electro optical sensors, it works more efficiently and accurately.

MAJOR APPLICATIONS OF PRECISION AGRICULTURE

Site specific application for nutrient management

Site specific nutrient application is one of the foremost components of precision agriculture. Research shows that fertilizers account for 30 to 70% of the yield. The variable rate application technology (VRT) for application of fertilizer is an important part of precision agriculture for applying fertilizers accurately as per demand of the crop and optimize productivity (Figure 1). Since, potato is a heavy nutrients feeder and these nutrients must be applied in such a précised manner and as per the crop demand that maximum nutrient reach to plant roots. This enables farmers to customize nutrient management according to the specific requirement of the crop and provides a framework for best management practices. Optimum fertilizer application plays a key role in improving the total factor productivity of various crops. Managing the land/crop within a field with different levels of input depending upon the yield potential of the crop and applying the fertilizer at required level results in reduced cost of production per unit area and reduced risk of environmental pollution.

Variable-rate fertilizer applications have been shown to improve efficiency and increase profits in many grower's fields. Profits is mainly in the form of increased yields without increasing total nutrient inputs or as sustained production at lower agriculture input quantity. Most of these systems consider both spatial and temporal variability, which can affect production. Most of the work in fertilizer application is focused on incorporating additional layers of real-time meteorological, soil and spatial information into the processes that can calculate fertilizer application rates. NDVI based N sensor is the most popular technique for on-the-go assessment of plant nitrogen. The NDVI is more concerned with the green index, basically reflectance of a crop canopy, which reflects the N variability in plants.

Koch *et al.* (2004) found that less total N fertilizer (6-46%) was used with the site-specific variable yield goal N management strategy when compared with uniform N management strategy and concluded that, variable rate N application utilizing site-specific management are more cost effective and viable than conventional practices.

Cambouris *et al.* (1999) conducted a 3-year trial to investigate the agronomic efficiency of VR application(VRA) of phosphorus(P) and potassium (K) fertilizers in potato production on a 2-ha field. In one year out of three, VRA of P and K significantly increased the total and marketable tuber yield compared with the uniform application of P and K. However, the effect of soil series on tuber yield was more significant and more consistent over growing seasons than the effect of application treatment. These more precise recommendations will help to ensure the success of PA approaches in potato production.

Chatterjee *et al.* (2015) in his work on management of soil nutrients with special emphasis on different forms of potassium considering their spatial variation in intensive cropping system of West Bengal, India concluded that the geostatistics-based mapping provided an opportunity to assess the variability in the distribution of native nutrients and other yield-limiting soil parameters across a large area. This could facilitate strategizing the appropriate management of nutrients leading to better yield, while ensuring a more effective environmental protection.

Site specific application for water management

Potato plants have sparse and shallow root system and hence very sensitive to water stress. Water requirement of the crop ranges from 350-550mm depending on the location, soil, variety *etc*. By using precision agriculture tools, site specific optimum supply of water as per the demand can be made efficiently.

A study conducted in province of Quebec, Canada, showed the spatial and temporal variability of soil water content and its effect on total yield (Allaire *et al.*, 2014). The authors found significant correlations between soil temperature, soil water content and total yield on a large scale and reported that low temperatures during the early growing season and lack of water late stage had a negative influence on yield. Precision agriculture in this case can be used to specifically supply water according to the initial water content of soil, which will help in improving input use efficiency and in turn total factor productivity.

Another emerging technology in the field of irrigation is variable rate irrigation. Modelling and soil sensors are being used to study water requirement in soil. Variable rate irrigation (VRI) is more advantageous as there is great variability found in soil texture in fields, excess or uniform irrigation may cause water and nutrient loss by runoff. In potato, it also deteriorates tuber quality. VRI includes different precision agriculture tools that facilitate a centre pivot irrigation system to optimize irrigation application. VRI technology allows growers to easily apply varying rates of irrigation water based on individual management zones within fields. Management zones are field areas possessing homogenous features for landscape and soil properties. These features lead to similar crop yield potential and input-use efficiency for seed, nutrients and water (Doerge, 1999). Control systems on center pivots allow the right amount of water to be applied to these management zones within the field.

Micro irrigation is commonly used technology for precise application of irrigation water. Water saving and productivity gains are higher in case of micro-irrigation as compared flood method of irrigation for the same crops. Micro-irrigation is also found to be reducing energy (electricity) requirement, weed problems, soil erosion and cost of cultivation. The on-farm irrigation efficiency of properly designed and managed drip and sprinkler irrigation system is estimated to be about 90 and 70 percent respectively. Studies conducted to compare micro-sprinkler, drip and furrow irrigation systems for potato has conclusively proven superiority of micro-irrigation. Economic analyses have also revealed microirrigation as a profitable alternative for potato production in semi-arid environment over the existing irrigation method.

Another approach for precise application of irrigation water is pulse irrigation. Pulsing irrigation refer to the practice of irrigating for a short period then waiting for another short period, and repeating this on-off cycle until the entire irrigation water is applied (Eric *et al.*, 2004). Under this technique water is applied to the plant in less amount with high frequency. Pulse irrigation helps in water saving by reducing runoff and leaching and by increasing water use efficiency in both heavy and sandy soil. Intermittent operation of sprinklers and foggers can provide evaporative cooling for temperature control. Smaller pipes and lower capacity pumps can be used which reduces the cost of irrigation.

PA for insect/pest/disease management

Potato crop is highly prone to incidence of insect and diseases and often heavy losses are inflicted depending upon growing situation and crop age. It is obvious that early pest and diseases can reduce yield loss considerably. If the symptoms are observed early, it might be corrected during the growing season. Since the objective is to reduce the spread of disease or pest to minimize production losses, this may be accomplished with monitoring and early detection. Application of agrochemicals is at present the most commonly used practice to protect plants from diseases, pests and weeds. In India, generally, manual or tractor operated uniform spraying systems are being used for insect/pest management. However, uniformity in these application systems depend on speed of operation and discharge rate of the liquid chemical. Therefore, speed based automatic flow control system could be a solution to maintain uniformity in application of the agro-chemicals. Different sensors are used for early crop disease detection, thus allowing a grower to adopt quick and timely crop protection (Table 1). As early as 1933, visible aerial photography was used to detect viral disease in potato. Seelan *et al.* (2003) used this imagery, flown low over potato, wheat and sugar beets to map numerous stresses on crops. In the high spatial resolution (70 cm) imagery wind damage, fertilizer skips and disease were visible.

Precautionary measures based on the meteorological conditions and precised application of pesticides and fungicides helped in reducing the one third of expenses. This reduction in input use improved the total factor productivity. Disease forecasting allows the prediction of probable outbreaks and decision support system can help in management of any further increase in disease intensity. This allows us to take strategic decisions about the disease management.

Late blight one of the most dreaded disease in the case of potato has been managed effectively using forecast models and DSS in terms of forecasting it appearance and management this has led to improve potato productivity and net return.

Table 1: Sensors for early detection of diseases and pests

Different methods of detection	Principle of Working
Pest detection sensors	Sensors provide real-time data from the field
Low-power Image Sensor	Wireless sensor which periodically capture images of the trap contents and sends them remotely to a control station. This images are used for determination of the number of pests found at each trap.
Acoustic sensor	Monitoring the noise level of the insect pests thus detecting the infestation at a very early stage.
Thermography Disease Detection Method	Captures infrared radiation emitted from the plant surface. It is based on detection of change in plant surface temperature.
Fluorescence Disease Detection Method	Measure changes in chlorophyll and photosynthetic activity, thus detecting the pathogen presence.
Hyperspectral Disease Detection Method	Measure the changes in reflectance that are the results of the biophysical and biochemical characteristic changes experienced upon infection.
Gas Chromatography Disease Detection Method	Determine volatile chemical compounds released by the infected plants.

INITIATIVES FOR DEVELOPMENT OF PA IN INDIA

In India, precision farming is still at a very budding stage. More than 86% of operational holdings are small or marginal size (< 2 ha) which is the biggest bottleneck in the practice of PF in India. Under such situations developing field specific recommendation of inputs is fairly good steps towards development of precision agriculture. According to a report, in the states like Punjab, Rajasthan, Haryana and Gujarat, more than 20% of agricultural lands have operational holding size of more than 4 ha, in such case intra-field variations can also be taken as input management options. Some discrete initiatives started towards the application of this technology. India–US Knowledge Initiative on Agriculture (KIA) has started working on PA in India in the year 2007. Tamil Nadu State Government sanctioned a scheme named "Tamil Nadu Precision Farming Project" implemented in Dharmapuri and Krishnagiri districts covering an area of 400 ha. High value crops such as hybrid tomatoes, capsicum, babycorn, white onion, cabbage, and cauliflower are cultivated under this scheme.

ICAR institutes Project Directorate for Cropping Systems Research (PDCSR) now ICAR-IIFSR, Modipuram, Meerut (Uttar Pradesh state) in collaboration with Central Institute of Agricultural Engineering (CIAE), Bhopal initiated variable rate input application in different cropping systems. In 2009, Space Application Centre (ISRO), Ahmedabad started experiments in the Central Potato Research Station farm at Jalandhar, Punjab, to study the role of remote sensing in mapping the variability with respect to space and time.

The Precision Farming Development Centres (PFDC) established in year 2001 play leading role in the development of regionally differentiated technology validation and dissemination. Presently, there are twenty-two (22) PFDC which have been established in India to promote precision farming and plasti-culture applications for hi-tech horticulture/ agriculture. The PFDC at IARI, New Delhi; University of Agriculture Sciences, Bangalore; Gujarat Agriculture University, Navsari; Indian Institute of Technology, Kharagpur and Central Institute of Sub-Tropical Horticulture (CISH), Lucknow function as Centres for excellence for Precision Farming (CEPF) and are equipped to take up research and development works on Precision Farming. The CEPFs function as mother centres for providing technical support to other PFDCs located in the region. ICAR Institutes and Institutes in the private sector are also involved in technology development. Recently, few start-ups have also come up to advise growers of cash crop on precision agriculture practices resulting into reduced cost of cultivation improvement in net income.

As an example of collaborative effort of private and Govt. agencies, precision farming centre has been established by MSSRF (M.S. Swaminathan Research Foundation – a non-profit trust) at Kannivadi in Tamil Nadu with financial support from the National Bank for Agriculture and Rural Development (NABARD). This Precision Farming Centre receives the help of Arava Research and Development Centre of Israel and works with an objective of poverty alleviation by applying PA technologies.

STATUS OF PRECISION AGRICULTURE IN POTATO CROP IN INDIA

When we talk in strict sense there is hardly any report on use of PA technique like VRT for the management of nutrient, water and pest in potato crop in India. However, there have been approaches to improve precision in recommendations based in field specific observation to optimise input application and yield enhancement. A decision support tool has been developed by ICAR-CPRI for providing information on the optimum time of planting and the likely consequences of early or late planting of potato in about 173 locations of Nilgiris region of Tamil Nadu state in India. This DSS developed for the purpose of potato crop scheduling is of great significance, which enables the farmers as well as extension functionaries for taking right decisions on timing the planting and harvesting of potato crop.

Spatial variability maps of available nutrients developed for potato growing pockets by ICAR-CPRI can also be tool to have nutrient recommendation specific to field than using a blanket recommendation for the region. These maps are useful for identifying specific locale of potato growing pockets with different nutrient management problems. Further, the DSS developed for giving soil test based nutrient recommendations based on QUEFT model is a step forward toward precision. The quantitative evaluation of fertility of tropical soils (QUEFTS) model was calibrated for the estimation of NPK requirements for different targeted yields of potato. The results of the study showed that observed yields of potato with different amount of nutrients were in agreement with the values predicted by the model. Therefore, the QUEFTS model based NPK fertilizer recommendations can be adopted for site-specific nutrient management of potato. DSS has also been developed based on QUEFTS model to give soil test based field specific nutrient recommendation for different potato growing regions. Nitrogen is one element in potato cultivation for which in season real time decision on quantity and time of application is very important. Low cost tools like chlorophyll meter (SPAD) and leaf color chart (LCC) are simple, portable diagnostic tools that can be used for in situ measurement of the crop N status and proved to be useful for small farms of developing countries. Singh *et al.* (2019) worked on SPAD for in situ real time nitrogen management increasing number of splits of nitrogen (N)application is based on SPAD value. Through this technique it could be possible to save 10 to 20 % of recommended N without compromising on yield, thereby, improving efficiency and precision.

The only most important disease having maximum economic importance and requiring in season real time decision making in potato is late blight. Precision in the management of this diseases increased immensely with the development a Pan India computerized forecasting model 'INDOBLIGHTCAST', which is web based, to forecast late blight and the model could forecast the disease well in advance in comparison to other forecasting models tested. This has led to decreased number of sprays in several potato growing pockets.

Kumar *et al.* (2016) conducted the study to develop a decision support system (DSS) and integrated with soil moisture sensor based on tensiometric principle, for real time irrigation scheduling either on time basis or soil moisture sensor basis. The designed system was successfully tested on potato crop under different methods of irrigation at Precision Farming Development Centre (PFDC), IARI, and New Delhi, India. The reference evapotranspiration, water requirement of potato crop estimated from DSS was verified and tested with CROPWAT model, and approximately similar results were obtained. The performance evaluation of developed system helped to control water application as per crop needs with its various functionality were found satisfactorily.

These works are adding to help development of PA revolution in India directly or most probably indirectly. The need-based nutrient application for potato and application of DSS in Indian agriculture have been initiated in some places. To utilize the full benefits of PA in Indian condition, an organized, well-planned, long-term policy suitable for Indian farming sector is required.

PRECISION AGRICULTURE FOR IMPROVING TOTAL FACTOR PRODUCTIVITY

In agriculture, people often misinterpret productivity as ability of a production system to produce more yield. But, the consideration of yield alone as a measure of productivity provides misleading indication of the degree of productivity improvement in agriculture (Chandel, 2007). There are many concepts of productivity but best-known concepts among them are partial productivity and total factor productivity. Partial productivity measures contribution of one factor/input (say land, labour or capital) to output growth when other factors are kept constant. *e.g.* land productivity is measure of the land needed to meet food demands, like yield of crop per unit of land *etc.* However, it does not reveal whether increase in productivity is due to increase in use of inputs or increase in input use efficiency or due to use of improved technology. Further, it also ignores time, secondary products and inputs. Therefore, the other concept of productivity *i.e.* total factor productivity (TFP) has been found more relevant. It measures the ratio of total agriculture output to total production input i.e. output with the combined use of all resources and higher TFP indicates efficient use of agriculture inputs. That's why, it is regarded as accurate productivity measure. The best measure of TFP is to compare output with the combined use of all resources. Using PA technologies in potato cultivation can lead to increase in input use efficiency. Evert *et al.* (2017) reported that PA can lead to reduction in use of pesticide by 23% (expressed as EIQ) and nitrogen fertilizer by 15% in potato production, these results suggests that PA can improve total factor productivity. Rana and Anwer (2018) studied potato production scenario in India and analysed total factor productivity of potato production over the years. They found that the TFP have shown tremendous improvement and this change was mainly contributed by growth in technology change. Hence by implementing precision agriculture technology obviously we can improve total factor productivity making potato cultivation more efficient and profitable.

CONCLUSION

Potato being input efficient important food crop with future promise has great potential for use of precision agriculture. PA provides the ability to utilize crop inputs more effectively including farm equipment, seeds/seedlings, fertilizers, pesticides and irrigation water. With the growing concern of economic viability and environmental sustainability use of PA tools like GPS, GIS, UAV, models and DSS are likely to increase. It is likely that efficient and effective use of inputs would lead to greater crop yield and/or product quality, without polluting the environment. However, the cost effectiveness of precision agriculture has not conclusively proven. Many challenges are there in the use of PA, like, lack of knowledge and technical know-how, poor connectivity in many areas, unavailability of high spatial resolution imagery for remote areas. In Indian scenario small landholdings appears to a major bottleneck to start with. Despite these challenges, with advancement in user friendly tools (apps, DSS, Models) and development of various types of cost effective sensors to capture minor variation of relevant abiotic and biotic parameters in soil and crop would lead to make this technology more usable and cost effective in coming days. Besides, with the improved availability of high resolution imagery from remote sensing and spatial analysis would provide valuable information by allowing complete understanding of the spatial complexity of the characteristics of a field and its crops, and providing information about the different parameters which in turn would be very useful in improving precision in input delivery leading to increased total factor productivity.

REFERENCES

Allaire SE, Cambouris AN, Lafond JA, Lange SF, Pelletier B and Dutilleul P (2014) Spatial variability of potato tuber yield and plant nitrogen uptake related to soil properties. *Agronomy Journal* **106**:851–859

Cambouris AN, Nolin MC and Simard RR (1999) Precision management of fertilizer (P and K) for potato crop in Quebec, Canada. In: Robert PC *et al.* (eds.) Proceeding fourth international conference precision agricul-ture. St. Paul, Minnesota 847–857

Chandel BS (2007) How Substantial is the Total Factor Productivity Growth in Oilseeds in India? *Indian Journal of Agricultural Economics.* **62**(2): 144–158

Chatterjee S, Santra P, Majumdar K, Ghosh D, Das I and Sanyal SK (2015) Geostatistical approach for management of soil nutrients with special emphasis on different forms of potassium considering their spatial variation in intensive cropping system of West Bengal, India. *Environment Monitoring Assessment* **187**:183

Doerge T (1999) Defining management zones for precision farming. *Crop Insights* **8**(21):1–5

Eric S, David S and Robert H (2004) To pulse or not to pulse drip irrigation that is the question UF/IFAS-Horticultural Sciences Department. Florida, USANFREC-SV-Vegetarian (04-05)

FAOSTAT (2017) food and agriculture organization statistical database: http://FAO.org/faostat/en/#search/potatoes

Koch B, Khosla R, Frasier WM, Westfall DG and Inman D (2004) Economic feasibility of variable-rate nitrogen application utilizing site-specific management zones. *Journal of Agronomy* **96**:1572-1580

Kumar J, Patel N and Raput, TBS (2016) Development and integration of soil moisture sensor with drip system for precise irrigation scheduling through mobile phone. *Journal of Applied and Natural Science* **8**(4): 1959-1965

Li B, Xu X and Han J (2019) The estimation of crop emergence in potatoes by UAV RGB imagery. *Plant Methods* **15**: 15p

Mogili UM and Deepak BBVL (2018) Review on application of drone systems in precision agriculture. *Procedia Computer Science* **133**: 502-509

Rana R and Anwer Md. (2018) Potato production scenario and analysis of its total factor productivity in India. *Indian Journal of Agricultural Sciences* **88**:1354-1361

Seelan SK, Laguette S, Casady GM and Seielstad GA (2003) Remote sensing applications for precision agriculture: A learning community approach. *Remote Sensing of Environment* **88**:157-169

Singh SP, Kumar M, Dua VK, Sharma SK, Sadawarti MJ and Roy S (2019) Leaf chlorophyll meter- A non-destructive method for scheduling nitrogen in potato crop. *Potato Journal* **46**(1): 73-80

Van Evert FK, Fountas S, Jakovetic D, Crnojevic V, Travlos I and Kempenaar C (2017) Big data for weed control and crop protection. *Weed Research* **57**: 218–233

10

Natural, Zero Budget, Organic Agriculture for Sustainability and Cost Effectiveness

SP Singh[1], Sanjay Rawal[2], VK Dua[3], Jagdev Sharma[3], YP Singh[1]
MJ Sadawarti[1] and S Katare[1]

[1]*ICAR-Central Potato Research Institute Regional Station, Gwalior-474020*
MP, India
[2]*ICAR- Central Potato Research Institute Regional Station*
Modipuram-250110, UP, India
[3]*ICAR- Central Potato Research Institute, Shimla-171001, HP, India*

INTRODUCTION

Agricultural biodiversity is the basis of global food security. Wide variety of crops, soil and cropping systems are nurtured by human cultural diversity since times immemorial. Potato is a major food crop which is widely acknowledged now. It is consumed by the world's largest population than any other vegetable, produces more calories and protein per unit of land and time than any other food crop. Potato has much potential for sustainable, non-chemical farming, fits well with many crop rotations, and does very well with natural fertilizers. Over the years chemical based farming has started experiencing reduced production and increased costs, or both (Singh *et al.*, 2011 and Sreenivasa *et al.*, 2010). Monoculture of crops such as rice, wheat and cotton etc., depletes topsoil of nutrients and reduces diversity of beneficial microbes leading to reduced soil productivity. It is finally making the crop plants vulnerable to parasites and pathogens. Environmental pollution caused by chemical fertilizers and pesticides is posing a serious threat worldwide (Doran *et al.*, 1996). Healthy soil is the foundation upon which sustainable agriculture is built. Farming practices differ mainly based on soil inputs and crop protection measures (Devarinti, 2016). Thus there is growing interest in organically produced products and farmers are shifting from conventional to organic farming. In India, Sikkim is the first state where completely organic farming is adopted. Farming practices differ mainly based on soil inputs and crop protection measures (Devarinti, 2016) in different regions. Different types of farming based on use of natural products are described in the chapter.

TYPES OF FARMING

Organic farming

Although considered more sustainable, it however is less productive than conventional farming. Organic farming results in improved soil structure with higher organic matter concentrations and higher soil aggregation, a reduction in groundwater nitrate concentrations, and plant-parasitic nematodes. Amongst different countries Australia has the highest area under organic farming (Figure 1).

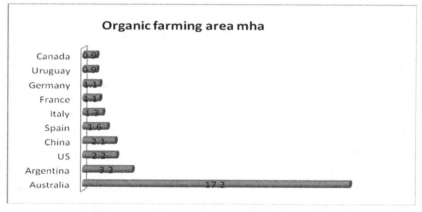

Fig. 1: The 10 countries with the largest areas (Million hectares) of organic agricultural land in 2011 (Source: FIBL-IFOAM survey 2012)

Natural farming

Natural farming, philosophy is to work with nature in order to produce healthy food, keep ourselves and the land healthy. Everything in nature is useful and serves a purpose in the web of life. Also termed 'Do Nothing Farming', because the farmer is considered only to be a facilitator - the real work is done by nature herself. No tillage and farming without the application of herbicides, inorganic fertilizers and pesticides is practiced. Here, actual physical work and labor has been found to reduce by up to 80% compared to other farming systems. Natural farming differs from organic farming by not using any organic manure like FYM and vermi-compost. In Japan, Fukuoka started Natural farming by experimenting with the Nature and following the natural ways of crop propagation.

He achieved yields similar to those of chemical farming but without soil erosion. The essence of natural farming is minimizing the external inputs to the farm land, which degenerate the soil nature. At first, because there was no habitat for many of the insects, he had to make natural insecticide like pyrethrum which comes from chrysanthemum roots in order to keep pests like cabbage worm and cabbage moths away. Zero- Budget Natural Farming (ZBNF) is proposed by Subash Palekar, in India, with the same philosophy but with the indigenous supplements. In ZBNF, soil is supplemented with the microbial inoculums like Beejamruth and Jeevamruth to accelerate the propagation of soil micro flora, beneficial to soil enrichment. Indigenous pesticide decoctions of leaves with cow urine Neemastram and Bramhastram etc., are introduced. The philosophy of the natural farming is to nurture the growth of these beneficial microorganisms without using external manure and chemical pesticides (Devarinti, 2016 and Palekar 2009).

Beejamruth

Application of Beejamruth is advocated in ZBNF. It is a seed treatment mixture prepared from cow dung, cow urine, lime and a handful of soil. Naturally occurring beneficial microorganisms are found in cow dung. These microorganisms are cultured in the form of Beejamruth and applied to the seeds as inoculum. It is reported that seed treatment with beejamruth protects the crop from harmful soil-borne pathogens (Palekar, 2009 and Sreenivasa *et al.,* 2010).

Jeevamruth

Soil microorganisms play an active role in soil fertility as they are involved in the cycle of nutrients like carbon and nitrogen, which are required for plant growth. They are responsible for the decomposition of the organic matter entering the soil and therefore in the recycling of nutrients in soil. PGPR, cyanobacteria and mycorrhiza constitute soil microorganisms. They participate in decomposition, mineralization and nutrient supply to the plants. Phosphate Solubilizing Bacteria (PSB) and mycorrhizal fungi can also increase the availability of mineral nutrients (phosphorus) to plants. Nitrogen-fixing bacteria can transform nitrogen in the atmosphere into soluble nitrogenous compounds useful for plant growth. These microorganisms, which improve the fertility status of the soil and contribute to plant growth. They may also show antagonism (biological control) to pathogens. Soil is saturated with all the nutrients, but these are in the non-available form to the roots of the plants. Beneficial micro-organisms in Jeevamrith convert the nutrients in non-available form into dissolved form, when it is inoculated to the soil (Aulakh *et al.,* 2013). Jeevamruth is either sprayed/sprinkled on the crop field or added to the irrigation tank in regular interval of 15 days until the soil is enriched. Composition of Jeevamruth water is 200 litre, cow dung 10 kg., cow urine 5-10 liter, Jaggary 1-2 kg, flour of the pulses 1 kg, a handful of soil (Palekar, 2009). This mixture is well stirred for few days and sprayed on crop for every fortnight. It is shown that this mixture facilitates the growth of beneficial microorganisms. Application of Jeevamruth facilitated the growth of beneficial soil microorganisms and improved crop yield.

ORGANIC MANURES

Organic matter consists of the substances originating from plant remnants and animal remains and contains nutrients beneficial to soil health. The need for organic matter is unlimited, the more added to the soil, the better the soil will become. Organic manures are traditional sources of nutrients, which help in maintaining the soil fertility. Among the organic manures, farmyard manure (FYM) is the principal source and is commonly available to the local farmers. They are relatively cheap soil amendments, rich in nitrogen, helping in sustaining the soil fertility and protection of the environment. Organic manures contain plant nutrients, though in small quantities in comparison to the chemical fertilizers. The presence of growth hormones and enzymes make them essential for improvement of soil fertility and productivity. The supply of essential micronutrients through organic manures has also improved plant metabolic activities especially in the early vigorous growth of plant. Application of farmyard manure up to 30 t/ha has significantly increased the potato growth and yield contributing traits as well as the tuber yield. Manures reduce soil pH, improve the soil fertility and water holding capacity of soil. The organic manure has low C-N ratio, supply 20–30% to the current potato crop and 40–60% is stored in the soil. Continuous application of organic manure for long periods results in an increased output of organic matter annually. The soil organic carbon has been improved with the application of FYM. Soil organic carbon was improved from 18 to 62% with organic sources compared to chemical fertilizers. At lower fertility, the organic manures showed the maximum response than at higher fertility levels (Canali et al., 2010).

Sources of organic matter

Composts: Compost results from the decomposition of various forms of organic matter. It plays an essential role in enriching nutrient content in the surface layer of the soil. Finished compost contains an abundance of nutrients. Application of fully decomposed manure is more beneficial and do not enhance soil and tuber borne diseases in potato.

Bio-fertilizers: Application of biofertilizers is gaining attention in recent times. These are alternative sources being eco-friendly, fuel independent and cost effective help in improving soil health and a better crop nutrient management. The ecological and agricultural importance of these organisms depends upon the ability of certain species to carry out both photosynthetic nitrogen fixation and proliferation in diverse habitats. An integrated nutrient management strategy i.e., organic manures and crop rotations with legumes will improve the soil fertility and sustainability of rice-based cropping systems.

Crop residues: On an average, 25% of the total nitrogen, 50% of total phosphorus and 75% of total potassium in the crop harvest are retained in the residues. An estimated 377 million tons of crop residues per year are available in India. With the incorporation of green manure or crop residues, the organic matter has been improved and soil physical conditions has been altered i.e., decrease in bulk density, increase in total pore space, water stable aggregates and hydraulic conductivity. Dhaincha (*Sesbania aculeata*), Sunnhemp (*Crotalaria juncea* Linn.), blackgram (*Vigna mungo* [L.]), cowpea (*Vigna unguiculata* [L.]) and greengram (*Vigna radiata* [L.]) are some of the important legumes used as green manure plants and they are adaptable to different cropping system. The impact of residue incorporation on succeeding crops depends on the produced quantity of residues and time and method of incorporation. Residue retention in mungbean (*Vigna radiata* [L.]) – potato rotation has increased yields of both crops and nitrogen balances of the crop rotation. Mungbean and lentil (*Lens culinaris*) residues returned to soil have fixed about 112 and 68 kg N/ha, respectively which has resulted in positive N balances (64 and 27 kg N/ha, respectively) of the cropping system and hence the fertilizer N requirement could be reduced (Goffart, *et al.*, 2008).

Legume cover crops: The legumes/pulses contribute to the sustainability of cropping systems through biological nitrogen fixation, which supplies nitrogen to the system, diversification of cropping system, which reduces the disease, pest and weed incidence and provide food and feed that are rich in protein. Soil fertility and the physical properties have been enhanced with use of the legumes/ green manure crops. Grain legumes shed their leaves near maturity and the above ground biomass after harvesting (seeds along with residues and roots) contains nitrogen, improving the soil nitrogen balance and productivity. The legume residues contain about 20–80 kg N/ha (about 70% of it is derived from biological nitrogen fixation) depending upon the type of crop and the full N benefits will be realized if all the residues are incorporated after harvesting the seed yield.

MULCHING

Mulch is used as a soil cover in potato crop. Crop residues that can be used as organic mulch are paddy-straw (Figure 2), wheat-straw, groundnut foliage, etc.

Fig. 2: Paddy straw mulch in potato crop

Mulching with straw improves soil moisture content and conducive to the growth of microorganisms and earthworms. It also improves tuber emergence without tillage. Growth of weeds holds back effectively. Growth of covering crops like legumes increases the nitrogen fixation in the soil. Harvesting weed before flowering and covering the open land reduces the area for the weeds and improves the organic matter content in the soil. With this practice usage of herbicides can be avoided (Doring *et al.*, 2005). The benefits of mulch are:

- Mulches provide a more stable microclimate in the soil by regulating temperature and moisture, which promotes plant growth.

- Increase availability of nutrients in the soil as a result of higher activity of soil microorganisms that decompose organic matter at more favorable soil temperature and moisture.

- Higher yields reduce crop weed competition.

ORGANIC/NATURAL CULTIVATION PRACTIES FOR POTATO

Organic farming and minimum tillage

Annual tillage chemical fertilization and pesticide use consistently affect populations of earthworms. When tillage is avoided, soil moisture content is increased, augment the propagation of earthworms. Earthworms are known to make the soil porous and enrich the soil with their castings. Seeds are scattered and covered by straw before harvesting the previous crop. Seeds are germinated by the arrival of next favorable season. In ZBNF, this practice is not given prominence.

Use of manures: Manure has the same function as other organic fertilizers, but is richer in nitrogen, because of its urine content. Different types of manure have different water and nutrient composition depending on the type of animal, their age providing the manure and the food it got (Table 1).

Table 1: Water and nutrient content of different organic manures from animals and birds

Sources of manure	Water content (%)	Nutrient content (kg/ton)		
		Nitrogen	Phosphorus	Potassium
Buffalo	85	26.2	4.5	13.0
Cow	85	22.0	2.6	13.7
Poultry (chicken, duck etc.)	62	65.3	13.7	12.8
Pig	85	28.4	6.8	19.9
Sheep	66	50.6	6.7	39.7
Horse	66	32.8	4.3	24.2

Organic manures should be applied at planting time by mixing it straight into the soil or by placing it to the left and right of the seeds. Organic manure requirements for potato crops are minimally 20 ton/ha. Using more than this will further improve soil structure and fertility.

Timely planting: Natural/organic agriculture is only economical if crop is timely planted. Early and late planting suffer most due to insect, pest, weed etc. Since, in natural/organic agriculture crop pest cannot be controlled but are managed by avoiding their critical period and enhancing natural enemies of pests, diversifying crops and enhancing resistance in crop plants by nutrition.

Hoeing and earthing up: Begin earthing up early in the growth of the crop, even when the plants are about six inches tall. Continue hilling up until one high ridge covering the developing tubers has replaced. The practice of earthing up protects the potatoes from light which will turn potato skins green and toxic. Also, spores of various diseases, which may be present on the leaves, will be harmlessly washed off onto the ground, where most will die, unable to come into contact with the tubers. Late blight is a very common disease that can be minimized this way. Late blight generally appears as brownish splotches on the leaves, especially between the veins, and as it spreads the leaves will turn brown, curl, disintegrate, and drop off: Prior to earthing up, cut off any infested leaves with a razor blade, and remove from the plot. After removal, hill up as normal, and you will reduce the spread of the disease, which unchecked could spread throughout your plot.

Growing resistant/efficient varieties: In the years of sever outbreak of insect, pests and diseases there is no effective bio-pesticide which can effectively control diseases and pests in organic farming crops. Hence, it is better to opt disease and pest resistant varieties suitable for the region so that crop can be saved from total failure. Details of diseased and insect, pest resistant varieties is mentioned in Table 2.

Table 2: Resistant/tolerant from insects, pests and nutrient/water efficient varieties

Name of variety	Disease/pest
Kufri Anand	Moderate resistant to late blight. Immune to wart disease. Tolerant to hopper burn, Tolerant to Gemini virus
Kufri Arun, Kufri Himalini[1]	Moderate resistant to late blight.
Kufri Badshah	Resistant to PVX, early and late blight
Kufri Chamatkar	Immune to wart disease; Resistant to early blight and charcoal rot.
Kufri Chipsona- 1	Resistant to late blight.
Kufri Chipsona- 2, Kufri Frysona	Resistant to late blight; Immune to wart disease
Kufri Chipsona- 3	Resistant to late blight.
Kufri Chipsona- 4	Field resistant to late blight.
Kufri Gaurav	Moderate resistant to late blight; High tolerance to nutrient stress.
Kufri Garima	Moderate resistant to late blight.
Kufri Girdhari	Highly resistant to late blight.
Kufri Jyoti, Kufri Kanchan, Kufri Kashigaro	Moderately resistant to early and late blight; Immune to wart disease
Kufri Khyati	Field resistant to early and late blight.
Kufri Kuber	Resistance to PLRV
Kufri Kundan	Moderate resistant to late blight and resistance charcoal rot.

Source: NIPHM, NCIPM and Directorate of Plant Protection, Quarantine & Storage

DISEASE MANAGEMENT

In organic systems, cultural practices form the basis of a disease management program. Promote plant health by maintaining a biologically active, well-structured, adequately drained and aerated soil that supplies the requisite amount and balance of nutrients. Choose varieties resistant to one or more important diseases whenever possible. Plant only clean, disease-free seed and maintain the best growing conditions possible. Rotation is an important management practice for pathogens that overwinter in soil or in crop debris. Rotating between crop families is useful for many diseases, but may not be effective for pathogens with a wide host range, such as *Sclerotinia*- white mold, *Rhizoctonia*- black scurf, *Colletotrichum*- black dot, *Verticillium*- wilt, common scab, or nematodes. Rotation with a grain crop, preferably a crop or crops that will be in place for one or more seasons, deprives many disease-causing organisms of a host, and also contributes to a healthy soil structure that promotes vigorous plant growth. The same practices are effective for preventing the buildup of root damaging nematodes in the soil, but keep in mind that certain grain crops are also hosts for some nematode species. Maximizing air movement and leaf drying is a common theme. Many plant diseases are favored by long periods of leaf wetness. Any practice that promotes faster leaf drying, such as orienting rows with the prevailing wind, or using a wider row or plant spacing, can slow disease development. Fields surrounded by trees or brush, that tend to hold moisture after rain or dew, should be avoided if possible, especially for a crop like potatoes, with a long list of potential disease problems. Insect damage can create susceptibility to disease. Feeding by the European corn borer (ECB) can create an avenue for disease infection by *Erwinia* spp., the pathogen that causes black leg and bacterial soft rot. Survival and establishment of ECB larvae vary depending on potato cultivar and field conditions. Scouting fields weekly is key to early detection and evaluation of control measures. Allowing pest populations to build past thresholds can leave few or no options for control. All currently available fungicides allowed for organic production are protectants meaning they must be present on the plant surface before disease inoculum arrives to effectively prevent infection. Biological products must be handled carefully to keep the microbes alive. In addition to disease control, fungicides containing copper may have antifeedant activity against some insect pests including the Colorado potato beetle.

Late blight management in organic potato production

Planting potato seed from unreliable sources increases late blight risk. So the seed should be purchased from certified seed potatoes sources. Certified seed is not guaranteed to be late-blight free, but in general, should be less risky because certified seed production is monitored by seed professionals and certified growers tend to be well informed about significant potato pests affecting seed quality. Inspect seed potatoes for late blight symptoms upon arrival, and if seed is stored before planting, inspect it again before planting as late blight can spread during storage. If late blight is suspected, diseased tubers should be removed. Do not plant infected tubers as they are an excellent source of primary (initial) inoculum to start an early epidemic. On organic farms, early epidemics have a high likelihood of destroying the entire crop.

Planting date: Manipulating planting dates to avoid high disease risk periods is an important and effective strategy for controlling disease. Manipulation of planting date will however, depend on local conditions and risk periods. In areas where the highest likelihood of late blight onset is late in the season, plant short season varieties early so foliage has died back before high risk periods begin.

Drip irrigation to manage late blight: Compared to overhead sprinkler drip irrigations dramatically reduces late blight risk by eliminating the contribution of rain and overhead irrigation to leaf wetness. If sprinklers are to be used then do not irrigate in the afternoon if the foliage cannot dry before dark; wet foliage at dark will typically remain wet all night long, resulting in a very long wetting period and increased disease risk. If applicable, use drip irrigation to minimize irrigation-applied leaf wetness.

Hilling to manage Late Blight: High hilling, and prevention of crack development in hills, can reduce the movement of late blight spores through the soil to tubers, thereby reducing tuber late blight risk.

PEST MANAGEMENT

Insect pests are a food source for natural enemies. If there are no pests, there will be no natural enemies. Hence a low level of insect pests should be acceptable. Natural enemies are more sensitive to insecticides. Tests have shown that natural enemies are more numerous in quantity and species diversity in unsprayed fields than in fields sprayed with insecticides. The overall result is smaller pest populations in sprayed fields. Adult insects that act as natural enemies (particularly parasitoids) feed on nectar. Therefore, planting nector producing flowers on the edges of fields will provide food to these parasitoid species and induce longevity. Predatory flies need compost as a habitat for its maggots. Hence composts and manures should be used regularly.

Biological control

Natural enemies have been discussed in the section above on farmer-friendly insects. Biological control will develop better if supported with another strategy, i.e. using natural pesticides, which originate from plants and are less likely to disturb or kill natural enemies. Plant materials that can be used as insecticides are the following:

Neem: Neem (*Azadirachta indica*) is a tree that is toxic and repellent to numerous pests, particularly insect larvae, aphids and thrips. All parts of this plant are toxic, but toxicity is highest in the seeds. Before using neem its plant parts should be pulverised until they are soft, and dilute them with clean water. Spray the mixture onto plants. Because the toxicity does not last long in direct sunlight, it is best to spray in the late afternoon. If the mixture is too concentrated, it will poison plants leaving them looking as if they have been burnt. The benefits and drawbacks of natural pesticides are given in Table 3. The presence of a wide variety of plant types, and especially a wide variety of flowering plants, will provide habitat for many beneficial insects. One of the major drawbacks of chemical pest management is that it destroys the natural balances of the insect world, particularly by killing beneficial predators and parasites. Growing three or four crops on same piece of land simultaneously will ensure crop diversity and reduce pest attack.

Table 3: Benefit and drawbacks of natural pesticides

	Benefits	Drawbacks
1	They are often cheap and easy for farmers to make	Materials are not always readily available to farmers
2	They are generally not toxic to humans or livestock	They must be applied appropriately and repeatedly
3	They do not pollute the environment, because their residues are easily broken down	Some natural pesticides also poison natural enemies
4	They rarely lead to insect immunity	It is difficult to determine the correct doses to apply. Bits of plants often block sprayer nozzles

Physical barriers for pests: Strategies include using traps and trap crops. Effective trap crops for leaf miner flies are all kinds of beans, as the flies prefer these plants to potatoes to lay eggs on. Trap crops contribute to increasing the role of parasitoids, and can be used as follows:

- Plant beans at the same time as potatoes on the edges of potato beds.

- One week after emergence conduct observations of these plants. Collect leaves affected by leaf miner flies and put them in the parasitoid release cage. Continue to do observations every other day.

Yellow sticky traps: These are only appropriate for insects which are attracted to the color yellow. Initially yellow traps were only used for observing the presence and quantity of these insect pests, but recently these are also being used as a means for reducing leaf miner fly populations. Traps can be made from a yellow coloured materials such as yellow plastic, yellow painted boards, oil bottles etc. can be used for making traps. Trap material can be smeared with something sticky such as glue, starch solution or old engine oil, then put them in the field about 10-20 cm above the tops of the plants. Bamboo stakes can be used for supporting the yellow boards. About 80 traps are required for one hectare. Position flat traps in line with the path of the sun (west-east). Traps that gleam in sunlight will be more effective. Many farmers have changed their insecticide use patterns as a result of the success obtained with yellow traps. When farmers find lots of leaf miner flies on their traps, they feel they are controlling them successfully and insecticide sprays are not needed. This supports the development of natural enemies, increasing their numbers, diversity and impact. Finally, natural enemies can control leaf miner flies by themselves. Drawbacks of yellow traps are:

- They only trap adult insects, while the actual pest is the larva. Trapped insects may have already laid their eggs on leaves. Hence, the effects of such traps on population regulation are limited.

- Some natural enemies in the parasitoid and predator groups are caught, as they are also attracted to yellow.

WEEDS AND THEIR CONTROL

Weeds

Weeds are both damaging and beneficial. Weeds are all the unwanted plants growing on farming land. They compete with the main crop and considered to produce nothing of any benefit. However, having weeds also have many beneficial effects (Table 4).

Table 4: Harmful and beneficial effects of weeds

S.N.	Harmful effects	Beneficial effects
1	They compete with potato plants for nutrients, sunlight, water and living space	They serve as green manure providing trace elements and improving soil structure. Weeds form a fundamental ingredient in making organic fertilizer
2	They require expenditure in their control	They form a covering layer protecting soil from sunlight or erosion damage
3	They become a food source for pests enabling them to thrive, even when there are no potato plants in the field	They become a food source for natural enemies. Weeds produce flowers and nectar that can be an alternative food source for parasitoids
4	They host diseases that affect potato plants. Crop rotation will not work if disease and pest supporting weeds are still present in the field	They become a food source for livestock

Weed management

Weeds can be managed in two ways:

- Before crop emergence: – The weeds incidence can be prevented to large extent by clearing away the before their seeds become mature weeds from the field during tilling, making raised seedbeds and planting. Collected weeds can be turned into manure by proper composting..

- Management when potato crop is growing – It can be done by pulling up weeds or burying them in the soil. It should be done twice in a season, at 30 and 50 DAP. Be careful when weeding at 50 DAP as tubers are starting to form, and any damage to potato plant root systems will affect yield and increase susceptibility to disease.

Most organic crop growers rely on hoeing (mechanical in large farms or hand hoeing in small farms) as a safe and available method for controlling weeds. However, hand hoeing for a long time would inadvertently damage or remove some of the vegetable plants, while missing some of the weeds. In addition, organic crop growers were unwilling to accept hoeing damage to their vegetable crops and to increase plants spacing because of yields losses. Also, the method is highly expensive if enough labor is used to remove weeds. Cultivation had been shown to reduce the yields of several crops, including potato (*Solanum tuberosum*) and asparagus (*Asparagus officinalis*) because of root pruning and crop damage. Hand weeding or hoeing is safe and very effective against annual and biennial weeds (Boydston, 2010). However, with rapid industrialization and urbanization in developing countries, human labor is rapidly becoming scarce and expensive. Mulching is quite effective in controlling the weeds. Both organic and inorganic materials are effective in controlling weeds (Figure 3).

Fig. 3: Use of paddy straw mulch and plastic mulch for weed control in Potato

Soil solarization

Soil solarization is a nonchemical method successfully used in many countries to control or reduce soil borne plant pathogens, weeds and mites. Solarization involves the use of transparent polyethylene sheeting to trap the heat from solar radiation to raise soil temperature to levels that are lethal to weed seeds and seedlings. Soil solarization for 2, 4 and 6 weeks with chicken manure has increased the average weight of cabbage plants by 55, 70 and 75%, respectively compared to the control with chicken manure. Soil solarization is a promising method to reduce the populations of soil borne pests and weeds without using pesticides. Weed control effectiveness is dependent on moist soil, sufficiently high air temperatures and solar radiation, and an adequate length of exposure. Moist soil is essential to heat conductivity and for keeping seeds in a more susceptible imbibed state. The effects of solarization on weed emergence were apparent for a short time after plastic was removed. Soil solarization increases temperature by up to 10 to 21°C in the upper soil layer, and increased levels of N, P, K, Na and EC in soil, but a slight effect was detected on OM (%) and pH of solarized soil comparing with an unsolarized one. There is a decrease in disease incidence, an increase in growth of various crops, and an improvement in crop yield (up to 437%) and crop quality as shown in Table 5.

Table 5: Effect of soil solarization on yield of different crops [Source: Satour (1997)]

Crop	Yield (t/acre)		% increase
	Non solar	Solar	
Onion	7	21	300
Tomato	8	35	437
Potato	7.5	12	160

Natural herbicides

The term 'natural-product' be defined as "ingredients extracted directly from plants or animal products as opposed to being produced synthetically and that are as good as or better than synthetic herbicides and that are likely to be much safer." Corn gluten meal, Vinegar (Acetic acid) and Citric acid have promises as non-synthetic herbicides for controlling weeds. Cinmethylin, a natural herbicide produced by species of sage, controls many annual grasses and suppresses some broadleaved weed species. Different classes of compounds have been known for the potential use as natural herbicides.

Haulm killing before harvesting

It is recommended to kill the haulms before harvesting when some of the produce will be used for seed. Under sub-tropical conditions, haulm killing can be done at 80 DAP (depending on the variety), by cutting them at the base of their stems or by pulling haulms. The foliage can be collected and composted.

Benefits of haulm killing

- To make tubers harden more quickly so they can be harvested sooner. Normally, harvesting can be done two weeks after haulm killing.

- To prevent diseases spreading from plant stems to tubers. Viral diseases in particular will spread to tubers if stems begin to wilt and dry out. The same occurs to other diseases such as late blight, stem rot and bacterial wilt.

Harvest time

Based on maturity period, potato varieties are early, medium or late maturing varieties. Kufri Chipsona need more time before harvesting (100-110 days) than Kufri Khyati, Kufri Pukhraj which is ready for harvest at 80-90 DAP (Days after planting).

Harvesting and estimating yield

Harvesting methods affect tuber quality. Potatoes can be harvested in two ways, directly by hand or by using a hoe. Harvesting by hand takes longer and is more labor intensive, but will produce good quality, undamaged tubers. Using a hoe is less time-consuming and labor-intensive, but some tubers will be damaged in the process. When soil is too hard and is covered with grass, dig up potato beds using a hoe. Be careful when loosening the soil so as not to damage potato tubers. If the whole produce is going to be sold, then harvest all the potatoes in one go. However, if some of the tubers are for seed, harvest them at a different time from the ware potatoes. After harvesting, you should sanitize the field, by gathering and destroying harvest remnants such as plant parts, rotten tubers etc. Post-harvest sanitation is an important part of controlling various pests and diseases, by removing sources of contamination for the next crop from the field. The vegetation of some varieties can be allowed to die back naturally, but for many types this practice may result in highly over-sized tubers. When a test digging indicates that a variety is at the size you prefer for harvest, it is time to prepare the potatoes for storage. The goal of this process is to cure and thicken the skin, and to slow the respiration of the tubers by closing the lenticels (the breathing openings on the potato skin). Most varieties, when cured properly, will store for months. New, uncured potatoes, with their delightfully thin skins, will not keep, but are delicious if eaten soon after harvesting. To begin curing the potatoes, give them a last complete watering. A day or two later, cut off all the vines at ground level, they will already very likely show signs of aging and decay. Remove the vines from the field, and let the potatoes begin curing. Removing the vines reduces the chances of spores or infection on the leaves coming in contact with the tubers; disease-causing organisms will generally dry out and die on the now-exposed soil surface. Test dig some potatoes eight to ten days after removing the vines; the skins should be quite tough by then. If not, give them a few more days.

Cost effectiveness

Presently conventional farming inputs are subsidized by governments. For organic farming to compete with highly subsidized conventional farming, government steps are needed for supporting of organic farmers. Since, conventional farming is disturbing ecology, that cost should also be included in cost of cultivation of conventional farming. Like- wise organic farming is in accordance with nature; hence, it should be given benefit of it. Since organic farming is localized, inputs are from wastes of farm produce and integrated with animals, external purchase cost is bare minimum. Though yields are less under organic farming but sustainability is highest in organic farming system (Table 6). Among cropping system, its stability, productivity, profitability revolves around potato crop. Higher the production of potato, higher is the stability and profitability etc.

Table 6: Mean combined yield in terms of rice equivalent yield of various nutrient management under different cropping system during four consecutive years (2004-05 to 2007-08)

Treatment	Rice-Equivalent		Yield (q/ha)		
	2004-05	2005-06	2006-07	2007-08	Mean
Nutrient Management					
100% organic	129.99	114.81	100.76	96.94	110.62
100% inorganic	157.85	131.77	112.30	107.86	127.44
Integrated (50% each of organic and inorganic)	149.77	124.19	108.71	107.11	122.44
Cropping system					
Green Manure-Rice-Durum wheat	78.93	78.91	58.32	60.16	69.08
Rice-Potato-Okra	226.69	178.52	168.35	175.08	187.16
Rice-Berseem (fodder and seed)	149.88	118.18	104.90	89.01	115.49
Rice-Vegetable Pea-Sorghum (fodder)	127.95	118.75	97.43	91.62	108.94
	Nutrient Management (M)		Cropping system (CS)		M x CS
SE m±	2.65		3.06		5.29
CD at 5%	7.62		8.79		NS

NS- Non significant

CONCLUSION

Indiscriminate use of chemical fertilizers and pesticides has posed a serious threat to the soil and environment. Many investigations have shown their adverse effects of change in soil nature, soil contamination, ground water pollution and decrease in soil micro flora etc. Farming practices like organic farming or natural farming utilizing natural products with the minimum external inputs and by application of supplements like Jeevamruth, improve the soil fertility by increasing the soil micro flora and available nutrients. This method of farming encourages multi cropping and biodiversity of micro and macro flora and improve soil health. Labor and production costs are minimized. The farming practices based on natural products are eco-friendly and crop by-products becomes input instead of pollutants in this system.

REFERENCES

Aulakh CS, Singh H, Walia SS, Phutela RP, and Singh G (2013) Evaluation of microbial culture (Jeevamrit) preparation and its effect on productivity of field crops. *Indian Journal of Agronomy* **58** (2): 182-186

Boydston RA (2010) Managing weeds in potato rotations without herbicides. *American Journal of Potato Research* **87**: 420–427

Canali S, Ciaccia C, Antichi D, Barberi P, Montemurro F and Tittarelli F (2010) Interactions between green manure and amendment type and rate: effects on organic potato and soil mineral N dynamic. *Journal of Food and Agricultural Environment* **8**:537–543

Devarinti SR (2016) Natural Farming: Eco-Friendly and Sustainable. *Agrotechnology* **5**(2):147

Doran JW, Sarrantonio M and Liebieg MA (1996) Soil health and sustainability. *Advance in Agronomy* **56**.

Doring TF, Brandt M, Heß J, Finckh MR and Saucke H (2005) Effects of straw mulch on soil nitrate dynamics, weeds, yield, and soil erosion in organically grown potatoes. *Field Crops Research* **94**:238–249

Goffart JP, Olivier M and Frankinet M (2008) Potato crop nitrogen status assessment to improve N fertilization management and efficiency: past–present–future. *Potato Research* **51**:353–381

Palekar S (2009) How to Practice Natural Farming? All India Pingalwara Charitable Society. 22–27

Singh JS, Pandey VC and Singh DP (2011) Efficient soil microorganisms: A new dimension for sustainable agriculture and environmental development. *Agriculiure, Ecocology and Environmemt* **140**: 339–353

Sreenivasa MN, Naik NM and Bhat SN (2010) Beejamruth: A source for beneficial bacteria. *Karnataka Journal of Agriculture Science* **17**: 72-77

11

Weed Management in Potato Crop

Sanjay Rawal[1], Pooja Mankar[2], S.P. Singh[3] and V.K. Dua[2]

[1]*ICAR-Central Potato Research Institute Regional Station*
Modipuram-250 110, Uttar Pradesh, India
[2]*ICAR-Central Potato Research Institute, Shimla-171 001*
Himachal Pradesh, India
[3]*ICAR-Central Potato Research Institute Regional Station*
Gwalior-474 006 Madhya Pradesh, India

INTRODUCTION

Control of weeds below economic threshold level in potato crop is a substantial component of crop management strategy for achieving optimum productivity. In our country, potatoes are raised either in assured irrigation or as rainfed crop with comparatively heavy doses of fertilizers. Therefore, weeds are bound to become menace during active crop growth period of this crop and these undesired plants needs to be controlled in early stages of crop growth by adopting several techniques. Current scenario of crop production is different from past due to many reasons. Farm holdings are becoming smaller day-by-day and majority of potato growers are small and marginal farmers. They have limited capacity for purchase of inputs and farm machines. They need economical, efficient and easier to adopt weed control technologies. Farm labour is also costly now in many regions as they are migrating towards cities and industrial townships. This scenario becomes more complex as weeds would be major challenge due to global warming, which is slowly changing climates of potato growing regions in India. The concern for environmental issues emphasize upon adopting systems and integrated approach for managing weeds in such a way that these do not reduce the potato and system productivity (CPRI, 2014).

WEEDS AND CROP LOSSES

Weeds, generally termed as unwanted or undesired plants, provide inter-plant competition for all the field crops. Potatoes are grown as short duration early or late crop (60-75 days), main crop for table and processing purpose (90-120 days) and as seed crop (80-90 days). Understanding weed-potato crop interaction is essential for devising a suitable strategy for effective weed control. When the crop gets less of resources than what it needs, the economic productivity is bound to decline. Magnitude of yield reduction depends on the density and competitiveness of various weeds among themselves and with the crop and the weed control methods adopted. Crop-weed competition also depends upon the season and time of planting of potato crop. Early crop suffers more due to weeds infestation as it has slow growth and lanky canopy compared to main crop. Similar is the case for *kharif* potatoes in plateaue regions due to rains and higher environmental temperature (Lal and Dua, 2003). Dry matter production of weeds is negatively correlated with tuber yield and up to 80 per cent reduction in the productivity of potato crop has been observed (Dash *et al.*, 2009).

Slow growth of early potato crop leaves ample time for weeds to become more aggressive. In main crop, which has faster growth rate from emergence to bulking, weeds are suppressed quickly. Competition between potato crop and weeds, for all macro and micronutrients, is certainly the most severe and is an important factor in determining the yield of this crop. Nitrogen, phosphorous and potash removal by weeds may vary between 21-43, 7-8 and 17-49 kg per ha, respectively, during whole crop duration. Soil water availability also becomes one of the limiting factors to influence crop yield under restricted soil moisture conditions in the presence of weeds. Weeds are generally better adapted to withstand adverse conditions in comparison to potatoes as they can deplete soil moisture rapidly, making the potato crop suffer in terms of growth, tuberisation and finally the yield. Competition for light and space between weeds and potato is another important component as weeds might over-power the potato plants because of their larger population, hardy nature, faster growth, proliferated and dense rooting system.

Several weeds harbour insects, pests, vectors and diseases, which adversely influence the growth and productivity of potato crop. These weed species serve both as an alternate and collateral host for pests and pathogens and help in their perpetuation and spread. That is why clean cultivation is one of the principles in integrated management for preventing pests and diseases (CPRI, 2012). Leafhopper, aphid, white flies and mites are severe pest in early crop, and aphids are important pest during main crop season in plains. In hills, white fly, tuber moth, cutworm and white grub are very serious pests. *Epilachna* beetle, a defoliator in hills and plateau region can cause 10-20% yield loss due to reduced photosynthesis. Major vectors of potato viral diseases are aphids, white flies, leafhoppers and thrips. Apart from spreading a large number of viruses in potatoes, these are also damaging crop canopy for considerable productivity losses (Table 1).

Weeds also harbour major fungal, bacterial and viral potato diseases and play a definite role in their multiplication and spread. Wart disease (under quarantine) thrives upon weeds *Solanum pimpinellifolium* and *S. sisymbrifolium,* and may contribute significantly towards maintaining high inoculation of this pathogen. Bacterial wilt has the potential for 30-70 per cent productivity decline in affected fields of plateau region. Several weed hosts (Table 2) harbor the bacterium of this disease, which leads to multiplication, and perpetuation of the disease. Most of the potato crop viruses have weed hosts widely distributed in fields and vicinity that serve as a reservoir for viruses like PVS, PVY, PVX, potato leafroll virus, potato top roll, potato phyllody, purple top roll, marginal flavescence and witch's broom etc (Table 3). Their share in yield losses may vary from 7 to 60 per cent depending upon their intensity and complexity of two or more strains. Abundance of weed species in and around potato crop will certainly accelerate disease epidemics and assist in rapid multiplication of specific vectors/ viruses.

Weed infestation is a nuisance in performing daily farm operations, which increases labour cost, breakages and depreciation of farm machineries during inter-culture and harvesting operations of the crop. It is also a problem in irrigation channels and drainage system of potato fields. Weeds like *Cyperus rotundus* L. may damage tubers physically by penetration of its vegetative parts (rhizome) thus deteriorating the tuber quality.

Table 1: Weed hosts of insects, pests and vectors of potato crop

Insects	Weed hosts
Epilachna beetle (*Henosepilachna vigintioctopunctata*)	*Datura stramonium* L., *Amaranthus caudatus* L.
Aphids (*Myzus persicae* Sulzer) and *Aphis gosspii* Glover)	*Ageratum conzoides* L., *Nicandra physaloides* Gaerten
White fly (*Bemisia tabaci* Gennadius),	Solanaceous & malvaceous weeds
Leaf hopper (*Alebroides nigroscutulatus* Dist and *Seriana equata* Singh)	Solanaceous & malvaceous weeds
Thrips (*Thrips hawaiensis, T. palmi, Megalurothrips distalis*)	*Amaranthus viridis*
Cutworm (*Agrotis ipsilon*)	*Chenopodium album, Solanum nigrum, Portulaca oleracea, Amaranthus viridis, Evolvulus alsinoides*
Leaf miner [*Phytomyza atricornis* (*Chromatomyia horticola*)]	*Cannabis sativa, Withania somnifera, Sonchus oleraeca* [*S. oleraceus*], *Chenopodium album, Parthenium* species

Table 2: Weed hosts of diseases of potato crop

Potato diseases	Weed hosts
Wart disease *(Synchytrium endobioticum)*	*Solanum pimpinellifolium, S. sisymbrifolium*
Late blight *(Phytophthora infestans)*	*Polygonum alatum, Buch-hum, Ipomea purpurea Lam., Sonchus oleraceus L., Datura stramonium*
Early blight (*Alternaria solani* Ell. & Mart)	*Nicandra physaloides*
Bacterial wilt (*Pseudomonus solanacearum* Smith)	*Ageratum conzoides, Amaranthus viridis, Datura metal* L., *Poygonum hydropiper, Phyllanthus niruri, Ranunculus schleratus*

Table 3: Weed hosts of viruses and virus like diseases of potato crop

Potato diseases	Weed hosts
Potato virus X	*Amaranthus caudatus, Amaranthus blitum, Chenopodium album* L., *Chenopodium murale* L., *Ipomea purpurea* Lam.
Potato virus S	*Amaranthus caudatus, Chenopodium murale* L.,*Chenopodium ficifolium* L.
Potato virus Y	*Ageratum conzoides* L., *Chenopodium murale* L.,*Convolvulus arvensis* L., *Datura metal* L.,*Nicandra physaloides* Gaerten, *Solanum nigrum* L.
Potato leaf roll virus	*Convolvulus arvensis* L., *Datura metal* L.,*Nicandra physaloides* Gaerten, *Solanum nigrum* L.
Potato top roll	*Convolvulus arvensis* L.
Potato phyllody	*Datura metal* L., *Datura stramonium,Nicandra physaloide* Gaerten, *Solanum nigrum* L.
Purple top roll Marginal Flavescence	*Datura metal* L., *Datura stramonium,Solanum nigrum* L.
Witch's broom	*Datura metal* L., *Datura stramonium,Nicandra physaloides* Gaerten, *Solanum nigrum* L.

MAJOR WEEDS OF POTATO CROP IN INDIA

Knowledge of dominant and invasive flora is very important for controlling weeds efficiently in different potato growing zones. Occurrence and intensity of weeds vary under various agro climatic regions, cropping systems as well as management conditions. The dominant weeds of potato crop in Indo-gangetic plains are *Cynodon dactylon* L. (Pers), *Cyperus rotundus* L., *Trianthema monogyna* L., *Chenopodium album* L., *Poa annua* L., *Anagallis arvensis* L., *Melilotus spp., Sonchus oleraceus* L. and *Vicia sativa* L. Major weeds of hills are *Amaranthus viridis* L., *Chenopodium spp., Oxalis spp., Digitaria sanguinalis* (L.) Scop., *Setaria glauca* (L). Beauv., *Spergula arvensis* L. and *Melilotus spp.* (Table 4). Potato crop is usually raised in wider crop geometry with liberal use of manures and fertilizers. Further, irrigations are frequent in plains and generally, rains are heavy in hills during crop season. All these practices are advantageous for early and faster growth of weeds even before the crop emerges out. It provides ample opportunity for weeds to flourish and dominate the crop, if not managed timely. This would culminate into reduction in tuber productivity particularly due to initial take over of weeds rather than late in the season. Initial period of 20-40 days after planting of potato crop is critical period for competition in plains, whereas, in hills it is upto 35-55 days after planting.

Table 4: Weeds of potato crop in India

Scientific name	Common name	Local name
Weed flora of the plains		
Amaranthis viridis L.	Pigweed	Jangali Chaulai
Anagallis arvensis L.	Pimpernel	Krishnaneel
Asphodelus tenuifolius Cavan	Wild onion	Piazi
Avena fatua L.	Wild oat	Jangali Jai
Chenopodium album L.	Lambs quarters	Bathua
Chenopodium murale L.	Goose foot	Kharthua
Cirsium arvense L. Scop.	Canada thistle	Kantaila
Convolvulus arvensis L.	Field bindweed	Hirankhuri
Coronopus didymus (L.) Sm.	Swinecress	Jangali halon
Cynodon dactylon (L.) Pers.	Bermuda grass	Dub
Cyperus iria L.	Yellow nutsedge	Motha
Cyperus rotundus L.	Purple nutsedge	Motha
Melilotus alba Desr.	White sweet clover	Safed senji
Melilotus indica L. All	Yellow sweet clover	Pilli senji
Oxalis corniculata L.	Indian sorrel	Khati-buti
Oxalis latifolia HBK	Wood sorrel	Khati-mithi ghas
Phalaris minor Retz.	Canary grass	Gulli-danda
Poa annua L.	Blue grass	Buin
Solanum nigrum L.	Blacknight shade	Makho
Sonchus oleraeeus L.	Sowthistle	Sow thistle
Setaria glauca L.Beauv.	Foxtails	Banra, Banari
Trianthema monogyna L.	Carpet weed	Patharchatta/ Its chit
Vicia sativa L.	Common vetch	Ruari, Ankari
Weed flora of the hills		
Amaranthus viridis L.	Pig weed	Jangali Chaulai
Bindens pilosa L.	Begger's sticks	Dipmal
Chenopodium album L.	Common lamb's quarters	Bathua
Chenopodium murale L.	Common lamb's quarters	Kharthua
Commelina benghalensis L.	Tropical spider wort	Kanchara/ Kanakaua
Cynodon dactylon L.Pers.	Bermuda grass	Dub
Digitaria sanguinalis L. Scop.	Crab grass	-
Echinochloa crusgalli (L.) Beauv	Bamyardgrass/ Watergrass	Savank
Melilotus indica L. All.	Annual yellow sweet clover	Pili senji
Oxalis corniculata L.	Wood sorrel	Khati-buti
Pennisetum clandestinun	Kikuya grass	Kikuya grass
Polygonum spp.	Black bird weed	-

(Contd.)

Rumex spp.	-	Jangali palak
Setaria glauca (L.) Beauv.	Foxtails	-
Spergula arvensis L.	Com spurry	Bundhania/ Matkan

WEED MANAGEMENT

Weed management is required to keep the intensity of undesirable plants below a limit where optimal tuber productivity and farm income are not impeded. Choice of a method of weed control depends upon severity of specific weeds, stage of weed growth, weather conditions and socio-economic condition of the farmers. Important weed management methods include cultural, mechanical and chemicals. Their proper integration into weed management strategy helps in sustaining crop yield.

Cultural Weed Control

Weeds are suppressed specifically in initial phases of crop growth by way of prevention and crop competition through adoption of best crop production practices. Weed emergence under better crop canopy are generally frail and will not be much harmful for tuber productivity. Further, these can be managed without much difficulty by adopting inter-cultivation or chemical methods. Adoption of suitable agronomic practices as outlined below can also reduce dependence on chemicals.

Crop Rotation

Well planned cropping systems can be quite useful in controlling weed density in long run. This is done following at least two-year crop rotation in a particular field or having green manure crops like *dhaincha*, cowpea etc. for smothering weeds. Two-year crop rotation will assist in reducing weed seed bank in field while in green manuring even for a shorter period (45-50 days), weeds get buried along with the green manure crops and are decomposed, which also add to the soil organic carbon. Thus, this operation also facilitates better potato growth, which will provide tough competition to weeds.

Hot weather cultivation

Hot and dry summer season should be harnessed in a cropping system for desiccating the weeds. Two-three deep field cultivations in the month of May and June in plains, where, maximum temperature is generally more than 40 °C, are very useful for the control of annual and perennial weeds like *Cynodon dactylon* L. Similarly, soil solarisation technique may be quite useful in specific situations. Soil solarization is done by using transparent polyethylene (TPE) film of 0.05 and 0.10 mm thickness for 30 and 40 days for killing weeds and the pathogens.

Field sanitation

This operation is done round the year on regular basis in potato fields or in the vicinity i.e. field channels and farm roads etc. to control weeds and herbaceous plants. Manual method, animal or tractor drawn implements are used for killing the weeds. This is very important for seed crop so as to avoid vector and disease infestation.

Crop planting

Seed bed is prepared thoroughly depending upon soil type of a region. Pre-plant tillage operations for making a proper soil tilth not only accelerate faster emergence of potato plant, but also destroy the weeds and give an edge to the crop. A competitive edge is given to potato crop by planting a variety at optimum date, where the soil bed is prepared properly and contains sufficient moisture. Well-sprouted seed tubers are planted at optimum crop geometry (60- 67.5 cm x 20-25 cm) at a proper depth (7.5-10 cm). This helps in faster emergence and growth of potato plant. At planting, manures and fertilizers are precisely placed in bands 5-6 cm below seed tubers, so that these inputs remain in root zone of the crop. Thus, the plants are able to harness nutrition efficiently in comparison to weeds, which will accelerate the vegetative growth. Faster coverage of fields by potato canopy deprives weeds from uptake of nutrients and thus assist potato crop in reducing their intensity particularly in initial phase of plant growth.

Mulching

Mulching is an efficient way for smothering the weed growth during crop season and more specifically for annuals. Main objective of mulching is to divest the weeds off solar radiation and thus inhibition of weeds. Germination of weeds is obstructed and this practice helps in conserving soil moisture, which facilitate quick emergence of potato plants. Crop residue, dry straw, dry grasses, pine needles and other vegetative material can be utilized for this purpose. Recently plastic mulching has also come up as a promising technology for weed control.

Inter-cultivation

Potato crop has critical period of crop-weed competition, so timely inter-cultivation and weeding are very relevant for maintaining better crop growth and higher tuber productivity. Inter-cultivation is better 20-25 days after planting when the plants are about 10-12 cm in height. Earthing up is done in morning of next day after inter-cultivation and weed removal for conserving soil moisture and proper ridging.

Haulm killing

Haulms are killed manually by using sickles or mechanically by using haulm cutters at chemical maturity of the crop for proper skin setting of tubers. This also helps in prevention of seed production of earlier weeds and mowing down of second flush of weeds that comes during senescence of potato crop.

Mechanical weed control

Mechanical methods involve the principle of eradication (uprooting and buring of weeds in soil layers), and control through destroying their vegetative parts (cutting, mowing or thermal treatment) and reducing the regenerative capacity of weed plants. These methods mainly include manual weeding, animal drawn implements and tractor-operated machines. The implements are designed according to the requirements of field operations. Mechanical weed management is very effective against annuals, but has limited utility for control of perennial weeds due to their deep rooted system or vegetative propagules. Another disadvantage is that this will not work, if soil moisture is at saturation level due to precipitation as mechanical traffic is obstructed. However, this approach is becoming more significant nowadays due to concerns for the environment and emphasis on avoidance of chemicals in food chain.

Manual method

Removal of weed plants by hand or by implements like *khurpi*, hand hoe, spades etc. is an old practice and still followed in many parts of India in potato crop by small and marginal farmers. This may be a feasible and efficient method for controlling the undesirable vegetation, provided manual labour is available. It is quite effective against annuals and biennials, as they do not re-generate from the pieces of vegetative parts left in soil after such operation. This method is particularly better as it destroys weeds within the rows, which are generally not controlled by the mechanical cultivation. Manual method has limitation of covering lesser area per unit time during crucial phase of rapid weed growth and is only practical for small farm size.

Animal drawn implements

Animal drawn three-tine cultivators are quite efficient and cost effective for inter-cultivation in potato crop. Narrow shovels are better for weeding operation as it will not damage roots and stolons of potato plants growing over ridges. One pair of bullocks per day can cover approximately one hectare of land. After inter-cultivation and desiccation of weeds, animal drawn single bottom ridger can be used for earthing up of the crop. These are as effective as tractor operated machines, but a pair of bullock add to cost of cultivation and coverage of field area per day is also limited.

Tractor drawn implements

Tractor operated machines are very efficient and can cover larger fields in a day. Spring tine cultivators consisting of spring tines with narrow reversible shovels fitted to a tractor tool bar may cultivate three or more potato rows at a time depending upon available brake horse power. As each tine is hinged at its base so the lateral position of shovels can be changed easily with a mild foot below, to reduce root and stolon damage if crop is cultivated at different stages of growth. Later on ridgers consisting of three or more bottoms are used for earthing up operation. Tractor based mechanization can cover 5-6 hectare of crop field per day easily, however, it requires good investment and economic support.

Chemical weed control

Research and development in herbicides for weed control in its present form is due to discovery of 'hormone' herbicides like IAA, NAA, MCPA and 2, 4 -D in early part of 20th century that revolutionized weed management in agriculture (Table 5). This works on the principle of selectivity of plant species and has several advantages over cultural or mechanical weed control. Chemicals can work where inter-cultivation is not feasible. It is very easy to control weeds during germination phase and early growth phase of a crop with chemicals. Weed control through weedicides is also faster and much less cumbersome as larger areas can be covered in a short span with less labour. Tillage operations are reduced considerably and selectivity of herbicides control weeds with in crop rows, where inter-cultivation is not effective. In potato, use of chemicals is highly efficient in controlling annual and perennial weeds, which are not so easily reduced by other methods (Table 5). Chemical weed management facilitates better growth of crop due to less root and shoot injuries, and results often in improved tuber productivity over other means of weed control. This method is very important in seed potato production programme as it minimizes the spread of mechanically transmitted viruses like X and S during inter-cultural operations. Work on herbicides has been carried out in potato crop in past six decades and several of them are used depending on their availability, cost effectiveness, type of weed flora and adopted cropping systems. Major disadvantage in use of chemicals is environmental concern and food safety. These chemicals have been divided in three categories as pre-planting, pre-emergence and post-emergence from application point of view due to their bio-chemical activities.

Pre-planting herbicides

Pre-planting herbicides are applied before planting of potatoes, and require soil incorporation to avoid photo-degradation. Their absorption is mainly by roots and translocation is throughout the plant system. Fluchloralin and pendimethalin are main herbicides and both compounds belong to dinitroanilines group available as emulisifiable concentrate. Fluchloralin (N- (2- chloroethyl)- 2, 6- dinitro- N-propyl- 4- (trifluoromethyl) aniline) is used for selective control of annual monocots and dicots. At 0.7-1.0 kg/ ha concentration, it provides efficient control of weeds with improved tuber yield. Pendimethalin (N- (1- ethylpropyl)- 2, 6-dinitro-3, 4 xylidine) is normally applied as pre-plant incorporation (1.0-1.5 kg/ ha) in the soil but it can also be applied as pre-emergence. Under normal crop conditions, it persists for less than a year at phytotoxic levels in soil. It kills annual broad-leaved weeds including some traditionally difficult species such as *Viola spp.*, *Veronica spp.*, and *Gallium aparine* L, and annual grasses.

Pre-emergence herbicides

Pre-emergence molecules are generally applied within 3-5 days after planting of potatoes and before the emergence of weeds and crop. These are selective, systemic, persist in top soil layer and remain in a very fine film on soil. Roots or shoots of emerging seedling absorb these chemicals. They penetrate into xylem and their translocation is apoplastic. Mode of action depends upon their specific bio-chemical properties affecting particular metabolic process (respiration, photosynthesis, protein and nucleic acid metabolism and enzymatic activities) in tissues of weed plant. A wide range of chemicals in this group is available for application in potato and growers may make their choice depending upon weed species and their intensity. Herbicide 2, 4-dichlorophenoxy acetic acid (2, 4-D) belongs to phenoxyacids group. It (0.5 kg/ ha) suppresses weed population in potato and gives at par tuber yield as conventional hand weeding. It is quite effective against weed species like *Chenopodium album* L., *Amaranthus viridis* L., *Cirsium arvensis* L., *Convolvulus arvensis* L. and several other broad leaf weeds.

Isoproturon (N'-(4-isopropylphenyl)-N, N-dimethylurea) a substituted urea compound is a selective soil active herbicide having limited shoot activity. Its application (0.5kg/ ha) decreases the grassy weed population and their dry matter accumulation resulting into higher tuber yield. Methabenzthiazuron (N-(benzothiazol-2-yl)-N, N'-dimethylurea) is a very promising herbicide, which belongs to substituted urea compound group, and is formulated as wettable powder. It also persists through out the crop growth period of potato and controls both monocot and dicot weeds. It is very effective herbicide against *Phalaris minor* Retz., *Echinochloa, crusgalli* L. Beauv., *Chenopodium album* L. and *Cirisium arvense* L. Scop at the rate of 1.0 kg/ ha. Oxyfluorfen (2-chloro-4-trifluoromethylphenyl 3-ethoxy-4-nitrophenyl either) is very important herbicide among diphenyl ether compounds and its main formulation is emulsifiable concentrate. It is absorbed by both leaves and roots, but its translocation is very little from these sites. So it is considered as a contact herbicide and exposure of weed plants to light is essential after its spray. It can control several annual weeds in potato crop efficiently like *Chenopodium album* L., *Anagallis arvensis* L., *Melilotus spp.*, *Trianthema monogyna* L. and *Vicia sativa* L. at a rate of 0.1-0.2 kg/ ha.

Triazine group of herbicides are extensively used for selective weed management in potato crop. These are adsorbed reversibly by clay and organic colloids in soil and thus have very little leaching. Atrazine (2-chloro-4-ethylamino-6-isopropylamino-1, 3, 5-triazine) is effective in controlling annual grasses and broad leaf weeds of potato crop and is absorbed through roots for further translocation to other plant parts. Atrazine is very potent for killing *Digitaria spp.*, *Echinochloa spp.*, *Cyperus rotundus* L., *Chenopodium album* L., and *Avena fatua* L. etc. Atrazine is recommended at 0.3-0.5 kg/ ha for better weed management in potato crop. Metribuzin (4-amino-6-t-butyl-3-(methylthio)-1, 2, 4-triazin-5(4H)-1) is the most effective herbicide among triazine group of chemicals and its persistence in soil is also less than other triazine molecules. Metribuzin is translocated through root and shoot, and has prolonged soil residual activity of 6-12 weeks. It is very effective against annual grasses and many broad leaf weeds in potato particularly *Chenopodium album* L., *Anagallis arvensis* L., *Melilotus* spp., *Vicia sativa* L., *Trianthema monogyna* L., *Poa annua* L. etc. at the rate of 0.7 to 1.0 kg/ ha.

Post-emergence herbicides

Chemicals applied after emergence of weeds and crop plants are termed as post- emergence herbicides. Paraquat (1, 1'-dimethyl 1- 4, 4'-bipyridylium), a contact herbicide of bipyridyliums group is commercially available as emulsifiable concentrate. Its spray on vegetative parts causes wilting and fast desiccation of foliage of weed plants within few hours. It is sprayed in the late afternoon or mid day as this gives some time for a little internal transport of molecules during night before development of acute phytotoxic symptoms for getting better results. It has no residual activity as it is rapidly adsorbed and inactivated in soil. Paraquat is the best herbicide used as directed spray for managing annual monocot and dicot weeds at a rate of 0.4 to 0.6 kg/ ha as early post-emergence in potato up to 5% emergence of crop.

New generation weedicides

Several molecules have been developed in past two decades that include rimsulfuron, sulfentrazone, clomazone, flumioxazin, ethalfluralin, prosulfocarb, triasulfuron, imazosulfuron, bentazone etc. used in potato crop. These need to be evaluated in Indian conditions as safer chemicals is a priority now for chemical weed management. Application of Bentazone (1.0-1.5 l/ ha) reduced total weed dry weight by 79.6-89.8 per cent over weedy check resulting into comparable tuber productivity to weed free plots. Mixing of compatible herbicides is a practical proposition for greater weed control efficiency of different weed species. Sulfentrazone (80 g/ ha) in combination with metribuzin (420 g/ha) improved control of *Amaranthus retroflexus*, *Chenopodium album*, *Solanum sarrachoides* and volunteer oat over their sole application. Flumioxazin (53 g/ha) with metribuzin, pendimethalin, S-metolachlor or ethalfluralin provided greater than 90% *Solanum saccharoides* control. Flumioxazin or rimsulfuron+ metribuzin provided more than 90% control of *Amaranthus retroflexus*, *Chenopodium album* and *Setaria viridis*. Similarly, Imazosulfuron+ S-metolachlor controlled yellow nutsedge more than 92 and 89% at 21 and 42 days after post emergence application, respectively.

Table 5: Major herbicides for weed management in potato

Herbicide	Dose (kg or l a.i./ha)	Type of weedflora controlled	Characteristics/mode of action
Pre-planting			
Fluchloralin	0.70-1.00	Annual grasses and broad leaf weeds	Systemic, selective, soil applied, absorbed by roots
Pendimethalin	1.00	Annual grasses and broad leaf weeds	Systemic, selective, soil applied, absorbed by roots
Pre-emergence			
Atrazine	0.50	Annual grasses and broad leaf weeds	Systemic, selective, soil applied, absorbed by roots
Isoproturon	0.50	Broad leaf weeds	Systemic, selective, soil active
Methabenzthiazuron	1.00	Annual grasses and broad leaf weeds	Selective, absorbed by roots, persists throughout potato crop growth
Metribuzin	0.75-1.00	Annual grasses and broad leaf weeds	Selective, root and shoot mobile, soil residual activity upto 6-12 months
Oxyfluorfen	0.10-0.20	Annual grasses and broad leaf weeds	Selective, absorbed by roots and shoots but translocation is very limited
2, 4-D	0.50	Broad leaf weeds	Selective, translocated, absorbed by roots and shoots
Post-emergence			
Paraquat	0.40-0.60	Annual grasses and broad leaf weeds	Contact, non-selective, absorption by leaves

Biological weed control

Use of parasites, predators and pathogens is getting attention to bring the weeds of a given crop or cropping system below economic threshold level by utilizing natural enemies of specific weed species. Attempts have been made to use combined effect of herbicide induced stress and arthropod herbivory to reduce the weed intensity. Fluroxypyr dose-response bioassays using volunteer potato were conducted in the presence and absence of Colorado potato beetle (*Leptinotarsa decemlineata*) herbivory in USA. Parameters like leaf area, shoot biomass, tuber number and weight were lower with herbivory. Season-long bioassays revealed that addition of herbivory reduced herbicide use by 65-85%. In another study, eighteen isolates belonging to nine genera have been obtained from the diseased parts of Eupatorium (*Chromolaena odorata*). Fungus *Aureobasidium pullulans* has potential for probable use on various target weed plants. Identification of fungi associated with the infected parts of *Lantana camara* for further exploitation as biological control agents is also in progress. Two fungi in pure culture (*Alternaria spp.* and *Fusarium spp.*) have been isolated for potential use in biological control strategies.

Allelopathy

Allelopathy deals with biochemical interactions among plants, algae and micro-organisms. It explains the mechanism of action of allelo-chemicals at molecular level and their influence on enemy plants and micro-organism for better survival of plants. It is novice field in agricultural research, which has latent potential for improvement and sustenance of potato productivity by replacing chemicals with nature's chemicals. Although fundamental work on this aspect has been carried out in our country, but no work has been done with reference to weed control in potato crop. Prospects of allelopathic weed management are better as it may provide a strategy for weed control in organic and sustainable potato farming. Organic substance like vinegar and clove oil based herbicide for weed control in potato have been found effective. Crop residue of faba bean (*Vicia faba*), vetch (*Vicia sativa*) and oleander (*Nerium oleander*) reduces germination of *Orobanche* spp. Rye acts as a very effective cover crop with triple planting density in potato crop and reduces dry weight of weeds considerably. Botanical pelargonic acid (30 per cent) has been found effective for weed management in potato crop at the rate of 20-25 l/ ha as post emergence, fast-acting, broad-spectrum and non-selective contact herbicide. Bio-herbicide agent *Dactylaria higginsii* (fungus) when produced on sorghum x sudangrass substrate is most virulent on nutsedge seedlings, and purple nutsedge was more susceptible than yellow nutsedge. Liquid derived from sweet potato shochu distillery waste (800 ml/ m^2) reduced *Digitaria ciliaris* and *Amaranthus patulus* germination significantly.

INTEGRATED WEED MANAGEMENT

This approach targets reduction in weed intensity below the economic threshold level with minimum damage to environment by combining all options in weed management strategy. This involves combined use of more than one weed management approaches comprising cultural, mechanical, chemical and other alternative methods. Combination of cultural and chemical methods of weed control is very effective in controlling weed infestation in early and later phase of crop growth. Weeding and hoeing followed by earthing up proved insufficient in controlling weeds, but combination of cultural and chemical weed control were more efficient in reducing weed intensity. Many combinations like oxyfluorfen (0.20 kg/ha) + one hand weeding was quite effective for tuber yield, and gross and net monetary returns (Borude *et al.*, 2001). Earthing up and mulching can also be very good component in integrated weed control.

Cropping system based weed management

Potato crop is grown mainly with food grain, vegetable and fodder crops in a sequence or as inter-crop in all potato-growing regions of the country. Proper development and adoption of cropping sequence also facilitates weed reduction. Among farmers, potato is a favourite crop as it adjusts well in various multiple and inter-cropping systems due to its flexibility in growth habits. However, suitable weed management strategy has to be worked out for promising potato based cropping and intercropping systems for efficient control of weeds without adversely affecting preceding or succeeding crops, especially when chemicals are used.

Inclusion of potato crop reduces *Phalaris minor* significantly in rice-wheat system (Chahal *et al.*, 2005). Similarly, the effects of catch crop (pea, toria or potato) in rice-wheat system is detrimental to weeds in rice-wheat system especially *Phalaris minor*. In North Eastern hills, inclusion of vegetables such as tomato, carrot and potato could be profitable and help in reducing weed intensity of the systems. In maize, potato and wheat based systems, herbicide atrazine in maize does not leave any residual effect on potato crop. Maize-potato-wheat crop sequence has lower weed density and higher system productivity, and also lower population of *Phalaris minor* (15-20 per cent) than in rice-wheat system. In potato and wheat/barley based systems, application of fluchloralin (0.5 kg/ ha) and metribuzin (0.25 kg/ ha) cut down the weed intensity distinctly and improve tuber productivity. These herbicides have no or little residual effect in soil for succeeding late sown wheat or barley crops. In pulses based systems, herbicides fluchloralin (1.0 kg/ ha), metribuzin (0.6 kg/ ha), methabenzthiazuron (1.0 kg/ ha), oxyfluorfen (0.15 kg/ ha) and paraquat (0.5 kg/ ha) in potato crop do not affect successive crops of green gram or blackgram adversely. These are efficient herbicide in controlling weeds in pulses based potato crop systems.

Weed control is difficult in intercropping systems unlike sole crop as choice of suitable mechanical or chemical method is limited. Selected mechanical operation or chemical should be able to reduce weed population in both main and component crop without having any adverse effect on any of them. In sugarcane and potato intercropping, pre-emergence application of metribuzin (0.8 kg/ ha) provides weed free environment throughout the crop season of both the crops due to their residual effect. Further, in sugarcane + potato intercropping system, weed growth is reduced and weed control efficiency improves (66 per cent) as cover provided by potatoes reduces the weed growth due to competition. In potato and maize intercropping, application of metribuzin (0.5 kg/ ha) as pre-emergence in potato crop controls weeds of this system effectively and economically. Pendimethalin is efficient in reducing weed growth in maize + potato inter-cropping. Although, methabenzthiazuron is better in managing weeds in intercropping of potato and maize, but had deleterious effect on subsequent maize crop to some extent. Weed control in potato and mustard intercropping specifically that of *Cyperus rotundus* is better due to allelopathic influence. This intercropping in 3:1 replacement series is remunerative and productive as compared to sole potato and *Brassica* spp.

Weed control under organic farming systems

Organic farming for eco-friendly crop raising is gaining momentum at international level and National Programme for Organic Production (NPOP) norms are adhered for weed management and use of products. Emphasis is given on balanced nutrition, suitable variety, fertile soils with high biological activity, crop rotations, green manuring, intercropping, mulching and land preparations etc. Products prepared at farm level from local plants, animals and microorganisms are permitted for weed control. Thermic weed control and physical methods are also permitted. All equipment from conventional farming systems should be properly cleaned and free from residues before being used in organic field. Use of synthetic herbicides is prohibited and farmers have to keep records for schedule of weed control at farm level for certification. Crop-weed competition is managed by optimum planting date, plant population, companion cropping, crop rotations and fertility manipulation. Weeds are thus suppressed out in initial phases of crop growth by making major components of crop growth in favour of potato crop. Weeds emerging out under better crop canopy are generally frail and will not be much harmful to tuber productivity. Potato cultivars having vigorous and rapid growing habits may prove better competitors for weeds as they cover fields quickly and overwhelm these undesirable plants. Crop rotation may be done following at least two-year crop rotation in a particular field or having green manure crops like *dhaincha*, cowpea etc. for smothering weeds. Two- three year crop rotation will assist in reducing weed seed bank in field, while green manuring (45-50 days) shall reduce their intensity as they get buried along with the green manure crops and are decomposed. Seed bed should be prepared thoroughly depending upon soil type of a region. Mulching of crop residue, dry straw, dry grasses, pine needles and other vegetative material may be utilized (Datta and Chakraborty, 1995). Plastic mulching is also permissible provided either the material is reused or disposed of properly.

Precision farming and weed management

Precise weed management is advantageous as it reduces chemical load in environment by way of their site-specific applications, and spatial and temporal information on weed occurrence and distribution helps in better understanding of weed biology and ecology for devising more efficient weed control approach (Aldrich and Kremer, 1997). Weed mapping of a field during the period of potato emergence, growth and maturity can be done and coordinates can be determined for each point using GPS kit. Values of distribution of all present weed species (g green mass m^2) can be computed to each coordinates and critical amount of present weed species per unit area in relation to yield reduction can be determined. Maps of weed distribution on a field plot can be developed for designing mode of spray and schedules. There is good possibility to obtain data on spatial variability of weed species using GPS and mapping software. Multiple, well timed shallow precision cultivations or flaming and application of nonselective organic herbicides coupled with new technologies for detecting crop rows and weeds hold promise in future precised weed control. Automated machine vision can be used to detect weeds or volunteer potato plants in real time for their effective control. Non-selective herbicides can be used with use of précised micro-sprayers for better biological efficacy.

Crop- weed modeling and decision support systems

Modeling of crop-weed interaction is another new area of interest for weed researchers to formulate précised weed management approach in different crops and cropping patterns (Kropff *et al.,* 1993). Different approaches are used in quantification of crop-weed competition and model development. Development of decision support system for optimisation of herbicides application is underway in many countries based on crop growth factors to optimize efficacy and selectivity of molecules. 'WeedCast', a weed emergence prediction model, has been developed in USA and it has been successfully used as a decision aid to schedule potato cultivation with and without herbicides in multi-location trials.

This has proved effective in decision making for cultivators at farm level trials. In India, Potato Weed Manager (PWM) software has been developed for weed control in field conditions. The software provides the information on the weed control method to be adopted based on the location of the field, type of major and minor weed flora and stage of crop growth.

Biotechnology and nanotechnology for weed management

Biotechnology can be used for achieving a breakthrough in weed management by developing genetically modified varieties resistant or tolerant to certain herbicides. Research groups are working on genetically modified potato lines for resistance to herbicides and success has been recorded in achieving minimal or no damage for herbicide sprays like chlorsulfuron. Scientists are also working on designing herbicide for its probable use in herbicide-resistant engineered plants, which could successfully catalytically be destroyed in such plants. These types of work are mainly going on in USA, Russian Federation and China for decreasing chemical load in the environment. Likewise, new vista of science 'Nanotechnology' can help a lot in revolutionizing future strategy of weed management in potato crop (Chinnamuthu and Boopathi, 2009). The work has already started on 'nano-herbicides', 'controlled release formulations' and 'smart delivery systems'. Herbicides are to be designed and formulated to release the active ingredient only when the soil receives moisture or irrigation, which is the right stage of weed flush. Control of parasitic weeds with nano-capsulated herbicides thereby reducing the phyto-toxicity of molecules on crops best explains the benefits of smart delivery system in agriculture. Properly functionalized nano-capsules provide better penetration through cuticle and allow slow and controlled release of active ingredients on reaching the target weeds. Nano encapsulation of chemicals with biodegradable materials makes it safer and easy to handle by the growers. In coming time, it will be possible to kill noxious weeds like *Cyperus rotundus* L. through smart delivery system.

Environmental safety, human health and Good Agricultural Practices (GAP)

The concept of GAP is based upon INDGAP (Indian Good Agricultural Practice) regulations framed by the Quality Council of India (QCI) and a systematic approach to ensure potato growers in identifying and managing the risks involved in process of potato production. This is almost in a protocol format which needs to be followed during crop season. Safe food production, environmental safeguard and quality hazards are priority. Studies show that isoproturon have been found in traces in soil or in potato tubers at harvest time. Applying weedicides carelessly can harm non-target organisms that are beneficial to agricultural ecologies and best way is to avoid injury to beneficial insects and microorganisms by judicious use of chemicals. Reduced microbial activity in soil under controlled environment or in field by use of herbicides like metribuzin, linuron and Pendimethalin have been reported. Control plots recorded higher populations of phosphate-solubilizing microorganisms and soil enzymatic activities while herbicides significantly reduced these. Therefore, herbicides should be used whenever their application is really necessary. Residues in soil can be avoided by use of safer herbicides, which degrade fast and do not persist for longer time. Integrated weed control strategy is adopted in GAP certification so that minimum herbicides are used. If recommended package for integrated weed control is followed then issue of MRL in GAP would be addressed properly. Integration of cultural and mechanical weed control measures in crop cycle and potato based cropping system should be strictly followed to avoid use of weedicides. After following all these practices, if chemical control is required then only recommended weedicides with the prescribed doses should be applied at right stage of crop growth. A record of weed control plan and schedule followed in crop season is to be maintained. Safe storage of herbicides has to be ensured and no spillage is permitted apart from in storage/ disposal site where adequate measures are adopted for avoiding contamination of surrounding environment.

Precautions in application and environmental protection

Information regarding direction for use should strictly be followed for better efficacy of herbicide and safety. Avoid drift as herbicide may damage sensitive crops in vicinity of potato crop, so do not spray in windy conditions. Care should be taken in avoiding spray overlap, as crop may be damaged due to over dose. Do not apply herbicide to potato crop if it is suffering from diseases, abiotic stress or nutrient deficiencies, or grown in acidic conditions. Herbicide should not be used if a previously applied residual herbicide is persisting in the soil and this is especially significant for high organic content soils. Pre or post emergence herbicides are applied to soil surfaces in such a way that they cover both sides of ridges uniformly. In case of post-emergence applications where crop is sheltering the weeds, it is essential that the spray penetrate weed canopy. Filters of 80 mesh size should be used in sprayers for avoiding nozzle clogging. It is important to clean sprayers after use, especially if they are used for more than one crop and for application of insecticides and fungicides. These chemicals are in general toxic to wildlife, so do not apply directly to water or to areas where surface water is present. Contamination of water bodies should be avoided when cleaning equipment, disposing equipment wash waters and containers. Do not allow direct spray from horizontal boom sprayers to fall within 5 m of top of bank of a static or flowing water body, unless a local environment risk assessment organization permits a statutory buffer zone, or within 1 m of the top of a ditch, which is dry at the time of application. Similarly, contamination of drainage is to be evaded. Care must be taken to ensure that off-target drift is minimized on food, forage and plantation crops. State laws, regulations, and guidelines should be followed for environmental safety.

IMPACT OF CLIMATE CHANGE ON WEEDS

It is envisaged that climate change will impact the geographical spread of agricultural pests, diseases and weeds. Although a number of weeds have C_4 photosynthesis and should theoretically not respond to increasing CO_2, however, a number of C_3 weeds can show a strong response. Many of the weeds associated with a crop are "wild" relatives and therefore have the same growth habits and photosynthetic pathway (e.g. rice and red rice, oat and wild oat). Although C_3 plants (e.g. potato) are more benefitted with increase in atmospheric CO_2 concentration, however except for the C_4 weed/C_3 crop combination in which CO_2 favour the crop, all other combinations favor the weed. It is also suggested that rising carbon dioxide changes weed populations with C_3 weeds selected preferentially. Parthenium, a C_3 weed may grow more rapidly under higher CO_2 levels and become more competitive. Thus, rising CO_2 may be a selection factor in weed species dominance. Nitrogen-fixing weeds, such as acacias may especially benefit because growth stimulated by CO_2 will not be constrained by low nitrogen levels. Some evidences have suggested that agronomic weeds may reduce crop yields further in a higher CO_2 environment. As far as weeds are concerned, these are going to be invasive in many potato growing regions of the country. Response of invasive weeds is about three times the average for recent CO_2 increases. Therefore, consideration of this aspect is important to devise suitable weed control strategy and tactics. Review of work done by climatologists and agronomists focusing on cropping patterns, and pests, diseases and weeds in Europe and North America suggest variability in weed infestation of north and southern potato growing zones. Initial evidence indicates that rising CO_2 may be a factor in establishment of invasive weeds. As in Greece, species like *Solanum elaeagnifolium* (Cavanilles) not only became invasive, but also host Colorado potato beetle in post harvest period of spring potatoes. Increasing CO_2 also reduces herbicide efficacy. In a number of studies, it has been found that increased CO_2 improved resistance to glyphosate. As carbon dioxide increases, glyphosate efficacy is reduced in controlling *Chenopodium album*. However, the basis for reduction is not entirely known. Thus the evidences indicates that climate change, particularly rise in CO_2 per se can potentially effect weed populations, crop losses due to weeds, species diversity and effect weed control efforts. As the impacts of climate change on ecosystem are complex, so useful analysis of climate-change impacts on crops and cropping systems will often require consideration of the wide array of other biota that interact with plants, including plant diseases, animal herbivores, and weeds.

WAY FORWARD

Weed biology is fundamental to better understanding of survival tactics of weeds in a specific agro-ecology. In this context, nasty and probable invasive weeds of potato crop should be given priority under changing climate scenario of a region. Work on economic threshold levels of weed competition in newly released potato varieties and potato based cropping systems would be beneficial for farming community in reducing cost of cultivation. Continuous efforts ought to be made to evaluate the new and promising molecules in potato crop and its systems by adopting a protocol having observations for type of weeds controlled, phyto-toxicity, productivity, tuber defects and any impact on storage etc. New frontiers of science like biological methods, plastic mulching, allelopathic products, organic approach, precision farming, modeling, decision support systems, biotechnology and nanotechnological tools, and smart agriculture have to be explored in context of weed management in potato and potato based cropping systems for incorporating them in future weed control strategy (Das, 2016). More précised integrated weed management modules are required to be developed for various potato growing regions involving minimum cost by combining suitable methods of weed control. Food bio-safety and environmental issues must be addressed in coming time through planned studies as herbicidal residues and persistence cannot be tolerated now.

REFERENCES

Aldrich RJ and Kremer RJ (1997) Principles in weed management. Iowa State University Press, Iowa: 455p, ISBN 0-8138-2023-5

Borude SS, Solanke AV and Raundal PU (2001) Integrated weed management in potato in Western Maharashtra. *Journal of the Indian Potato Association* 28(2-4): 271-273

Chahal PS, Brar HS and Walia US (2005) Soil seed bank dynamics of *Phalaris minor* in relation to different cropping systems. *PAU Agricultural Research Journal* 42(1): 13-18

Chinnamuthu CR and Boopathi PM (2009) Nanotechnology and agroecosystem. *Madras Agricultural Journal* 96 (1-6): 17-31

CPRI (2012) Integrated management of potato pest. Central Potato Research Institute, Shimla, India. Technical bulletin-96: 52p

CPRI (2014) Weed management in potato. Central Potato Research Institute, Shimla, India. Technical bulletin-33: 78p

Das TK (2016) Weed science basics and applications. Jain Brothers, ISBN8183600964, 901p

Dash SN, Jena SN, Nayak A, Barik T and Pati P (2009) Weed management in potato. *Environment and Ecology* 27(2A): 940-941.

Datta T and Chakraborty T (1995) Effect of organic manures and subabul (*Leucaena leucocephala*) leaf mulching under varying levels of fertility on growth and yield of potato (*Solanum tuberosum*) and weed biomass. *Indian Journal of Agronomy* 40 (1): 140-142

Kropff MJ, Lotz LAP and Weaver SE (1993) Practical applications. In, Modelling crop- weed interactions. Kropff MJ and van Laar HH (eds), CAB International and IRRI, Wallingford: 149-186, ISBN 0 85198 745 1 (CABI)

Lal SS and Dua VK (2003) Weeds and their management. In, The Potato: Production and Utilization in Sub-tropics. Khurana SMP, Minhas JS and Pandey SK (eds), Mehta Publishers, New Delhi: 121-129, ISBN 81-88039-18-7

12

ICT Applications in Potato Cultivation

Shashi Rawat[1], VK Dua[1] and PM Govindakrishnan[1]

[1]ICAR-Central Potato Research Institute, Shimla-171 001, Himachal Pradesh India

INTRODUCTION

Information is power in the present age of rapid technological innovations; since information technology (IT) empowers people. IT emerged as the major driver of growth in all walks of life during last quarter of 20th century and is currently in a steady growth trajectory. It encompasses almost all human endeavors including agriculture. IT can facilitate agriculture in two broad ways; (i) as a tool for direct contribution to agricultural productivity and (ii) as an indirect tool for empowering farmers to take informed and quality decisions which will have positive impact on the way agriculture and allied activities are conducted. Precision farming extensively uses IT to make direct contribution to agricultural productivity. The indirect benefits of IT in empowering Indian farmer are significant and remains to be exploited. The emerging scenario of a deregulated agriculture under WTO has brought about a greater 'need' and urgency to make IT an integral part of decision making by agriculture sector. The changing environment faced by Indian farmers makes information not merely useful, but necessary to remain competitive. The farmer urgently requires timely and reliable sources of information inputs for taking decisions. At present, the farmer depends on trickling down of decision inputs from conventional sources, which are slow and unreliable. Besides, the personnel who work for the welfare of Indian farmers, such as extension workers, do not have access to latest information, which hinders their ability to serve the farming community effectively. IT will constitute the primary resource in agri-clinics, agri-business centers, *kisan* call centers, agri-*nukkads*, etc. that have been launched by Government of India and other organizations to ensure inclusive growth. Potato is an input-intensive crop and IT will have a special role in making its cultivation more productive and profitable.

In the pre-green revolution era, the input levels in agriculture were low and were primarily generated within the household, hence, dependence on external inputs was minimum. With the release of high yielding varieties during the green revolution period, the need for increasing input use was felt since those varieties generally yielded well under high input levels only. Therefore, farmers were encouraged to increase their input levels through external sources. Uniform recommendations for different inputs were developed which were based on average conditions. This led to over use of inputs at some situations while in others there was under use. However, this was not a very serious problem since the input levels used by farmers were often lower than the recommended doses; therefore, situations of over use of inputs were rare. The input use has however, increased over the years while the response to applied inputs is decreasing. Moreover, with the increased awareness and concern of environmental impact of inappropriate use of resources in agriculture, high cost of inputs and decreasing profit margins, the need for tailoring input use and management decisions to critically defined target situations is being increasingly realized. Thus, management of agricultural production needs to undergo a sea change. Decisions for specific situations would need detailed information about the target areas based on which relevant decisions can be made. This necessitates the need to collect and analyze a huge amount of data as well as develop user friendly delivery mechanism viz. decision support systems (DSS) and this is possible only through greater use of information technology (IT) tools. Geographical information system (GIS), remote sensing (RS) and crop modeling are some of the other tools being used to address the above concerns. Central Potato Research Institute has so far developed various decision support systems, viz. Indo-Blightcast: a potato late blight forecasting system, The Potato Pest Manager (PPM), Computer Aided Advisory System for Crop Scheduling (CAASPS), Potato Growing Season Descriptor (PGSD), Potato Growing Season Estimator, Advisory System for Nitrogen management in potato (ASNMP), Potato Weed Manager (PWM), VarTRAC: a bioinformatics tool for identifying potato varieties and Plausible Potato Growing Season Estimator. Details of these DSS in decision making are discussed in this chapter.

DECISION SUPPORT SYSTEMS

Computer Aided Advisory System for Potato Crop Scheduling

The optimum time of planting, the most suitable variety and the expected yield at different dates of harvest are vital information required by farmers for scheduling their planting and harvesting times as well as for choosing the variety to be grown. Obtaining such information through field experimentation in the diverse agro-climatic conditions in which potato is grown in India is an uphill task, but this information can be derived from crop models which can simulate crop growth, development and yield with reasonable accuracy under diverse situations.

However, use of crop models requires extensive data inputs as well as technical expertise to handle the model (Magarey *et al.*, 2002). Therefore, world over, models are handled by researchers hence take of models by field level workers is not very satisfactory. Decision Support Systems (DSS) on the other hand provide a method for delivery of information in a user friendly and simple way. Therefore, this DSS "Computer Aided Advisory System for Potato Crop Scheduling" (CAASPS) has been developed (Govindakrishnan *et al.*, 2011) with the following purposes:

a) To provide information on the expected yields of different varieties planted at different times to enable farmers to decide on the most suitable one for their respective locations.

b) To help decide the time of harvest based on yield accrued at 60, 70, 80 and 90 days after planting.

c) To indicate the varietal performance under different dates of planting and crop durations and thus help choose the appropriate variety.

This DSS consists of a database and a user interface. The database consists of state, district and location names along with Infocrop-Potato model (Aggarwal *et al.*, 2004) derived yield outputs.

The model outputs were derived as follows:

1) Weather database were created for important locations in India using MARKSIM weather generator.

2) Suitable thermal window was delineated for each location by defining screening rules for maximum temperature ($< 35°C$) and minimum temperature ($>2°C$ & $< 21°C$) at least three weeks after the minimum temperature criteria was fulfilled.

3) Infocrop-potato model was run for five planting situations starting from ten days earlier to the beginning of the suitable thermal window identified by the screening rules and staggered at 10 days interval.

4) For each date of planting, the model was run for 10 varieties under potential situations and 80% of the potential yield was taken as attainable yield.

5) Yield output of each variety at 60,70,80 and 90 days after planting were linked to corresponding spatial attributes viz. state, district and location names in MS Access.

User interface

A simple query system was designed for querying the database. The user first selects the State, and then the Districts of the state. The locations within the district for which information is available are then displayed for selection of one of them.

Once the location is selected, the five dates of planting for which model has been run for the selected location is displayed and the user is required to select one of them. When any of the dates is selected, the attainable yield data for all the ten varieties at four durations, corresponding to 60, 70, 80 and 90 days after start is displayed in tabular format (Figure 1).

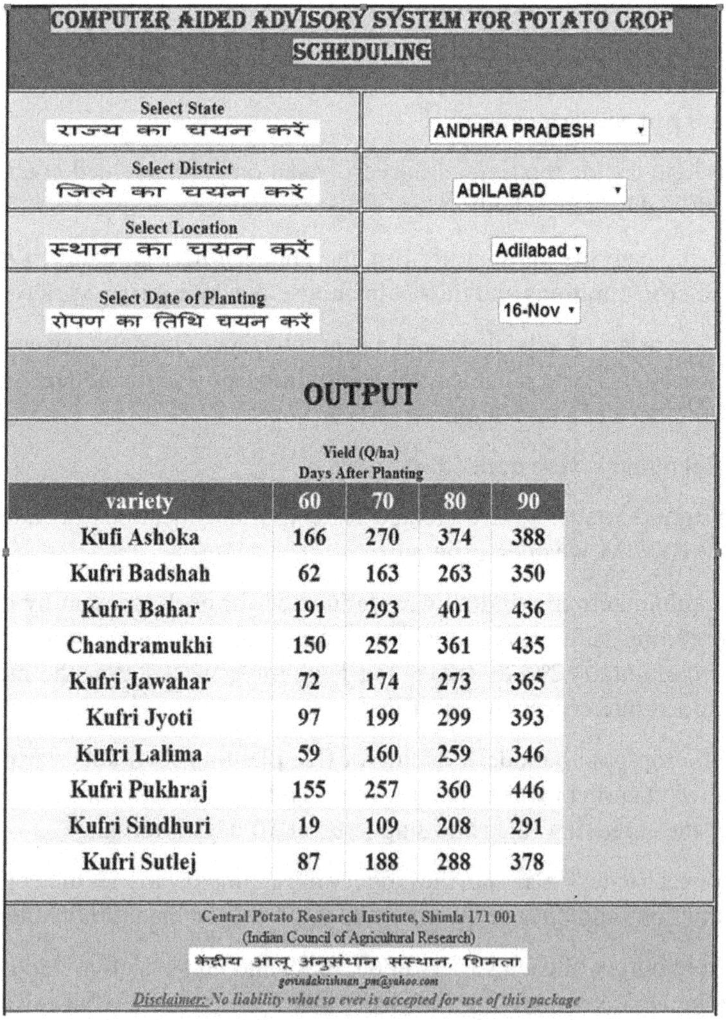

COMPUTER AIDED ADVISORY SYSTEM FOR POTATO CROP SCHEDULING

Select State राज्य का चयन करें	ANDHRA PRADESH ▾
Select District जिले का चयन करें	ADILABAD ▾
Select Location स्थान का चयन करें	Adilabad ▾
Select Date of Planting रोपण का तिथि चयन करें	16-Nov ▾

OUTPUT

variety	Yield (Q/ha) Days After Planting			
	60	70	80	90
Kufi Ashoka	166	270	374	388
Kufri Badshah	62	163	263	350
Kufri Bahar	191	293	401	436
Chandramukhi	150	252	361	435
Kufri Jawahar	72	174	273	365
Kufri Jyoti	97	199	299	393
Kufri Lalima	59	160	259	346
Kufri Pukhraj	155	257	360	446
Kufri Sindhuri	19	109	208	291
Kufri Sutlej	87	188	288	378

Central Potato Research Institute, Shimla 171 001
(Indian Council of Agricultural Research)
केंद्रीय आलू अनुसंधान संस्थान, शिमला
govindakrishnan_pm@yahoo.com
Disclaimer: No liability what so ever is accepted for use of this package

Fig. 1: Snap shot of CAASPS tool

Indo-blightcast

Late blight is the most dreaded disease of potato causing annual crop loss of about 12 billion • globally. Its appearance and spread is highly dependent on environmental factors. Under favourable conditions its spread is so fast that it can wipe out the crop within a weeks' time. In India it is very serious in the hills where it occurs regularly but in the plains it may or may not appear and even if it appears its time of occurrence would vary. The time of its occurrence and severity determines the yield loss which may exceed 40% country wide in some years. Prevention through prophylactic sprays of recommended chemicals is the best option since once it appears, it is very difficult to controls. This, however, requires information on the likely time of appearance of the disease and hence the importance of disease forecasting. INDO-BLIGHTCAST- was developed to predict the first appearance of late blight disease using daily weather data of meteorological stations. This model requires only daily weather data on maximum & minimum temperature and maximum & minimum RH and does not need local calibration for different regions. Hence it is robust and its predictions are broad based.

INDO-BLIGHTCAST has two modules one for data entry and the other for the general users to see the status of late blight forecast.

Data entry

The data entry module is user and password protected. The registered users can "Load data"(for viewing already entered data), "Add data" (to save entered data), "Edit data" (to change entered data values) and "Delete data" (to remove data) if required, in addition to running the model. However, any user (requires no registration) can know the late blight forecast at any location for selected date. The forecast is indicated by coloured buttons, Green colour indicates that late blight is not likely to appear soon; yellow colour indicates that late blight would appear very soon; and red colour indicates that the weather conditions have become suitable for late blight and it can appear any time within fifteen days. Thus depending upon the time required for taking control measures, the user may start preventive measures at yellow or red colour indication (Figure 2). The model has been developed and tested using the data on late blight appearance monitored at Central Potato Research Institute regional stations and All India Coordinated Research Project (Potato) centres over the past several years (Govindakrishnan, 2014).

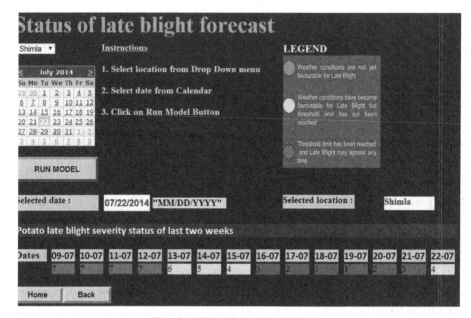

Fig. 2: INDO-BLIGHTCAST Tool

Potato pest manager (PPM)

For the management of diseases and pests two aspects are involved. First is to establish the identity of the disease/ pest and second is to recommend appropriate preventive and management practices to control them. With these objectives in view a DSS called PPM was developed (Govindakrishnan *et al.*, 2005) as detailed below:

Step 1

The photographs showing the symptoms of the diseases/pests are arranged in a photo gallery and displayed in sequence. The user is asked to match the symptoms in the photographs with those he has seen in the field and select the most closely matching one.

Step 2

The appropriateness of the selected photograph needs to be confirmed, because the user may not be fully conversant with the symptoms of different diseases or damage by pests. Information about the prevailing biotic/ abiotic factors is also necessary for a correct diagnosis. This is done through a set of confirmatory questions.

These are questions about the symptoms of the disease/damage by the pest, or conditions which need to be satisfied for the disease/pest occurrence. This information is arranged in a linear fashion. This arrangement allows insertion/ deletion of questions/an option to a question at any level without disturbing the overall structure. Furthermore, this information is presented in a format of questions/statements to the user, while answers are given as options to these questions/statements.

Step 3

Once, all the confirmatory questions are answered, the name of the disease/ pest corresponding to the selected photograph is displayed along with confidence percentage. The confidence percentage is calculated based on answers given to the confirmatory questions relevant to disease symptoms/ pest damage. Each confirmatory question/statement is assigned a certain value such that for all the questions, if the option corroborating the disease/ pest, whose photograph is selected, is chosen as the answer, the value adds up to 100.

However, the value allotted to each question may vary depending upon its significance.

Step 4

Many potato diseases/pests can only be controlled through preventive measures taken over a period of time before planting the crop and control is not possible once the disease/ pest appears. This is especially the case with diseases/ pests where symptoms are seen at/ after harvest. Therefore, the preventive measures applicable to the disease/ pest identified is displayed in this step which are the set of practices which would have prevented/mitigated the disease/ pest occurrence.

Step 5

In this step information required for suggesting control measures on the standing crop is obtained through a further series of questions. For example, information regarding severity of disease/pest damage, age of the crop, etc. is invariably required for deciding the chemicals to be used, their dosage, number of sprays etc. This information is again obtained from the user by presenting the questions or statements with various options. The questions or statements are arranged in tree structure and depending upon the answers given to each question a path is followed leading to a recommendation which is attached at the end node.

Step 6

This step displays the recommendation based on the options chosen (Figure 3).

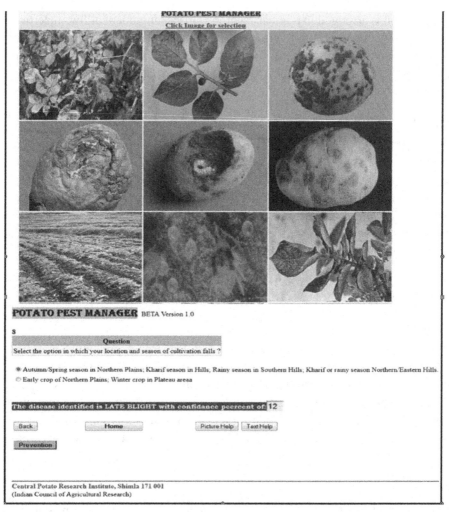

Fig. 3: Snap shot of PPM tool

Potato weed manager (PWM) Weeds cause enormous loss to potato production. They not only compete for moisture, nutrients, space and light but also harbour several pests and diseases as alternate hosts. A number of cultural, mechanical as well as chemical methods are available for controlling weeds in potato crop. Herbicides are available for control of different types of weeds at different stages. The selection of proper herbicide depends upon the type of weed flora and the stage of crop growth. However, weeds prevalent in potato crop vary from region to region and season to season and in the absence of knowledge about weed flora, it is difficult to give precise recommendation for their control. The knowledge of farmer about the weeds is limited to its local name and the extent of damage it may cause. Moreover, availability of proper guidance about weed control in the absence of technical advice may lead to improper control method leading to inefficient weed control. To alleviate this problem, a decision support tool "Potato Weed Manager" has been developed. The user inputs information on the major weed and associated weeds as well as the stage of the crop and based on this information the recommendations are provided (Figure 4). Potato Weed Manager is also converted to Mobile App which is available at Google Play Store from where it is freely available for download.

Potato Weed Manager
Select location of your farm.field?

Hills   Plans

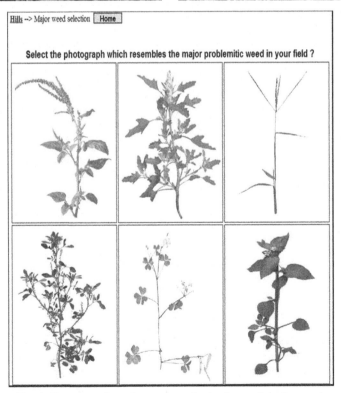

Fig. 4: Snapshot of weed manager and selection options for weeds

VarTRAC: A bioinformatics tool for identifying potato varieties

Authentic identification of potato cultivars is important for plant breeders, the variety registration and certification agencies, seed producers, merchants, farmers, growers, processors, and other end-users. Currently morphological descriptors are being used internationally for variety identification. However, there is a possibility of utilizing DNA fingerprint data to supplement morphological characters in near future. Central Potato Research Institute, Shimla is, therefore, developing both morphological and DNA fingerprint databases for potato cultivars' identification. Data on 50 different morphological attributes and DNA fingerprints based on 127 alleles from 4 micro-satellite markers are currently being used at CPRI for varietal identification. Manual analysis of such huge data is not easy. Therefore, computer software named "VarTRAC" was developed at CPRI (*Rawat et al.*, 2004) for speedy identification of a variety based on the morphological and DNA fingerprint data (Figure 5).

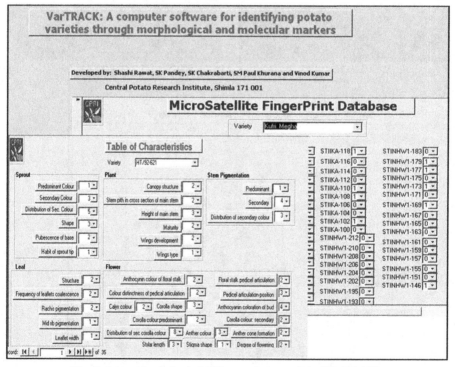

Fig. 5: A combined view of different windows of the "VarTRAC"

The database was created in MS Access with each morphological character taken as a field. All the characters necessary for the identification of a potato variety have been included. Scores are given for each character in a drop-down menu format and the users have only to select the appropriate score for each character. Further help has also been provided for proper scoring. As regards DNA fingerprints, the data on 127 alleles have been recorded by giving a score of one for those alleles which are present and zero for the absent ones.

The software can make generalized abstraction even from the minimum available information. For example, if only 5 morphological attributes of any unknown variety are known, the software can identify the group of varieties having similarity in respect of those 5 attributes.

Plausible potato growing seasons estimator

Potato growing season in India may be a single crop season or more than on thermally suitable period of varying duration which can be exploited wherever water is not a constraint. The continuous horizontal line at 1 shows the growing season while the vertical lines shows the breaks in the growing season due to unsuitable temperature conditions (Figure 6). Plausible Potato Growing Seasons Estimator (PPGSE) tool has been developed for estimating the number of growing seasons of potato crop and their duration for any location. The tool screens the daily maximum and minimum temperature of any location according to maximum and minimum threshold limits set by the user or user can use the default values and extracts those periods which meets the criteria for a period of more than 70 days continuously. The default rules for screening the database for estimating the start and end of growing period are as below:

i)   Maximum temperature should be less than a threshold value (35°C).

ii) Minimum temperature should be more than a threshold value (2°C) and less than another threshold value (21°C) up to about three weeks after planting.

iii) The end of the growing period is fixed at 120 days from the start of the growing period or when the crop growth is terminated when the threshold temperature values are exceeded whichever is early.

A spreadsheet with macros was created to process the raw data and extract the information required by the model. Two fields each with 365 records (maximum and minimum temperature) were created in spreadsheet and macros for screening the day as suitable or unsuitable for potato based on threshold limits for maximum and minimum temperature was written. The starting day number, ending day number and the total number of suitable days of the period more than 70 days were derived as outputs.

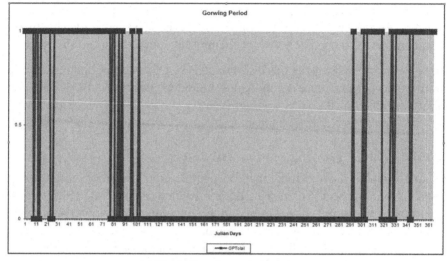

Fig. 6: Growing period graph

Potato growing season descriptor

Potato is one of the most sensitive crops to the environment. It has specific temperature and photoperiod requirements for growth and development. Apart from the phenology, growth and yield, the weather conditions during the growing season also affect the size of the tubers as well as its quality. There is also a wide variation in the pest and disease scenario affecting potato primarily due to differences in the growing season environmental factors at different locations/ seasons. Therefore, there is need for careful planning of the production strategy by analyzing the growing season.

The Potato Growing Season Descriptor (PGSD) consists of a database of daily meteorological data generated by MARKSIM weather generator for many locations (Rawat *et al.*, 2014). The daily meteorological data is analyzed using algorithms to determine the length of the plausible growing season, mean daily temperature, mean night temperature, accumulated growing degree days and accumulated P days during the growing season (Figure 7). The tool also gives the expected yield using a summary model. It is expected that these information would be useful for production managers, extension workers and farmers in planning their production strategy.

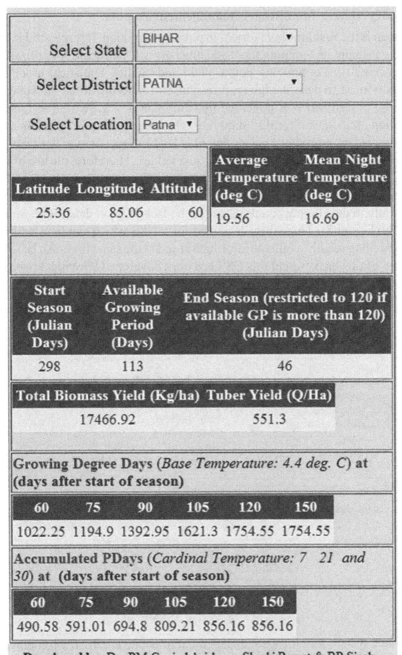

Fig. 7: Potato growing season descriptor

Advisory system for nitrogen management in potato

Nitrogen is the first limiting nutrient in potato production. It is required in large amount to maintain optimum shoot and tuber growth. Adequate N fertilization is also essential for optimizing potato yield and quality. Statistical models are generally used to describe the crop response to N fertilization. However, the response to fertilizer N is governed by many factors e.g. season, soil texture, irrigation method, soil fertility status etc. Thus, the average N optimum across sites derived through statistical models may lead to under or over fertilization in many situations due to a range of N optima values. Therefore, the use of more efficient approaches such as simulation models to develop site specific N recommendations is desirable. However, simulation models are not widely used especially in developing countries due to the lack of input data for running the models (Figure 8). Thus, there is a need for a DSS which can be used by the farmers using readily available information and at the same time also be robust. In view of this background this DSS has been developed (Govindakrishnan *et al.*, 2014).

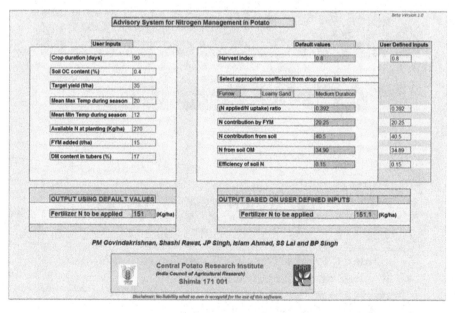

Fig. 8: Snap short of ASNMP tool

Advisory system for Nitrogen management in potato (ASNMP) computes the N requirement based on the soil texture, soil test values, crop duration, target yield, seasonal temperature, quantity of FYM added, harvest index and dry matter content in tubers. Information on the target yield is invariably available with the farmer based on his experience. The other inputs are also easily available.

Based on the inputs, the contribution of different components of the inputs and the efficiency of utilization are generated by the tool using equations reported in literature. Options to change the derived coefficients have also been provided.

REFERENCES

Aggarwal PK, Kalara N, Chander S and Pathak H (2004) Infocrop-a generic simulation model for annual crops in tropical environments. Indian Agricultural Research Institute, New Delhi: 132p

Govindakrishnan PM (2014) Disease forecasting in potato. In, Book of Abstracts: National seminar on emerging problems of potato. 120p

Govindakrishnan PM, Khurana SMP, Singh Sarjeet, Chandla VK, Ahmad I, Rawat S, Lal SS and Dua VK (2005) Development of a database for the identification and control of diseases and pests of potato in India. *International Pest Control* **42**(2): 78-81

Govindakrishnan PM, Rawat S, Singh JP, Ahmad I, Lal SS and Singh BP (2014) Advisory system for N management in potato-a framework for site specific nitrogen recommendation. In, Book of Abstracts: National seminar on emerging problems of potato. 208p

Govindakrishnan PM, Rawat Shashi, Pandey SK, Singh JP, Lal SS and Dua VK (2011) Computer Aided Advisory System for Potato Crop Scheduling (CAASPS)-A decision support tool for potato planting strategies. *International Journal of Agricultural and Statistical Sciences* **7**: 571-577

Magarey RD, Travis JW, Russo JM, Seem RC and Magarey PA (2002) Decision Support Systems: quenching the thirst. *Plant Disease* **86**: 1-14

Rawat S, Pandey SK, Chakrabarti SK, Khurana, SMP and Kumar Vinod (2004) VarTRAC: A computer software for identifying potato varieties through morphological and molecular markers. CPRI Newsletter-30: 1-2

Rawat S, Govindakrishnan PM and Singh BP (2014) Potato growing season descriptor. In, Book of Abstracts: National seminar on emerging problems of potato. 204p

13

Role of Mechanization in Potato Crop Management

Brajesh Nare[1] and Sukhwinder Singh[1]

[1]*ICAR-Central Potato Research Station, Jalandhar-144001, Punjab, India*

INTRODUCTION

Potato ranks third among the major food crops after wheat and rice in the world. It is most important and versatile food of the world. India is second largest potato producer next to china. Potato is bulky and watery in nature and requires lot of energy for production and processing. It is a short duration crop of 90-120 days, having highest yield per ay per unit area as compared to other major crops. Being a relatively short duration crop in the plains of India, its mechanization assumes a special importance in order to accomplish various farm operations in very limited time for its higher production. Manual production of potato is time consuming and tedius job, it requires about 1600-2000 man-hours per hectare. With the rising cost of production and depleting energy reserves, increase in crop productivity with minimum input of energy and cost one of the paramount concerns (Garg and Singh 2002). Sustenance of agricultural productivity goes hand in hand with mechanization of different farm operations, which aims at achieving timeliness of operations, efficient use of inputs safety and comfort of farmers, reduction in cost of produce and drudgery of farmers (Verma *et. al*, 1992).With the emergence of new crop rotations and enhanced cropping intensity, many farmers experience scarcity of time and labour. Mechanization plays an important role in timely completing farm operations involved in potato production. It implies a resource management strategy to increase productivity and economic returns. Efforts have been made at various levels to mechanise both production and processing. Various mechanical tools, implements and machinery have been developed for land development, fertilizer application, planting, weeding, harvesting, preparation for storage and on-farm processing of potato (Singh and Gulati 2014).

IMPLEMENTS FOR POTATO PRODUCTION TECHNOLOGY

Timely field operations and handling of crop are important factors for improving quality and quantity in potato production. Mechanization reduces time, labour and cost of production and also reduces drudgery to the farm workers/operators. Variety of machinery and equipments has been developed for various farm operations involved in potato crop cultivation like land leveling, seed bed preparation, planting, weeding and interculture, plant protection, digging, grading and seed treatment.

Seedbed preparation

After burying green manure crop (usually *dhaincha*) with disc plow, mould board plow or disc harrow (Figure 1) or rotavator one leveling makes field conditions favorable for decomposition of green manure. Potato requires good tilth of soils to ensure covering of stolons/tubers and also to provide adequate aeration for respiration of roots and tubers. To get the good yield, soil should be loose and friable with good drainage and aeration. Subsoiler is an important primary tillage implement used to till the soil at 20 to 50 cm depth. It breaks the hard pan below the soil surface and provide good drainage system. Pre-sowing irrigation has to be done before field preparation to have uniform germination. After 5-7 days of pre-sowing irrigation, 2-3 cross harrowing or 3-4 tiller operations followed by two leveling make the required seedbed for a good crop of potato. Among the primary and secondary tillage implements, the rotavator is most popular for field preparation as it provides fine level of soil tilth. All these implements are easily available in the market and are being used for other crops also.

Fig. 1: Green manuring with disc harrow and deep tillage with subsoiler improves infiltration and potato yield

Fertilizer application

Fertilizer type, quantity and placement are the important factors for good production of potato crop. Precise and judicious use of fertilizer is of prime importance to ensure its efficient utilization and to reduce the cost of farm input and environmental risk. After seed bed preparation, a fertilizer drill-cum-marker (Figure 2) or fertilizer broadcaster can be used for fertilizer application. Fertilizer drill –cum- line marker is a machine which ensures uniform placement of fertilizer in furrows and marks the line impressions at 600 or 650 mm fixed row spacings for tuber placement. With this machine fertilizer rate can be adjusted between 50-300 kg/ha with a field capacity of 4.0 ha/day. In case, fertilizer unit is attached with the planter than the use of fertilizer drill can be avoided. In potato crop, fertilizer should be placed 5-7 cms below the seed tubers and should not come in direct contact with the seed. Band placement helps in improving efficient utilization of fertilizer which results in improved production. Genrally fertilizers are hygroscopic in nature, hence it is recommended to properly clean the applicator after use.

Fig. 2: Fertilizer applicator cum marker

Planting

About 80-85 thousand seed potato has to be planted per hectare for good yield of the crop. This can be achieved by planting 35-55 gm of seed potato tubers per hectare. Uniform placement of tubers in prepared soil bed at specified spacing and depth is also important for better emergence as well as production of the crop. Uniform plant emergence can be achieved by planting of seed material with tractor operated potato planters (Figure 3). Function of the planter is to open a furrow, meter the seed tubers with a suitable metering mechanism, place the tubers in to the furrow and cover them by a soil layer of about 6-8 cm. These machines maintain uniformity of planting in terms of row to row (60 cms), plant to plant spacing and depth (6-8 cms) of planting. Various types of machinery are available in India for planting of potato viz.

(A)   Rotating drum type

(B)   Revolving magazine type

(C)   Belt and cup type and

(D)   Picker wheel type

First two are semi-automatic planters in which one person per row puts seed tubers in to the seed metering cells of the machines. Function of these machines is to open a furrows drop the seeds in to the soil through seed delivery tubes. Seed tubers are dropped in lines and a ridge of soil mass is formed over the tubers. Automatic planters are becoming popular now a days due to their distinct advantages over the other types. In this type of machine labour is required only for filling of hopper. All other activities like lifting individual tubers, dropping them into rows and than covering with soil mass are carried out by the machine itself.

Fig. 3: Potato planting with semi automatic and automatic planters

Weeding, top dressing and earthing up

Mechanical and chemical weed control methods are used in most of the crops. Potato being a tuber crop, is highly responsive to inter-row-cultivation. Three row tractor operated inter-row-cultivator is operated in 22-25 day crop when plant height is 8-10 cms. After operating this weeding machine, three or five row ridger (Figure 4) is used for earthing up and remaking the ridges. Ridger makes uniform ridges and accumulates sufficient soil mass around the plants. For light soils and low weed intensity areas another multipurpose machine is available which perform weeding, furrow opening, fertilizer application and re-earthing operation simultaneously. For chemical weed control, tractor operated multi-nozzle sprayer or spray gun are used to achieve fast and uniform coverage.

Fig. 4: Intercultural and earthing operations in potato crop

Plant protection

Potato crop needs to be protected from different disease carriers, insects and pests and mostly liquid chemical are sprayed on crop canopy to protect the crop. Many times delay in chemical spray can cause huge crop losses. Particularly, in case of late blight disease. Delay of 1-2 days can result in total loss of crop. Therefore, timely and uniform application of plant protection chemicals is important, which is possible with high capacity tractor operated sprayers. Multinozzle boom sprayer, having 12-16 nozzles, is also useful for potato crop. However spray gun is especially effective in case of late blight in potato plants as it covers lower surface of leaves more effectively (Figure 5). For small holdings, manually or battery operated knapsack sprayer can be used which has field capacity of 0.5 ha/hr.

Fig. 5: A view of spraying operation with power sprayers

Potato digging

Before starting harvesting operation haulms are removed manually sickles or with chemical methods. Potato digging is a cumbersome process and involves a lot of manpower. About 600-700 man-hours are required for manual digging of 300-400 quintals of potato from huge soil mass of around 10,000 quintals. In manual harvesting spade or *khurpa* is used which results up to 10% tuber cut or bruise which is a huge loss to the farmers and nation. In manual harvesting some portion is always left behind in the field. Animal drawn plow is another tool for potato digging. This method is faster compared to manual digging but in case of large scale farming harvesting is delayed with this method and fields become dry. Tractor operated diggers are fast, economical and cause least damage to the produce (Figure 6).

Fig. 6: Potato harvesting with tractor operated digger elevator

Digger elevator is an efficient machine which exposes 90-95% tubers in optimum field conditions. This machine has to be maintained properly as there are many parts which move in soil and cause more wear and tear. Multipurpose passive blade potato digger (Figure 7) is another less expensive equipment. It exposes about 80-85% potatoes and is easy to maintain. It can be

Fig. 7: Multipurpose potato digger

used in early crop (60 days without cutting haulms) and also can successfully work in dry field conditions.

Potato combine harvester

A two row potato combine harvesters has been developed at CRPI (Gulati and Singh 2019) to perform potato digging and picking operation combinely (Figure 8). Machine is to dig the potato tubers from 15-20 cm deep soil, separates the tubers from soil clods, conveys the tubers to the sorting platform and collects them in to a hopper. Further it can be unloaded in to a transporting trolley or on a heap.

Fig. 8: Potato combine harvester

Four workers have to be employed with this machine to sort the rotten tubers or soil clods passing over the sorting table. The machine is powered by a 50 to 60 hp tractor. The effective field capacity of the prototype is 0.26 ha/hr.

GRADING

Graded, sorted and properly packed potatoes always fetch good price in the market. Manual grading is not uniform and it requires a lot of time and energy which bring down the overall returns. Various manual as well as power operated graders have been developed for size grading of the potato tubers (Figure 9). The grader can grade the potatoes in four size grades e.g. < 20 mm, 20-32, 32 -52 and > 52 mm. The output capacity of this grader is 5-6 t/h and it would reduce the cost of grading by 68.18% (Gulati *et. al*, 2018). Rubber belts having different size of round or square shaped perforation are generally used in the grading system (Atwal and Gulati 2002). Lot of labour is still required in these high capacity graders for feeding sorting and packaging of the potatoes.

Fig. 9: Rubber belt sreen type potato sorting and grading machine

SEED TREATMENT

Disease free seed is the most important requirement to grow healthy potato crop. Seed potatoes needs to be treated before stroge to protect agrainst diseases. For controlling tuber and soil born diseases viz. common scab, black scurf and powdery scab, 3% boric acid treatment can be given with the help of an electric powered seed treatment machine. In this machine boric acid is sprayed on the potatoes moving and rolling over a conveyor. All the surfaces are coated with the chemical and the used solution is re-circulated. This machine can treat about 300-350 quintals of potatoes per day (Figure 10).

Fig. 10: Potato seed treatment system

CONLUSION

Potato is a crop with huge production per unit area. For production, handling and processing of potato it takes lot of time and energy if all the operations are completed manually or with the help of animal drawn equipments. A number of efficient farm machinery and equipment available in India to perform timely field operations like seed bed preparation, fertilizer application, planting, weeding, earthing, digging, grading and seed treatment. Therefore, in this era of competitive agriculture, for efficient use of expensive inputs like seed and fertilizer, farmers should opt for precise application of the farm inputs through mechanization and should use potato planters, fertilizer applicators, weeders, diggers, graders and other machinery. Of course some of these machines are expensive and are not with in the reach of small farmers. In that case group of farmers can purchase the machine or an individual farmer can have it and can go for custom hiring. Procuring of expensive machines by cooperatives can be a useful idea for having processing machineries at village or cluster levels.

REFRENCES

Atwal JS and Gulati S (2002) Design and development of a rubber rollers type of potato grader. *Journal of Indian Potato Association* **28** (1): 70-71

Garg, IK and Singh S (2002) Farm equipment for Punjab agriculture. Deparment of farm power and machinery, Punjab Agriculture University, Ludhiana, India. 187p

Gulati S, Sukhwinder S and Nare B (2018) Development and evaluation of a square wire mesh type of potato grader. *Potato Journal* **45**(1): 74-80

Gulati S and Singh M (2019) Design and development of two row tractor operated potato combine harvester. *Potato Journal* **46**(1): 81-85

Verma SR, Kalkat HA, Singh CP and Garg IK 1992. Development of potato harvesting equipment and the challenges ahead. In: Proceedings of National Colloquium on Potato mechanization in India held at PAU, Ludhiana, India, November, 11-12, 1992. 59-79

Singh S and Sunil Gulati (2014). Mechanization solutions for Indian potato cultivation. In: Souvenir, National Seminar on Emerging Problems of Potato, Nov 1-2, 2014, CPRI and IPA Shimla: 70-76

14

Potato Late Blight and Its Management

Sanjeev Sharma[1], Mehi Lal[2] and Sundaresha Sidappa[1]

[1]ICAR-Central Potato Research Institute, Shimla 171 001, India
[2]ICAR-Central Potato Research Institute Regional Station, Modipuram
Meerut-250 110, India

INTRODUCTION

Potato is an important horticultural food crop which has the potential to meet food demand of the fast growing human population across the world, including India. This is going to be the future food crop for the millions especially in the third world countries. However, this crop is highly prone to wide range of diseases and pests that are considered to be the main bottle neck in potato production across the globe. Amongst the biotic stresses, late blight caused by oomycete *Phytophthora infestans* (Mont.) de Bary has historically been an important disease of potatoes and tomatoes worldwide. In the mid 1800, late blight caused widespread crop failures throughout Northern Europe including Ireland where it was responsible for the Irish famine. Since then, it has spread far and wide and now occurs wherever potatoes are grown. Losses due to *P. infestans* have been estimated to • 12 billion per annum of which the losses in developing countries have been estimated around • 10 billion per annum (Haverkort *et al.*, 2009). In India, average annual losses to the tune of 15% have been estimated which amounts to 7.95 million tons of potatoes (Current production is 53 MT). Disease appears every year in epiphytotic form in hilly regions whereas in the plains, although it usually appears every year but its intensity is low (traces to 25%). It is only in few years that it assumes epiphytotic form leading to crop losses to the tune of 40%. Recently, late blight has become a serious problem in *kharif* grown potatoes and tomatoes in Karnataka state (Sharma *et al.*, 2015). *Phytophthora infestans* is considered as re-emerging pathogen due to regular emergence of its novel strains with increased virulence and its appearance in new locations with surprising intensity. Management of this devastating pathogen is challenged by its remarkable speed of adaptation to control strategies such as genetically resistant cultivars.

SYMPTOMS

Late blight affects leaves, stems and tubers. It appears on leaves as water-soaked irregular pale green lesions mostly near tip and margins which rapidly grow into large brown to purplish black necrotic spots. A white mildew, which consists of sporangiophores and spores of the pathogen, can be seen on lower surface of the infected leaves especially around the edges of the necrotic lesions. In dry weather the water-soaked areas dry up and turn brown. On stems and petioles light to dark brown lesions encircle the stems as a result the affected stems and petioles become weak at such locations and may collapse. Entire crop gives blackened blighted appearance especially under disease favourable conditions and may be destroyed within a week. Tubers in soil become infected by rain borne sporangia coming from the diseased foliage and show irregular reddish brown to purplish areas which extend into internal tissues of the tubers (Figure 1). The infected tubers usually are hard, dry and firm but may get attacked by soft rot causing bacteria and rot in field and stores. Sepals, flowers and berries may also get infected and show white mildew under humid and cloudy weather.

Fig. 1: Late blight symptoms (A) Foliar blight on upper surface (B) On lower surface (C) On stem (D On tubers

THE PATHOGEN

The disease is caused by an oomycete *Phytophthora infestans* (Mont.) de Bary, a diverse group of eukaryotic microorganisms in a group called the Stramenopiles, clustering together with others in a super group, the Chromalveaolata (Adl *et al.*, 2005). The position of oomycetes as a unique lineage of eukaryotes unrelated to true fungi but closely related to heterokont (brown) algae and diatoms is well established through molecular phylogenies and biochemical studies. *P. infestans* is heterothallic in nature and requires two mating types A_1 and A_2 for sexual reproduction. The pathogen is characterized by lemon shaped detachable, papillate sporangia produced on sympodially branched sporangiophores of indeterminate growth. The sporangiophores exhibit a characterized swelling at junction where sporangia are attached with the sporangiophores.

P. infestans is diploid and the genome is considerably larger (240 Mb) and by far the largest and most complex genome sequenced so far in the chromalveolates and even in true fungi. Its expansion results from a proliferation of repetitive DNA accounting for ~ 74% of the genome. Comparison with two other *Phytophthora* genomes showed rapid turnover and extensive expansion of specific families of secreted disease effector proteins, including many genes that are induced during infection or are predicted to have activities that alter host physiology. These fast-evolving effector genes are localized to highly dynamic and expanded regions of the *P. infestans* genome. This probably plays a crucial part in the rapid adaptability of the pathogen to host plants and underpins its evolutionary potential. A total of 17,797 protein-coding genes have been detected within the *P. infestans* genome. It also contains a diverse variety of transposons and many gene families encoding for effector proteins that are involved in causing pathogenicity. Overall the genome is having an extremely high repeat content (~ 74%) and to have an unusual gene distribution in that some areas contain many genes whereas others contain very few, which is thought to contribute to *P. infestans* evolutionary potential by promoting genome plasticity, thus enhancing genetic variation of effector genes leading to host adaptation (Haas *et al.*, 2009).

Phytophthora species, like many pathogens, secrete effector proteins that alter host physiology and facilitate colonization. The genome of *P. infestans* revealed large complex families of effector genes encoding secreted proteins that are implicated in pathogenesis. These fall into two broad categories: apoplastic effectors that accumulate in the plant intercellular space (apoplast) and cytoplasmic effectors that are translocated directly into the plant cell by a specialized infection structure called the haustorium. Apoplastic effectors include secreted hydrolytic enzymes such as proteases, lipases and glycosylases that probably degrade plant tissue; enzyme inhibitors to protect against host defence enzymes; and necrotizing toxins such as the Nep1-like proteins (NLPs) and PcF-like small cysteine-rich proteins (SCRs). Most notable among these are the RXLR and Crinkler (CRN) cytoplasmic effectors. At least 563 RXLR genes have been predicted in the *P. infestans* genome. All oomycete avirulence genes (encoding products recognized by plant hosts and resulting in host immunity) discovered so far encode RXLR effectors, modular secreted proteins containing the amino-terminal motif Arg-X-Leu-Arg (in which X represents any amino acid) that defines a domain required for delivery inside plant cells, followed by diverse, rapidly evolving carboxy-terminal effector domains. RXLR effector genes typically occupy a genomic environment that is gene sparse and repeat-rich. The mobile elements contributing to the dynamic nature of these repetitive regions may enable recombination events resulting in the higher rates of gene gain and gene loss observed for these effectors.

CRN cytoplasmic effectors were originally identified from *P. infestans* transcripts encoding putative secreted peptides that elicit necrosis *in planta*, a characteristic of plant innate immunity. Analysis of the *P. infestans* genome sequence revealed an enormous family of 196 CRN genes of unexpected complexity and diversity that is heavily expanded in *P. infestans*. Like RXLRs, CRNs are modular proteins. CRNs are defined by a highly conserved N-terminal 50-amino-acid LFLAK domain and an adjacent diversified DWL domain. Most (60%) possess a predicted signal peptide. Those lacking predicted signal peptides are typically found in CRN families containing members with secretion signals. CRN C-terminal regions exhibit a wide variety of domain structures, with 36 conserved domains and a further eight unique C termini identified among the 315 *Phytophthora* CRN proteins. A further 255 CRN genes are fragmented or otherwise disrupted and presumably non-functional. Both CRN and RXLR genes typically occur in repeat-rich, gene-sparse regions of the genome (Chakrabarti *et al.*, 2010).

DISEASE PERPETUATION AND PRE-DISPOSING FACTORS

P. infestans survives poorly in nature apart from its plant hosts. Under most conditions, the hyphae and asexual sporangia can survive for only brief periods in plant debris or soil, and are generally killed off during frosts or very warm weather. Persistence of *P. infestans* from year to year is a critical component of late blight epidemics. Infected seed tubers serve as the primary source of inoculum. The pathogen overwinters as mycelium in infected tubers, in refuse piles and volunteer plants or over summer in subtropical zones through tubers kept in cold stores. Potato tubers left in the field after harvest and cull potato tubers can produce volunteer plants which can carry over the pathogen to the next season. Besides, the pathogen is carried over latently by asymptomatic tubers. Survival of pathogen as oospores in soil serves as another source of primary inoculum. Tubers in soil get infected by contact with sporangia coming from infected haulms through rain water (Sharma *et al.*, 2017).

Appearance of late blight and its subsequent build-up and spread depends on several factors. Weather conditions such as temperature, relative humidity, rainfall, dew, sunshine hours etc. have a direct effect on *P. infestans* and are given as below.

Temperature

Pathogen growth	: 16-20°C
Spore production	: 18-22°C
Spore germination	: 10-20°C
Infection and disease development	:7.2-26.7°C (18-22±1°C)

Humidity

Spores are formed in moisture saturated atmosphere.

Spore germination and infection requires 100 per cent humidity.

Spores get killed under low humidity (less than 75% humidity)

Light

Spores are produced during the night and are sensitive to light.

Cloudiness favours disease development.

MANAGEMENT

Cultural practices

Sanitation: Reduction of the primary sources of inoculum is the first step in management of late blight. Control of contaminated sources such as waste heaps, infected tubers, volunteer plants, disease in neighbouring fields and re-growth after haulms destruction can help in management of the disease. The onset of epidemic can be delayed by 3 to 6 weeks if all primary infection from early potato is eliminated. Covering of dumps with black plastic sheet throughout the season and preventing seed tubers from becoming infected is an important step in reducing the primary inoculum. The sheeting must be kept in place and remain intact until the tubers are no longer viable (Cooke *et al.*, 2011). This prevents re-growth and proliferation of spores on piles thereby reducing the risk to nearby crops. Only good quality seed potatoes obtained from certified suppliers should be planted. Often discarded potatoes from the previous season and self-sown tubers can act as sources of inoculum.

Crop rotation: Importance of oospores as soil-borne inoculum is determined both by their formation in plant tissue and their survival in soil. There is a correlation between crop rotation and early blight infections. Infection usually start early in fields which are not subjected to crop rotations while decline in early infection in fields subjected for crop rotations (3 or more years) has been observed. The oospores can remain viable for 48 and 34 months in sandy and clay soils, respectively (Turkensteen *et al.*, 2000). This indicates that a sound crop rotation is important and is an effective way of reducing the risk of soil-borne infections of *P. infestans*.

Plant nutrition: Avoiding excess nitrogen and use of moderate nitrogen fertilization is often recommended as a cultural practice to delay the development of late blight as excess nitrogen increases succulency in plants and make them susceptible. Higher dose of phosphorus and potassium has been found to give a higher yield in a late blight year.

Crop geometry: Use of host density as a tool for management of late blight has also been used to control late blight. Tuber yield from both resistant and susceptible cultivar increase when these were grown in mixture as compared to the single genotype stands. Strip cropping of potatoes significantly reduced late blight severity in organic production when planted perpendicular to the wind neighboured by grass clover (Bounes and Finckh, 2008).

Control of tuber blight: High ridging is often used to reduce tuber contamination by blight. This normally involves piling soil or mulch around the stems of the potato blight meaning the pathogen has farther to travel to get to the tuber. Another approach is to destroy the canopy when blight reaches to 75% severity. By eliminating infected foliage, this reduces the likelihood of tuber infection (Sharma *et al.,* 2015).

Host resistance

Management of late blight through host resistance will remain the most environmentally and economically preferred option globally despite the fact that none of the variety could sustain the blight onslaught for more than 5-7 years. Development of resistant cultivars has played an important role in the control of late blight. *Solanum demissum,* a hexaploid wild species, has extensively been used to confer resistance against *P. infestans*. Field resistance is polygenic and more durable. *Solanum bulbocastanum, S. microdontum, S. verrucosum* and *S. chacoense* have been used as a source of field resistance in breeding programs. Since the pathogen is quite plastic, mutable matching races against major R genes develop readily and overcome the resistance of the new cultivars. However, major genes which have evolved naturally in *S. demmisum* population for thousands of years where late blight occurs annually still hold their importance. A multilineal combination of 11 resistant genes (R genes) identified so far, into commercial varieties has significant potential in management of late blight. Disease resistance in potato varieties together with use of fungicides can slow down the development of late blight. A variety with field resistance to late blight in tubers and a medium to high resistance in the foliage can help in reducing the use of fungicides. A large number of varieties have been developed by ICAR-Central Potato Research Institute, Shimla having resistance to late blight (Singh *et al.*, 2018). Large number of exotic potato genotypes possessing late blight resistance have been identified which can be used to develop varieties resistant to late blight.

Biotechnological approach

Genetic engineering may also provide options for generating resistant cultivars. A resistance gene effective against most known strains of blight has been identified from a wild relative of the potato, *Solanum bulbocastanum*, and introduced by genetic engineering into cultivated varieties of potato. Introgression of RB gene in Indian popular potato cultivars has demonstrated enhanced late blight resistance and generation of valuable genetic material for resistance breeding (Figure 2). It has also been demonstrated that genotypes with variable level of late blight resistance can be developed by crossing a specific RB-transgenic event with well adapted Indian potato cultivars (Shandil *et al.*, 2017).

Fig. 2: Performance of selected RB hybrids vs control

RNA interference (RNAi) has proved a powerful genetic tool for silencing genes in plants. Host-induced gene silencing of pathogen genes has provided a gene knockout strategy for a wide range of biotechnological applications. The RXLR effector Avr3a gene is largely responsible for virulence of oomycete plant pathogen *Phytophthora infestans*. Using this approach, we attempted to silence the Avr3a gene of *P. infestans* through siRNAi technology which could impart moderate resistance to *P. infestans* (Sharma, 2016).

Biocontrol

Heavy dependence on fungicides could pose threat to environment and human population. Biocontrol agents and biopesticides could be a safe option to the use of synthetic fungicides. Antagonism to *P. infestans* by some naturally occurring microorganisms such as *Trichoderma viride, Penicillium virdicatum, P. aurantiogiseum, Chetomium brasilense, Acremonium strictum, Myrothecium varrucaria, Penicillium aurantiogriseum, Epiccocum purpuranscens, Stachybotrys coccodes, Pseudomonas syringae, Fusarium graminearum* and *Pythium ultimum* have been observed in laboratory and field studies (Sharma *et al*, 2017).

Biosurfactants can be used as alternatives to chemical surfactants as their capability of reducing surface and interfacial tension with low toxicity, high specificity and biodegradability make them important for inhibiting pathogens. Biosurfactants are generally known to be less toxic compared with synthetic surfactants and rapidly degraded in the environment. Fluorescent pseudomonas capable of producing biosurfactants has recently received increased attention for biocontrol of zoospore producing pathogens. Several studies have shown lyses of zoospores within a minute after exposure to biosurfactants. Significant reduction in late blight development was observed when plants were treated with biosurfactant (*Pseudomonas koreansis* 2.74) and biosurfactants have the potential to induce resistance in potato to late blight (Bengtson *et al*., 2015). The metabolite of biosurfactant producing microorganism (*Pseudomonas aeruginosa*) has shown high efficacy against *P. infestans* under *in vitro* conditions (Tomar *et al*., 2013, 2019). The biocontrol agents in general have been found to be very effective under laboratory and glasshouse conditions but less effective under field conditions. However, an integrated use of biocontrol agents along with low dose of fungicides could help to reduce the quantity of fungicides used in the management of late blight (Lal *et al*., 2017a).

Use of fungicides

Spraying with fungicides has been a standard practice for control of late blight. Bordeaux mixture, which consists of copper sulfate, hydrated lime and water was a standard fungicide for many years. Subsequently organic fungicides especially carbamates which controlled both early and late blight replaced Bordeaux mixture. Metalaxyl – a phenylamide group of fungicides specific to oomycetes however, revolutionized late blight control. Since it was most effective its use increased rapidly and this became one of the major fungicide used world over but strains of *P. infestans* which do not respond to metalaxyl appeared worldwide. Since then efforts are on to minimize the fungicide resistant strains in *P. infestans*. Use of pure systemic fungicides over a long period leads to the development of fungicide resistant strains. Therefore, systemic fungicides should be used in combination with contact fungicides and ultimately check the appearance of resistant strains. Metalaxyl in mixture with unrelated contact fungicide however, could retard development of resistance in the pathogen. Cymoxanil mixtures have been found effective for managing metalaxyl resistant strains. Studies have shown that use of cyazofamid and mandipropamid could be reduced by 30% by adjusting the dose according to resistance level in a variety and used according to the infection pressure (Cooke *et al.*, 2011). Mixture of ametoctradin + dimethomorph has been found effective against late blight (Lal *et al.*, 2017b). Spray schedules have also been developed for effective management of late blight. Fluazinam, cyazofamid and mandipropamid have been used for the late blight management. Oxathiapiprolin has also shown efficacy against late blight (Sharma and Maheshwari, 2018).

Various substances other than fungicides have also been tested for management of late blight. Ammonium molybdate, cupric sulphate and potassium metabisulfate at 1mM partially inhibited the growth and spore germination of *P. infestans*, whereas ferric chloride, ferrous ammonium sulphate and $ZnSO_4$ at 10mM completely inhibited growth and spore germination. The foliar spray of $ZnSO_4$ and $CuSO_4$ (0.2%) micronutrients, delayed the onset of late blight by 12 days when used with host resistance, subsequently reduced disease severity with higher yield. Sub-phytotoixc dose of boron with reduced rate of propineb+iprovalidicab has been found more effective than treated with fungicides alone (Frenkel *et al.*, 2010). Similarly, application of potassium phosphate in combination with reduced doses of fungicides provided same level of protection as treatments with recommended full dose of fungicides. Thus, combined treatments could help to reduce the quantity of traditional fungicides and may also decrease the selection pressure for fungicide resistance development in the pathogen. β-aminobutyric acid (BABA) has been known as an inducer of disease-resistance. Plant activators *viz*, BABA and Phosphoric acid have been evaluated against late blight with combination of fungicides or alone (Tsai *et al.*, 2009). A 20–25% reduction of the fungicide dose in combination with BABA gave the same result on late blight development as full dose Shirlan alone in field condition, while reduced dose of Shirlan alone sometimes resulted in less effective protection. However, *in vitro* results indicated that the efficacy was lasted for only 4–5 days after BABA treatment and subsequently efficacy was lowered. The partially resistant cultivars Ovatio and Superb reacted to lower concentrations of BABA where no effect was found in susceptible cv. Bintje (Liljeroth *et al.*, 2010). Two SAR activators (BABA and phosphorous acid) were found effective against late blight of potato with significantly reduced disease severity (40- 60%). The expression of the defence related genes and *P. infestans* effecter proteins β-1,3 glucanase, PR-1 protein, phytophthora inhibitor, protease inhibitor, xyloglucanase, thaumatin protein, steroid binding proteins, proline, endochitinase and cyclophilin genes were up regulated with the SAR activator treatment compared to unsprayed (CPRI, 2014).

Late blight forecasting

Disease forecasting allows the prediction of probable outbreaks and decision support system can help in management of any further increase in disease intensity. This allows us to take strategic decisions about the disease management. Various concepts have been developed and utilized over the years for predicting late blight across the globe (Sharma and Singh, 2018). They include 'Dutch rules', Beaumont's periods, Irish rules, moving days concept, severity value accumulation, negative prognosis, mathematical models etc. Van Everdingen (1926) was the pioneer in using weather conditions for forecasting potato late blight under Holland conditions. He used dew periods, night temperature, cloudiness and rainfall, known as the *"Dutch rules"*, to predict initial appearance of late blight in Holland. Dutch rules in general were found satisfactory but sometimes the blight would appear even when the 'Dutch rules' were not fulfilled. After the development of more powerful computers, a large number of biotic and abiotic factors and their interrelations have been included in forecasting systems and decision support systems have been developed to manage the disease. Although not all these systems have been introduced in practice, still farmers have the option to use these systems in supporting their decision to workout spray schedules. Some of the prominent DSSs are BLITECAST, ProPhy, NegFry, PhytoPRE, Web-blight, Plant-Plus, PhytoPRE+2000, Chine-blight, Bio-PhytoPRE etc (Singh and Sharma, 2013). However, the forecasting models are region specific and have to be validated for each local situation. None of the models developed in Europe and elsewhere could forecast late blight appearance in India, hence a forecasting model JHULSACAST was developed which uses hourly temperature and RH data as input. This model also has to be calibrated to local situations and presently has been calibrated and validated for western Uttar Pradesh, Punjab, West Bengal and Tarai region of Uttarakhand. It has also been used to develop a DSS which has three components i) prediction of first appearance of disease, ii) decision rules for need based fungicide application, and iii) yield loss assessment model (Arora *et al.*, 2014). Similarly, a web-based DSS to economise on fungicide applications has also been developed. Recently, INDO-BLIGHTCAST, a pan India model, has been developed to predict appearance of late blight (Figure 3) which is generic and is applicable over a wide range of environments (Singh *et al.*, 2016).

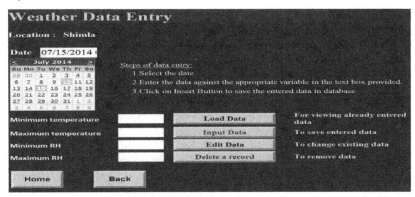

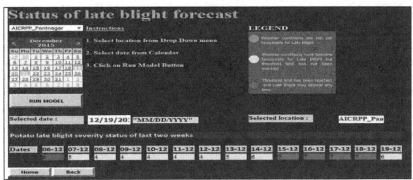

Fig. 3: Web shot of Indo-Blightcast model

REFERENCES

Adl SM *et al.* (2005) The new higher level classification of eukaryotes with emphasis on the taxonomy of protists. *J Eurkaryot Microbiol* 50: 399-451

Arora RK, Sharma Sanjeev and Singh BP (2014) Late blight disease of potato and its management. *Potato Journal* **41**(1): 16-40

Bengtson T, Holefors A, Liljeroth E, Hultberg M and Andreasson E (2015) Biosurfactants have the potential to induce defence against *Phytophthora infestans* in potato. *Potato Research* **58**: 83-90

Bounes H and Finckh MR (2008) Effects of strip intercropping of potatoes with non-hosts on late blight severity and tuber yield in organic production. *Plant Pathology* 57: 916-927

Chakrabarti SK, Singh BP and Sharma Sanjeev (2010) Genomics for understanding pathogenesis, epidemiology and management of diseases caused by fungi and fungi-like organisms. In: Bioinformatics- An Agricultural Perspective Hortinformatics 2010 Paper presented in

National Consultative Meet on Bioinformatics in Horticulture held on October 11-12, 2010 at Indian Institute of Spices Research, Calicut, Kerala: 29-45 p

Cooke L R, Schepers HTAM, Hermansen A, Bain RA, Bradshaw NJ, Ritchie F, Shaw DS, Evenhuis A, Kessel GJT, Wander JGN, Andersson B, Hansen JG, Hannukkala A, Naerstad R and Nielsen BJ (2011) Epidemiology and integrated control of potato late blight in Europe. *Potato Research* **54**: 183-222

Frenkel O, Yermiyahu U, Forbes GA Fry, WE and Shtienberg D (2010) Restriction of potato and tomato late blight development by sub-phytotoxic concentrations of boron. *Plant Pathology.* **59**: 626-33

Haas BJ *et al.* (2009) Genome sequence and analysis of the Irish potato famine pathogen *Phytophthora infestans. Nature* **461**(7262): 393-398

Haverkort AJ, Struik PC, Visser RGF and Jacobsen E (2009) Applied biotechnology to control late blight in potato caused by *Phytophthora infestans. Potato Research* **52**: 249-264.

Lal M, Yadav S, Sharma S, Singh BP and Kaushik SK (2017a) Integrated management of late blight of potato. *Journal of Appllied and Natural Science* 9 (3): 1821-1824

Lal Mehi, Singh, BP, Yadav, Saurabh and Sharma Sanjeev (2017b) Ametoctradin 27% + Dimethomorph 2027% (W/W) SC: A new molecule for management of late blight of potato in India. *Journal of Experimental Zoology India* **20** (2): 1119-1123

Liljeroth E, Bengtsson T, Wiik L and Anderson E (2010) Induced resistance in potato to *Phytophthora infestans* effects of BABA in greenhouse and field tests with different potato varieties. *European journal of Plant Pathology* **127**:171–183

Shandil RK, Chakrabarti SK, Singh BP, Sharma Sanjeev, Sundaresha S, Kaushik SK, Bhat AK and Sharma NN (2017) Genotypic background of the recipient plant is crucial for conferring RB gene mediated late blight resistance in potato. *BMC Genetics* DOI 101186/s12863-017-0490-x

Sharma S (2016) Advances in the management of potato diseases. In, Diseases of Commercial Crops in India. Gautam HR and Gupta SK (eds.) Neoti Book Agency PVT Ltd, New Delhi, India: 211-215

Sharma Sanjeev and Maheshwari Uma (2018) Bio-efficacy of QGU42 10%OD (oxathiapiprolin): A new molecule for management of late blight of potato in India. *Potato Journal* **45**(2): 93-98

Sharma Sanjeev, Sundaresha S and Singh BP (2015) Late blight of potato. In, A manual on diseases and pests of potato (Singh BP, Magesh M, Sharma Sanjeev, Sagar Vinay, Jeevalatha A and Sridhar J (eds.) ICAR-Central Potato Research Institute, Shimla, Technical Bulletin-101: 1-6

Sharma Sanjeev, Sagar Vinay and Singh BP (2017) Important diseases of potato and their management. In, Diseases of Commercial Crops in India.Gautam HR and Gupta SK (eds.) Neoti Book Agency PVT Ltd, New Delhi, India:433-456

Sharma Sanjeev and Singh, BP (2018) Forecasting for management of late blight of potato In, Alternative Approaches in Plant Disease Management. (In, Diseases of Commercial Crops in India. Gautam HR and Gupta SK (eds.) Neoti Book Agency PVT Ltd, New Delhi, India:439-458

Singh AK, Janakiram T, Chakrabarti SK, Bhardwaj V and Tiwari JK (2018) Indian potato varieties. ICAR-Central Potato Research Institute, Shimla India: 179p

Singh BP and Sharma Sanjeev (2013) Forecasting of potato late blight. *International Journal of Innovative Horticulture* **2**(2): 1-11

Singh BP, Govindakrishnan PM, Ahmad Islam, Rawat Shashi, Sharma Sanjeev and Sreekumar J (2016) INDO-BLIGHTCAST-A model for forecasting late blight across agroecologies. *International Journal of Pest Managemnt* **62**(4): 360-367

Tomar S, Singh BP, Khan MA, Kumar S, Sharma S and Lal M (2013) Identification of *Pseudomonas aeruginosa* strain producing biosurfactant with antifungal activity against *Phytophthora infestans*. *Potato Journal* **40**: 155-163

Tomar S, Lal M, Khan MA, Singh BP and Sharma S (2019) Characterization of glycolipid biosurfactant from *Pseudomonas aeruginosa* PA1 and its efficacy against *Phytophthora infestans*. *Journal of Environmental Biology* **40**(4): 725-730

Tsai J, Ann P, Wang IT, Wang SY and Hu CY (2009) Control of *Phytophthora* late blight of potato and tomato with neutralized phosphorous acid. *Journal of Taiwan Agricultural Reserach* **58**: 185-195

Turkensteen J, Flier WG, Wanningen R and Mulder A (2000) Production survival and infectivity of oospores of *Phytophthora infestans*. *Plant Pathology* **49**: 688-696.

15

Soil and Tuber Borne Diseases of Potato and Their Management

Vinay Sagar and Sanjeev Sharma

Division of Plant Protection, ICAR-CPRI, Shimla 171 001, India

INTRODUCTION

Potato is the world's most important non-grain food commodity that ranks fourth as main food crop in the world after rice, wheat and maize. The crop is grown in more than 100 countries, mainly in Asia (195.67 million tons) and Europe (121.76 million tons) (yield in 2017 as per FAOSTAT, 2019). Because of its efficiency in producing high dry matter, energy and edible protein per unit area per unit time, it holds promise for food security in the scenario of ever growing world population. The full potential of this crop however, can only be realized if diseases and pests are kept under control, especially in a subtropical country like India, where the weather is highly conducive for a number of pathogens.

A number of soil and tuber borne diseases are known to affect the potato crop causing heavy yield losses. These diseases not only reduce quality and market value of the produce but also spread to new un-infested fields through infected seed potato tubers. Diseases such as black scurf, common scab and powdery scab disfigure potato tubers and reduce their market value whereas dry rot, charcoal rot and soft rot spoil the harvested produce during transit and storage. Potato wart is a serious disease which once established is difficult to eradicate and thus require quarantine measures. Bacterial wilt or brown rot of potato is another serious disease which is likely to threaten potato cultivation with the rise in global temperature and requires immediate attention.

Potato is becoming a more and more important foodstuff in the world and therefore, its cultivation is expanding fast especially in developing countries. This intensive cultivation, in fact, has led to the emergence of diseases in certain areas where they were of less importance in the past. For instance common scab, known to be a minor disease about 2-3 decades ago, is now emerging as serious problem in many potato growing areas of the country including Indo Gangetic plains. Similarly brown rot/ bacterial wilt which was endemic in mid hills and plateau region, has now become a limiting factor for potato cultivation in many areas in Madhya Pradesh, West Bengal, north eastern region and in Kumoan hills of Uttrakhand. *Sclerotium* wilt on the other hand has shown upward trend in plateau region especially in Karnataka. This is because the dynamics of pathogens keeps on changing over time and space and also with the changing farming practices of a region. Many such new diseases are likely to emerge in a big way in future also, hence warrants regular and careful monitoring. In the following pages, the major soil and tuber borne diseases are discussed with special reference to India.

BLACK SCURF AND STEM CANKER

Black scurf is perhaps the oldest known disease of potato with worldwide distribution. The disease kills potato sprouts, causes damping off of seedlings, delay crop emergence and reduce crop stand (Figure 1). The tuber quality is also affected by the development of sclerotia on progeny tubers, thus their market value is reduced. Yield losses up to 35% primarily due to reduced crop stand have been reported. The most common symptoms of the disease appear on tubers as black irregular lumpy encrustations of fungal sclerotia which stick to the surface of tubers (black scurf). Reddish brown lesions develop on stems and often girdle them. Partial or complete girdling of the stems results in rosetting of plant tops, purple pigmentation, upward chlorosis or rolling of leaves. Formation of aerial tubers in axis of leaves due to interference with starch translocation, are also observed in the infected plants. Other symptoms on tubers could be cracking, malformation, pitting and stem end necrosis (Dutt, 1979). The sexual or perfect stage of the pathogen appears on stem just above soil line as whitish grey mat or mycelial felt. These mats are often located above a lesion on the below ground portion of stem and are generally visible later in the growing season under favorable weather conditions (Arora and Khurana, 2004).

Fig. 1: Formation of aerial tubers (a), purple pigmentation and rosetting of top leaves (b) and black scurf symptoms (c) due to *R. solani*.

The disease is prevalent in most potato growing regions of the world including India, North-West, Central Volga-Vyatka, Ural regions of Russian Federation, West and East Siberia, Far East, Baltic countries, Belarus, woodlands of Ukraine, north and central regions of Kazakhstan and many other potato growing countries. In India, it has spread widely in most potato growing regions.

The disease is caused by a fungal pathogen *Rhizoctonia solani* Kuhn (perfect basidial stage: *Thenatephorus cucumeris* (Frank) Donk). *R. solani* consists of genetically defined populations that are distinguished by anastomosis between hyphae of the isolates belonging to the same 'anastomosis group' (AG). Fourteen different anastomosis groups *viz.,* AG-1 to AG-10, AG-BI, AG-11, AG-12, and AG-13 have been recognized. Isolates of *R. solani* from potato mostly belong to anastomosis group 3 (AG-3). However, strains belonging to other anastomosis groups (AG-1, AG-2-1, AG-2-2, AG-4, AG-5, AG-7 and AG-9) have also been isolated from potato stems, stolons, roots and tubers, as well as from soils in which potatoes were grown. In Central Mexico, in addition to AG-3, isolates of AG-2, 4, -5, and -7 have been collected from potato plants and/or tubers from fields. In India, only AG-3 (PT) isolates of *R. solani* have been found responsible for black scurf and stem canker disease of potato. However, since pathogen is tuber-borne in nature, there is every possibility of introduction of other pathogenic AGs in India with import of seed/ ware potato.

The pathogen is both soil and seed borne but the disease spreads to new growing areas primarily through sclerotia-covered seed tubers. Soil-borne infection emerges later in the season since the fungus needs some time to grow into proximity with its potato host. Sclerotia of the pathogen germinate between 8 to 30°C and invade emerging sprouts or potato stems. Optimum temperature for germination of sclerotia is 23°C and for development of stem lesions is 18°C. Very high summer temperatures are not conducive for production of sclerotia and their survival. High soil temperature (28-32°C) and high soil moisture favour development of sclerotia. Sclerotial development on tubers is initiated depending on environmental conditions. Maximum development of sclerotia takes place in the period between dehaulming and harvest of the crop. Late harvested crops show more black scurf incidence. The host range of *R. solani* is very wide and the pathogen causes diseases of many economically important plants belonging to Solanaceae, Fabaceae, Asteraceae, Poaceae and Brassicaceae as well as other ornamental plants and forest trees.

The use of healthy seed tubers free from sclerotia, seed tuber treatment with 3% boric acid (dip for 30 minutes/ spray) before cold storage, planting in relatively dry and warm soil with shallow covering of seed tubers to achieve rapid crop emergence with less opportunity for the fungus to attack, two to four year crop rotation with cereals and legumes, soil solarization with transparent polyethylene mulching during hot summer months especially in Indian subtropical plains has been found very effective for control of the disease. Bio-control agents such as *Trichoderma viride, T. harzianum* and fungicides such as benomyl, thiabendazole, carboxin, pencycuron and azoxystrobin have been reported effective for control of the disease. Boric acid and pencycuron are the two chemicals that are frequently used by Indian farmers to control black scurf. Seed tubers treatment with boric acid is recommended before sprouting usually prior to cold storage whereas pencycuron can be applied to the sprouted tubers at planting. More than 97% disease control by penflufen (0.062, 0.083%) dip treatment of scurf infected tubers for 10 minutes has also been achieved.

FUSARIUM DRY ROT

Fusarium dry rot is an important post-harvest disease worldwide which causes significant losses during storage and transit of both seed and table potatoes. The affected tubers if used as seed may cause wilt of plants in field. Planting un-suberized cut pieces of potato tubers can result in *Fusarium* seed piece decay. The cut seed pieces rot from surface inward eventually destroying growing buds and result in poor emergence of the crop. Under such conditions losses by *Fusarium* rots may go up to 50 percent. In general, losses associated with dry rot have been estimated to range from 6% to 25%, and occasionally losses as great as 60% have been reported during long-term storage (Arora and Sagar, 2014).

The initial symptoms on potato tubers appear as small brown lesions on surface. These lesions subsequently enlarge, appear dark, sunken and wrinkle producing white, pink, or brown pustules. In later stages, a cavity often develop in the centre of the concentric ring and whitish, pinkish or dark brown growth of fungal mycelium may become visible. Rotten tubers may shrivel and get mummified (Figure 2).

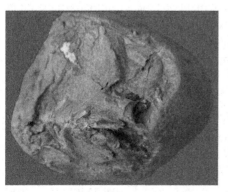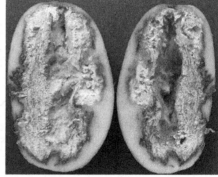

Fig. 2: Symptoms of Fusarium dry rot of potato tubers

The disease is distributed world-wide and occurs where ever potatoes are grown including India. Infection of potato tubers by dry rot pathogens occurs through wounds inflicted during harvesting, grading, cutting, and handling of tubers. As many as thirteen species of *Fusarium* have been reported to cause dry rot of potatoes. *F. sulphureum* Schlechtend. (syn. *F. sambucinum* Fuckel) is the most common pathogenic species in North America and some regions of Europe, whereas *F. coeruleum* (Libert) Sacc. (syn. *F. solani* var. *coeruleum*) is considered to be the predominant causal agent in the United Kingdom and *F. oxysporum* Schlechtend in plains of India. Other pathogenic species associated with dry rot include *F. avenaceum* (Fr.:Fr.) Sacc., *F. culmorum* (Wm. G. Sm.) Sacc., *F. acuminatum* Ellis & Everh. *F. crookwellense* L. W. Burgess, P. E. Nelson & T. A. Toussoun, *F. equiseti* (Corda) Sacc., *F. graminearum* Schwabe, *F. scirpi* Lambotte & Fautrey, *F. semitectum* Berk. & Ravenel, *F. sporotrichioides* Sherb., and *F. tricintum* (Corda) Sacc. In India, at least nine species of *Fusarium* have been reported to cause dry rot of potato. Adding to it, *F. sambucinum* was reported for the first time to cause dry rot of potatoes in cold stores in India.

Infected or rotting tubers are main source of spread of the inoculum of *Fusarium* spp. and results in soil infestation. The pathogen may also get introduced to new locations through contaminated soil, farm implements, through wind and irrigation water etc. The pathogen cannot infect intact periderm and lenticels of tubers however, cuts and wounds created during harvest, grading, transport and storage predispose them to infection. An increase in interval between haulm cutting and the harvest increases strength of tuber skin and is generally believed to reduce dry rots but the contrary view also exist. Development of disease is also affected by moisture and temperature. The fungus grows well between 15 to 28°C, however, disease development continues at low temperature in cold stores especially dry rot due to *F. sambucinum*. Storage period and relative humidity have been found to be positively correlated with dry rot whereas maximum temperature was negatively correlated. *Fusarium spp.* can cause a number of diseases of wild and economically important plants grown as food or fibre.

Planting of healthy seed, seed tuber treatment with 3% boric acid, adopting sanitation measures to avoid soil contamination through farm implements, irrigation water and reducing soil inoculum through crop rotation and eliminating volunteer potatoes are some of the measures which can reduce the risk of dry rot. Delaying harvesting for about two weeks after haulm destruction when skin of the tubers have matured and avoiding injury to tubers during harvest, handling and transportation minimize the dry rots. Harvested potatoes should be stored at around 13 to 18°C and moderate humidity for two to three weeks for bruises to heal before putting the potato in cold stores.

WART

Potato wart is an important and serious disease worldwide. The disease once established is difficult to eradicate since the resting sporangia can survive inter-host periods for up to 20 years. Rough warty mostly spherical outgrowths or protuberances appear on buds and eyes of tubers, stolons, or underground stems or at stem base. Wart may appear occasionally on above ground stem, leaf or flowers. Underground galls are white to light pink when young and become brown or light black with age. Above ground galls are green to brown or black. The wart tissues are soft and spongy (Figure 3). Tubers may get completely replaced by warts which desiccate or decay at harvest (Phadtare and Singh, 1993).

Potato wart has been reported in Asia, Africa, Europe, Oceania, North America, and South America. In India, the disease occurs in Darjeeling hills and has been managed by enforcing strict quarantine legislation.

The disease is caused by *Synchytriun endobioticum* (schilberszky) Percival, a member of *Chytridales*. Numerous pathotypes of the fungus exist and at least 43 pathotypes have been described from Europe. Wart is favoured by periodic flooding followed by drainage and aeration since free water is required for germination of sporangia and

Fig. 3: Warty outgrowths on stem and tubers of potato

dispersal of zoospores. Mean temperature below 18°C and annual precipitation of about 70 cm favour disease development.

The pathogen spreads from one locality to another through infected seed tubers, infested soil adhering tubers, machinery and other carriers of contaminated soil. Resting sporangia survive passage through the digestive track of animals fed with the infected potatoes, and contaminate manure, therefore, can disperse the inoculum. Earthworms have been found to serve as means of inoculum dispersal. The resting sporangia can also be dispersed by wind-blown soil or by flowing surface water. Resting sporangia of *S. endobioticum* are endogenously dormant and can remain viable for 40 to 50 years at depths of up to 50 cm in soil. Potato is the principal host of the pathogen, although, experimentally a number of species of *Solanaceae* takes infection upon artificial inoculation.

The disease has been successfully managed by sanitation, long crop rotation (e.g. maize/cabbage/pea followed by bean or radish), growing resistant and immune varieties (e.g. Kufri Jyoti, Kufri Sherpa, Kufri Bahar, Kufri Kanchan, Kufri Khasigaro etc.) and by enforcing strict quarantine legislation in India (Singh and Shekhawat, 2000). Periodic surveys are however, required to monitor viability of the pathogen in soil and efficiency of the quarantine measures.

CHARCOAL ROT

The disease is favoured by high soil moisture coupled with temperature exceeding 28°C. The affected tubers rot both in field and during storage and can cause severe losses under unusually warm wet weather. The fungus attacks the growing plants and tubers at harvest and storage. Affected plants in field exhibit stem blight or shallow rot similar to black leg and cause the foliage to wilt and turn yellow. Early symptoms on tubers develop around eyes, lenticels and stolon end where a dark light grey, soft, water soaked lesion develop on the surface of the tuber (Figure 4). Subsequently, the lesions become filled with black mycelium and sclerotia of the pathogen. Secondary organisms may develop in such lesions especially under wet conditions causing significant losses.

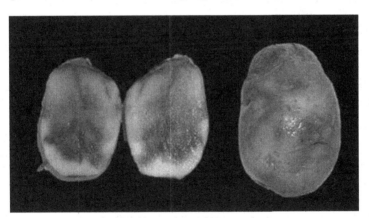

Fig. 4: Charcoal rot of potato tubers

The disease is prevalent in the Mediterranean region, warmer areas of India and Peru, Hawaii and Southern United States of America. The disease is incited by a fungus *Macrophomina phaseolina* (Tassi) Goidanich. Black, smooth, hard 0.1 to 1.0 mm sized sclerotia of the fungus develop within roots, stems, tubers and leaves. *M. phaseolina* is a weakly parasitic soil fungus and over-winters in soil as micro-sclerotia in plant debris, weeds and alternate host crops. The pathogen spread through the infected seed tubers and through the infested soil carried along with the implements. Poor plant nutrition and wounds predispose the plants to charcoal rot. Temperature around 30°C or above are very favourable for infection, the rot is slow at 20 to 25°C and stops at 10°C or below. Fungal growth stops in tubers placed in cold stores but it resumes the growth after cold storage. The fungus is found on underground parts of an extremely wide range of plants, both cultivated and wild. At least 284 hosts have been recorded (Hooker, 1981).

Charcoal rot used to be a number one problem in Bihar and eastern Uttar Pradesh, was eliminated by adjusting the dates of planting and harvesting. The disease is active only when soil temperature goes beyond 28°C. Therefore, planting early maturing cultivars, frequent irrigations to keep down the soil temperature and harvesting before the soil temperature exceeds 28°C can reduce the disease incidence in eastern plains of India. Rotation with non-host crops and use of seed from disease free area, avoiding cuts and bruises at harvest can also be followed to reduce disease incidence.

BACTERIAL WILT

Bacterial wilt or brown rot caused by *Ralstonia solanacearum* is one of the most damaging diseases of potato worldwide estimated to affect potato crop in 3.75 million acres in approximately 80 countries with global damage estimates exceeding $950 million per year. Strains of this pathogen affect more than 450 plant species in over 54 botanical families throughout the world, including a wide range of crop plants, ornamentals and weeds. In India, Losses up to 75 per cent have been recorded under extreme conditions. With increase in global temperature, the disease is likely to spread to new areas and affect potato cultivation there.

The disease causes wilting of plants in standing crop and also causes rot of infected tubers in field, storage and transit. Another indirect loss results from the spread of the disease through latently infected tubers (infected tubers without exhibiting visible symptoms) when used as seed. Potato breeder seed production cannot be undertaken in fields having even slightest bacterial wilt incidence. There is zero tolerance to this disease in most international seed certification systems.

The earliest symptom of the disease is slight wilting in leaves of top branches during hot sunny days. The leaves show drooping due to loss of turgidity followed by total unrecoverable wilt. In well-established infections, cross-sections of stems reveal brown discoloration of infected tissues. In advanced stages of wilt, cut end of base of the stem may show dull white ooze on squeezing (Figure 5). Bacterial wilt in field can be distinguished from other fungal wilts by placing the stem cut sections in clear water. Within a few minutes, a whitish thread like streaming can be observed coming out from cut end into water. This streaming represents the bacterial ooze exuding from the cut ends of colonized vascular bundles. The same test can also be carried out to see infection in tuber.

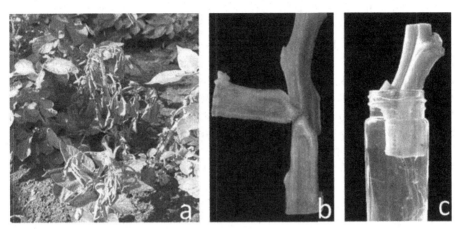

Fig. 5: Symptoms of bacterial wilt (a), brown discoloration of stem tissues (b), and bacterial streaming in clear water from cut section of potato stem (c) infected with *R. solanacearum*

In tubers, two types of symptoms are produced; they are vascular rot and pitted lesion on surface. In vascular rot, the vascular tissues of a transversely cut tuber show water soaked brown circles where dirty white sticky drops appear in about 2-3 minutes. In advanced stages of wilt, bacterial mass may ooze out from eyes. Such eyes may carry soil glued with the bacterial ooze (Figure 6). The other kinds of symptom are observed as lesions on tuber. The lesions are produced due to infection through lenticels (skin pore). Initially, water soaked spots develop which enlarge in the form of pitted lesion. The tubers may not rot in storage and also may not show vascular browning. These symptoms on tuber surface are more common in north eastern region of India.

Fig. 6: Vascular browning of tubers (left) and emergence of bacterial mass through eye due to *R. solanacearum*

Ralstonia solanacearum is a Gram-negative, rod-shaped, strictly aerobic bacterium that measures 0.5-0.7 x 1.5-2.5 μm in size. For most strains, the optimal growth temperature is between 28 and 32°C; however some strains have a lower optimal growth temperature of 27°C. Strains of *R. solanacearum* have conventionally been classified into five races (related to the ability to wilt members of the family *Solanaceae* (r1), banana (r2), potato and tomato in temperate conditions (r3), ginger (r4) and mulberry (r5) and six biovars (metabolic profiles related to the ability to metabolize a panel of three sugar alcohols and three disaccharides). Based on this classification, potatoes are known to be affected by either r1 (bv 1, 3 and 4), frequent at warmer areas and lower elevations in the tropics, and r3 (bv 2), more common in higher elevations or latitudes. A new classification scheme was described for strains of *R. solanacearum*, based on variation of DNA sequences. Four phylotypes were identified within the species that broadly reflect the ancestral relationships and geographical origin of the strains. Phylotype I contains strains of Asiatic origin which belong to bv 3, 4, and 5. Phylotype II (American origin) contains r1bv1, r2bv1 (Moko disease causing strains), r3bv2 and bv2T strains. Phylotype III contains strains from Africa and Indian Ocean, which belong to bv1 and bv2T. Phylotype IV contains strains from Indonesia, Japan and a single strain from Australia. Each phylotype can further be subdivided into sequevars based on differences in the sequence of a portion of the *endoglucanase (egl)* gene. In India, the bacterial wilt of potato is known to be caused by strains of phylotype I, IIB and IV of *R. solanacearum*.

Recently using a polyphasic taxonomic approach on an extensive set of strains of *Ralstonia solanacearum* species complex (RSSC) representing all four phylotypes, divided the RSSC into three genospecies. According to this study, the *R. solanacearum* is restricted to strains of *R. solanacearum* phylotype II only. The second genospecies includes the type strain of *R. syzygii* and contains only phylotype IV strains. This genospecies is subdivided into three distinct groups, namely *R. syzygii* subsp. *syzygii* (the causal agent of Sumatra disease on clove trees in Indonesia), *R. syzygii* subsp. *celebesensis* (the causal agent of the banana blood disease) and *R. syzygii* subsp. *indonesiensis* (phylotype IV strains isolated from potato, tomato, chilli pepper, clove). The third genospecies is designated as *R. pseudosolanacearum* and include *R. solanacearum* strains belonging to phylotype I and III. This division has been in the meantime supported by the outcome of proteomic and genomic data.

Bacterial wilt or brown rot has a worldwide distribution. It is a destructive disease of potato especially in tropical and subtropical parts of Asia, Africa, South and Central America and in some soils and waterways in Europe and Australia. In India, the disease is endemic in Karnataka, Western Maharashtra,

Madhya Pradesh, eastern plains of Assam, Orissa and West Bengal, Chhota Nagpur plateau, north-western Kumaon hills, eastern hills of West Bengal, Meghalaya, Manipur, Tripura, Mizoram, Arunachal Pradesh and in Nilgiris, Annamalai and Palani hills of Tamil Nadu (Shekhawat *et al.*, 2000). Bacterial wilt is a serious problem in Malwa region and adjoining areas in Madhya Pradesh where potato is grown for processing industry. However, it has not been noticed in the north-western high hills (excluding Kumaon hills) and in the North-western and North-central plains which are major seed producing zones of the country and need to be protected from the introduction of the disease.

Infected tubers and plant debris in soil are two major sources of inoculum. The pathogen infects roots of healthy plants through wounds. Nematodes such as *Meloidogyne incognita* which affect potato roots and tubers increase wilt incidence. Inoculum potential of about 10^7 cfu/g soil favours infection which however is dependent on other predisposing factors. Race 1 has greater ability to survive in soil than race 3 because of the better competitiveness; wide host range and higher aggressiveness of race 1. Mean soil temperature below 15°C and above 35°C do not favour the disease development. Soil moisture influences the disease in at least four ways; (i) increasing survival of the bacterium in the soil, (ii) increasing infection (iii) increasing disease development after infection, and (iv) increasing exit of the bacterium from host and spread through the soil. *Ralstonia solanacearum* is capable of causing brown rot in a wide range of soil types and levels of acidity. In majority of the cases, the disease has been reported in acidic soils (pH 4.3 to 6.8) and only in a few cases in alkaline soils.

Several other factors that affect pathogen survival in soil and water also affect disease development. The soil type and physicochemical properties have significant influence on survival of the pathogen. Soils having high clay and silt content with higher water holding capacity are favourable for long survival while high sand contents disfavor its survival. Also, soil moisture and temperature exert a combined effect on survival of the pathogen. The congenial conditions for slow decline of population and virulence for race 1 and 3 are temperature between 10-30°C, soil moisture between 20-60 WHC, heavy soils and aerobic conditions (Shekhawat *et al.*, 1993).

The pathogen survives through infected seed tubers and in plant debris in soil. Symptomless plants may harbour the bacterium and transmit it to progeny tubers as latent infection. This could lead to severe disease outbreaks when the tubers are grown at disease free sites. High soil moisture, temperature, oxygen stress and soil type affect the survival of the pathogen. The pathogen population declines gradually in soil devoid of host plants and their debris.

Transmission of *R. solanacearum* from one area to another occurs through infected seed, irrigation water and farm implements. Under favorable conditions, potato plants infected with *R. solanacearum* may not show any disease symptoms. In this case, latently infected tubers used for potato seed production may play a major role in spread of the bacterium from infected potato seed production sites to healthy potato-growing sites.

Management

The control of bacterial wilt has proved to be very difficult because of both the seed and soil borne nature of the pathogen and especially in the case of race 1 due to its broad host range. Chemical control is nearly impossible. Soil fumigants have shown either slight or no effects. Antibiotics such as streptomycin, ampicillin, tetracycline and penicillin hardly have any effect; in fact, streptomycin application increased the incidence of bacterial wilt in Egypt. Biological control has been investigated, but is still in its infancy. Potato cultivars developed in Colombia with a *Solanum phureja* and *S. demissum* background showed resistance to *R. solanacearum*, but the race and strain diversity of the pathogen made it difficult to utilize these in other countries. The absolute control of bacterial wilt, at present, is difficult to achieve, however, economic losses can be brought down considerably using the following eco-friendly package of practices:

Healthy seed

Use of healthy planting material can take care of almost 80% of bacterial wilt problem. Fortunately, bacterial wilt free areas in western and central Indo-Gangetic plains can be the source of disease free seed in India. Tubers should not be cut since the cutting knife spreads the disease and also cut tubers can contact disease from soil easily.

Field sanitation and cultural management

Where the field is already infested, the disease can be minimized by the following agronomic practices:

- Two to three years' crop rotation with non-host crops like cereals, garlic, onion, cabbage and *sanai* (Indian Senna) reduces the wilt infestation. Solanaceous crops like tomato, brinjal, capsicum, chilli, tobacco etc. should not be cultivated in rotation with potato. Paddy, sugarcane and soybean, though are not hosts of *R. solanacearum*, still they carry the pathogen and contribute to the disease perpetuation and therefore, should be avoided.

- Pathogen enters in plant through root or stolen injuries. Such injuries cannot be avoided during intercultural operations. Therefore, tillage should be restricted to the minimum and it is advisable to follow full earthing-up at planting.

- The pathogen perpetuates in the root system of many weeds and crops. Field should be cleaned from weeds and root/foliage remnants and burn them. The pathogen in remnants can be exposed to high temperature above 40°C in summer in plains and plateau and low temperature below 5°C in hills by giving deep ploughing. This may cause extinction of pathogen from the field.

Chemical control

Incidence of bacterial wilt declines by application of bleaching powder @ 12 kg/h mixed with fertilizer or soil drenching after first earthing up together with the use of healthy seed. Application of copra and pea nut meal has been reported to reduce the disease. Biocontrol of bacterial wilt by use of antagonists such as *Pseudomonas flourescens*, *Bacillus* spp, avirulent *P. solanacearum* and actinomycetes have been found to be effective in some countries. In India, biocontrol of bacterial wilt by use of antagonists such as *Pseudomonas flourescens*, *Bacillus* spp, have been found to be effective at some locations (Shekawat *et al*, 1993). Similarly, *Bacillus subtilis* (strain B5) recovered from rhizosphere soil of potato plants from bacterial wilt infested fields of Bhowali, Uttarakhand controlled tuber borne *Ralstonia solanacearum* under different argo-climate conditions and enhanced the crop yield. A bio-formulation of *Bacillus subtilis* (strain B5) has also been developed. More such biocontrol agents effective against bacterial wilt need be identified and their deployment in potato field needs to be standardized for management of the disease in an eco-friendly manner.

Based on intensive ecological and epidemiological studies at ICAR-Central Potato Research Institute, Shimla, the following practices are recommended for checking the bacterial wilt in different agro-climatic zones of the country.

Zone I: It comprises of non-endemic areas like Gujarat, Maharashtra, north-western and north-central plains. This zone is characterised by hot and dry summer with scanty vegetation (April-June); temperature may go up to 40-43°C. The bacterial wilt is no more a major problem. Therefore, deep ploughing in summer and use of disease free seed is adequate for the disease control.

Zone II: It includes north-western mid hills (up to 2200 masl), north-eastern hills and the Nilgiris. The zone is characterised by mild summer, profuse vegetation with a maximum temperature range of 26-30°C. Winter temperature may go as low as 3-6°C. Many weed hosts can provide perpetual niche for colonisation and survival of the bacteria. The use of disease free seed plus application of stable bleaching powder @ 12kg/ha mixed with fertiliser at planting, ploughing the field in September- October and exposing the soil to winter temperature are adequate for disease control. The application of bleaching powder can be substituted by 2 year crop rotation with crops like wheat, barley, finger millet, cabbage, cauliflower, knol-khol, carrot, onion, garlic etc. Early planting preferably in February and early harvesting are recommended to minimise the exposure of the crop to high temperature which favours the disease.

Zone III: It includes eastern plains and Deccan plateau. The area is relatively rich in vegetation. Day temperature sometimes reaches 38°C. Heavy precipitation occurs due to western disturbances. Eastern plains and Deccan plateau have many symptomless carriers of the pathogen. Therefore, management of the disease is most difficult. However, the disease can be kept under check with practices like use of disease free seed, application of bleaching powder, blind earthing-up and ploughing in March and leaving the soil exposed to summer temperatures during April- May and crop rotations along with clean cultivation.

Zone IV: It includes north western high hills (above 2200 masl excluding Kumaon hills). This zone has a temperate climate with severe winters; daily temperature ranges from −10 to 5°C during December–January. Snow is common during these months. Bacterial wilt is not endemic and the use of disease free seed alone is adequate.

COMMON SCAB OF POTATO

Common scab causes superficial lesions on surface of potato tubers. The disease seldom affects yield but economic losses may be high as the affected tubers fetch low price in market due to their poor appearance. Seed lots exceeding 5 per cent incidence are rejected by seed certification agencies (in India) causing huge loss to seed industry. This disease was first recorded in Patna during 1958. Since then, it has become endemic in various potato growing states. Scab begins as small reddish or brownish spot on the surface of the potato tubers and its initial infection takes place during juvenile period of tubers. Infection takes place mainly through lenticels and surrounding periderm turns brown and rough. Lesion becomes corky due to elongation and division of invaded cells. Under Indian conditions multiple kinds of symptoms have been recorded and they are grouped as (1) a mere brownish roughening or abrasion of tuber skin (2) proliferated lenticels with hard corky deposition, might lead to star shaped lesion (3) raised rough and corky pustules (4) 3-4 mm deep pits surrounded by hard corky tissue (5) concentric series of wrinkled layers of cork around central black core (Figure 7).

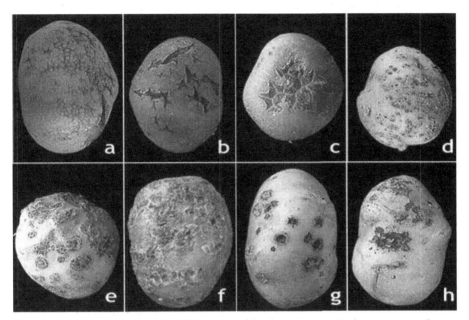

Fig. 7: Various types of scab symptoms caused by *Streptomyces* species on potato tubers

Common scab occurs in most potato producing areas in Africa, Asia, Europe, North and South America. In India, it was known to occur in Lahaul Valley (Himachal Pradesh) in severe form since 1969; its frequent occurrence in plains was reported in 1979-1980. Afterwards, it became a major problem in almost all agro climatic zones of India. Now, the disease has covered almost all the potato growing areas of the country and is posing a serious threat to successful potato cultivation. In Eastern Uttar Pradesh, common scab on potato has been reported every year in moderate to severe form. Its real impact is felt in states like Punjab, Uttar Pradesh and Lahaul valley of Himachal Pradesh where potato production is for seed industry. The disease is spreading fast in some areas in Indo Gangetic plains due to cultivation of potato year after year in the same land.

At least 13 different *Streptomyces* spp. have been found to cause common scab on potato worldwide. The prominent among them are *Streptomyces scabies* (Thaxter) Lambert and Loria, *S. acidiscabies* Bambert and Loria, *S. turgidscabis* Takeunchi, *S. collinus* Lindenbein; *S. griseus* (Krainsky), *S. longisporoflavus*, *S. cinereus* , *S. violanceoruber*, *S. alborgriseolus*, *S. griseoflavus*, *S. catenulae* and others. *Streptomyces* spp. may be pathogenic or non-pathogenic. The pathogenic isolates (*S. scabies, S. europaescabies, S. bottropensis*) and non-pathogenic (*S. griseus, S. coelicolor*) strains from common scab infected tubers collected from northern, eastern and north-eastern parts of India. The pathogenic species produce thaxtomins which are phytotoxins and cause hypertrophy and cell death. The identification and taxonomy of *Streptomyces* spp. has been based on morphological and physiological characteristics combined with thaxtomin production and pathogenicity tests *in vitro* and *in vivo*. Ability to produce thaxtomin toxin is strongly correlated with the pathogen's pathogenicity (Wanner and Kirk, 2015).

Potato is physiologically most susceptible to *Streptomyces* spp. in the period following tuber initiation. *Streptomyces* species infects the newly formed tubers through stomata and immature lenticels. Once the periderm has differentiated, tubers are no longer susceptible to the pathogen. The pathogen is both seed and soil borne. It can survive in soil for several years in plant debris and infested soil. Soil conditions greatly influence the pathogen. Favourable conditions include pH between 5.2 to 8.0 or more, temperature in the range of 20 to 30°C and low soil moisture. The pathogen is aerobic in nature and maintaining high soil moisture for 10 to 20 days after tuber initiation can help in reducing the common scab. However, scab outbreaks have been reported in irrigated or wet soil conditions in northern Europe, Israel, and Canada. The organism is a tuber-borne and is well-adapted saprophyte that persists in soil on decaying organic matter and manure for several years. Infected tubers serve as source of inoculum in the field, giving rise to infected progeny tubers. The pathogenic *Streptomyces* species are both soil and tuber-borne. Tuber-borne inoculum is likely to be involved in the distribution of new strains or species.

The pathogen is difficult to eradicate because of long survival both on seed tubers and in soils. As common scab pathogen is a bacterium, it is not controlled by seed-applied fungicides. Earlier, formaldehyde, urea formaldehyde, and manganese sulphate were used for control of common scab, but are no longer applied in fields. In a 2-year Canadian study, seed-applied fludioxonil resulted in a 57.8% reduction of common scab severity, and use of a seed-applied biopesticide containing *Bacillus subtilis* resulted in a 56.1% reduction.

A perusal through literature provides information on effectiveness of various soil amendments, seed-applied, in-furrow applied, and foliar-applied treatments from many locations, over many years. Unfortunately, few to no treatments provide highly effective and reliable control of the disease across locations.

In general, common scab can be managed by use of disease free seed tubers, tuber treatment with boric acid (3% for 30 min.) before or after cold storage (before sprouting), keeping the moisture near to field capacity right from tuber initiation until the tubers measure 1 cm in diameter, following 3-4 year crop rotation with wheat, pea, oats, barley, soybean, sorghum, bajra, green manuring and deep ploughing the potato fields in April and leaving the soil exposed to high temperatures during summer (May to June) in the North Indian plains.

SOFT ROT OR BLACK LEG

Bacterial soft rot can cause significant loss of potato tubers at harvest, transit and storage. Losses due to poor handling of the produce, poorly ventilated storage or transit may go up to 100 per cent (Somani and Shekhawat, 1990). Soft rot bacteria usually infect potato tubers which have been damaged by mechanical injury or in the presence of other tuber borne pathogens. On tubers, the disease appears as small soft water soaked spots around lenticels which enlarge under high humidity. A brown to black pigment may develop around the lesion. The soft rot affected tubers become slimy and foul smelling and brown liquid oozes out from the affected tubers (Figure 8). The tuber skin remains intact and sometimes the rotten tubers are swollen due to gas formation. At harvest, many small rotten tubers with intact skin can be seen. In cooler regions black leg phase develops from soft rot infected seed tubers. A soft black lesion appears at base of stem which extend from the decaying mother seed tuber in soil to a little above ground level. The affected plants become stunted, exhibit yellow chlorotic foliage, wilt and die without producing fresh tubers (Somani and Shekhawat, 1990).

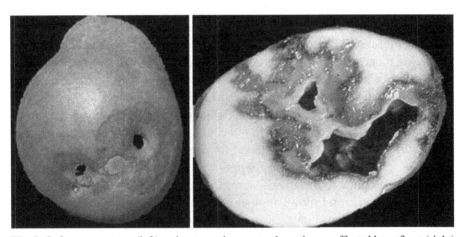

Fig. 8: Soft rot symptoms (left) and potato tuber cut to show tissues affected by soft rot (right)

Bacterial soft rot of potato is found wherever potatoes are grown. The disease affect the crop at all stages of growth but it is more serious on potato tubers under poor storage conditions especially in warm and wet climate. Black-leg (*Pectobacterium atrosepticum*) phase of the disease is not common in India. It occurs only rarely in the Shimla hills in HP, the Kumaon hills in Uttarakhand, Ootacamund in Nilgiris and also in Bihar plains. Stem and petiole rot due to *Pectobacterium carotovora* sub sp. *carotovora* has been observed in Shimla, Jalandhar, Ambala, Panipat, Meerut, Agra, Kanpur, Allahabad, and Burdwan.

Pectobacterium atrosepticum (van Hall) (syn. *Erwinia carotovora* sub sp. *atroseptica*), *Pectobacterium carotovorum* sub sp. *carotovorum* (Jones) (syn. *Erwinia carotovora* subsp. *carotovora*), *Dickeya* spp. (including *D. dianthicola, D. dadantii, D. zeae*) (syn. *Erwinia chrysanthemi*), *Bacillus polymyxa, B. subtilis, B. mesentericus, B. megaterium* de Bary, *Pseudomonas marginalis* (Brown) Stevens, *P. viridiflava* (Burkholder) Dowson, *Clostridium* spp., *Micrococcus* spp., and *Flavobacterium* have been found to cause soft rot. *P. atrosepticum*, the primary enterobacteria causing soft rots, produce pectolytic enzymes and degrade pectin in middle lamella of host cells, breakdown tissues and cause soft rot and the decay (Czajkowski *et al.*, 2015). The decaying tissue become slimy and foul smelling and brown liquid oozes out from the soft rot affected tubers. About 1500 strains of pectinolytic *Erwinia* have been isolated from infected plants and tubers. The pathogen produce certain volatile compounds such as ammonia, trimethylamine and several volatile sulfides and early detection of such volatile compounds in storage could be used as a method to detect the disease at initial stage.

Soft rot bacteria may be carried latently in lenticels, wounds and on surface of tubers without any visible symptoms and spread to healthy tubers in stores, during seed cutting, handling and planting. Water film on surface of tuber which cause proliferation of lenticels and creates anaerobic conditions and injury on surface of tuber predispose potatoes to soft rot. From soft rot infected seed tubers bacteria may enter vascular tissues of developing stems and can develop black leg under favourable conditions. From black leg infected plants the pathogen can reach daughter tubers through stolons and initiate tuber decay at the site of tuber attachment. Decaying tubers in soil could serve as source of contamination for healthy tubers. The threshold level for disease development is about 103 cells of *E. carotovora* sub sp. *atroseptica* per tuber. Tubers harvested in wet soil, poor ventilation in transit and storage promotes the rot. In warm climates, where one potato crop follows another or where only short rotation cycles are applied, the bacteria can pass easily from one crop to the next, especially in poorly drained soil. The bacteria can survive at places where rotten potatoes and vegetables are dumped.

An integrated approach involving practices like planting of whole seed potato or well suberized seed pieces in well-drained soil, tuber treatment with 3 percent boric acid (Somani and Shekhawat, 1990), restricting nitrogen dose to minimum (150 kg/ha), application of stable bleaching powder before planting and during last irrigation, crop rotation following green manure- potato – wheat reduces soft rot, avoiding bruises and cuts to potato tubers during harvest, handling, and proper aeration during storage and transit can minimize soft rots. Adjusting planting time to avoid hot weather during plant emergence and harvesting the crop before soil temperature rises above 28°C are also recommended to minimize the losses due to soft rot.

REFERENCES

Arora RK and Khurana SMP (2004). Major fungal and bacterial diseases of potato and their management. In, Fruit and Vegetable Diseases. K.G. Mukerji (ed), Kluwer Academic Publishers, the Netherlands: 189-231

Arora RK and Sagar V (2014) Tuber borne diseases of potato. In, Diseases of Vegetable Crops, Diagnosis and Management. Singh Dinesh, Chowadappa P and Sharma Pratibha (eds), pp Indian Phytopathological Society, New Delhi, India:1-58

Czajkowski R, Perombelon MCM, Jafra S, Lojkowska E, Potrykus M, van der Wolf JM and Sledz W (2015) Detection, identification and differentiation of *Pectobacterium* and *Dickeya* species causing potato blackleg and tuber soft rot: a review. *Annals of Applied Biology* 166: 18-38

Dutt BL (1979) Bacterial and fungal diseases of potato. ICAR, New Delhi, Technical bulletin: 199p

Hooker WJ (1981) *Compendium of potato diseases. American Phytopathological Society,* St. Paul, Minnesota, 125p

Phadtare SG and Singh PH (1993) Wart disease of potato. In, Advances in Horticulture. Vol.7. Potato. K.L. Chadha and JS Grewal (eds), Malhotra Publishing House, New Delhi: 443-462

Shekhawat GS, Chakrabarti SK and Gadevar AV (2000) Potato bacterial wilt in India. Central Potato Research Institute, Shimla, India. Technical Bulletin-38(Revised 2000):

Shekhawat GS, Singh BP and Jeswani MD (1993) *Soil and tuber borne diseases of potato.* Central Potato Research Institute, Shimla, India. Technical bulletin- **41**:47p

Singh PH and Shekhawat GS (2000) Wart disease of potato in Darjeeling Hills. Central Potato Research Institute, Shimla, India. Technical bulletin- 19 (Revised): 73p

Somani AK and Shekhawat GS (1990) Bacterial soft rot of potato in India. Central Potato Research Institute, Shimla, India. Technical bulletin-21: 32p

Wanner LA and Kirk WW (2015) *Streptomyces*- from basic microbiology to role as a plant pathogen. *American Journal of Potato Research* **92**: 236-242.

16

Viral and Viroid Diseases of Potato and Their Management

Ravinder Kumar[1], A. Jeevalatha [2], Baswaraj R.[1] and Rahul Kumar Tiwari[1]

[1]*ICAR-Central Potato Research Institute, Shimla- 171001, Himachal Pradesh India*
[2]*ICAR - Indian Institute of Spices Research, Kozhikode-673012, Kerala, India*

INTRODUCTION

Potato (*Solanum tuberosum* L.) is afflicted by many kinds of viruses and their various strains belonging to different taxonomic groups (Table 1) throughout the world resulting in decline in yield and tuber quality. Exact estimation of yield and economic losses caused by potato viruses depends on a number of factors mainly strain of the virus, variety, season and growing conditions etc. Generally crop losses ranges between 5 to15 per cent if all plants are secondarily infected with PVX and PVS; 15-30 per cent for 100 per cent secondary infection of PVY^n and 40-70 per cent due to PLRV. Besides, in Europe and North America *Potato spindle tuber viroid* (PSTVd) is well known to reduce yields significantly (16-64 per cent) depending on the viroid strain/potato variety and warm weather (Singh, 1988). The severe strains of *Potato virus Y* (PVY) and *Potato leafroll virus* (PLRV) have the potential to reduce yield up to 80 per cent, while mild viruses like *Potato virus X* (PVX), *Potato virus S* (PVS) and *Potato virus M* (PVM) can cause up to 30 per cent yield loss in infected plants. A tospovirus *Groundnut bud necrosis virus* (GBNV) causing severe stem/leaf necrosis disease in plains/plateaux of central/western India heavily infects the early crop of potato. Similarly, a whitefly transmitted begomovirus *Tomato leaf curl New Delhi virus-potato* (ToLCNDV) known to cause apical leafcurl disease in India, has become a serious problem in North Indian plains.

The viruses can be mechanically transmitted with plant sap, by contact and with seeds. Infected tubers, whitefly and aphid vectors are important factors for the transmission of the viruses in nature. Potato viruses are transmitted by aphids in two ways. The viruses are either non-persistent (stylet-borne) or persistent and circulative (they are ingested and persist in the aphid throughout its life cycle). The most common persistent and circulative virus affecting potatoes in India is PLRV. Common non-persistent or stylet-borne viruses are PVY, PVA, PVS and PVM. Control of circulative viruses by insecticides is highly effective, which is not the ease with stylet borne viruses. Sustainable potato production is possible only if the virus incidence is kept under check especially in sub tropics where the weather is highly conducive both for vectors and the common viruses. In this chapter the major viruses and viroid of potato including symptomatology, diagnostics and their management have been discussed.

GENOME STRUCTURE AND REPLICATION OF POTATO VIRUSES

Genome is entire sequence of DNA or RNA or total genetic material in an organism. Viruses are small obligate intracellular pathogens containing either a RNA or DNA genome surrounded by a protective, virus-coded protein coat. Viruses are mobile genetic elements, most probably of cellular origin and characterized by a long co-evolution of virus and the host. A complete virus particle is known as virion, the main function of which is to deliver its DNA or RNA genome into the host cell, so that the genome can be expressed (transcribed and translated) by the host cell. Of late, knowledge of the genome has become most important part of biological as well as biotechnological research. A wealth of information on potato viruses has been generated that provides detailed understanding of the viral structure. Virus infected cells have proved extremely useful as model systems for the study of basic aspects of cell biology. In many cases, DNA viruses utilize cellular enzymes for synthesis of their DNA genomes and mRNAs; all viruses utilize normal cellular ribosomes, tRNAs, and translation factors for synthesis of their proteins. The taxonomy and genome organization of potato viruses are explained and illustrated at the official home page (http://www.ictvonline.org/virusTaxInfo.asp) of the International Committee on Taxonomy of Viruses (ICTV). The infection cycle comprises of 1) entry of the virus into the cell and disassembly of the virus capsids, 2) translation of the viral RNA, genome replication and transcription, and 3) encapsidation and cell-to-cell movement. The majority of potato viruses have a single-stranded RNA (ssRNA) genome (Table 1) that is replicated in cytoplasm by the viral RNA-dependent RNA polymerase. However, tospoviruses exemplified by TSWV contain a negative-sense ssRNA genome of an ambisense nature: open reading frames (genes) exist in both reading directions. A third type of RNA genome is exemplified by nucleorhabdoviruses such as PYDW that has a negative-sense ssRNA genome replicated in the nucleus. Tospo- and nucleorhabdoviruses differ from most RNA viruses as their virions are enveloped by a lipid membrane. A few viruses such as geminiviruses have a single-stranded, small ambisense DNA genome which encodes only a few structural and non-structural proteins and replicated in the nucleus by the host DNA polymerase. Their virions are not enveloped (Vreugdenhil *et al.*, 2007). PSTVd is one of the smallest known viroid infecting potato worldwide.

MAJOR VIRUSES AND VIROID OF POTATO

Potato virus Y (PVY)

Potato Virus Y (PVY) is an important virus causing serious damage in potato (10-100 per cent yield losses) and other solanaceous crops worlwide. The severity of this disease depends on the PVY strain involved, host tolerance, time of infection and environmental factors. The distribution of PVY is global, although some virus strains are restricted to certain continents. The common strain PVY^O occurs worldwide, while PVY^N occurs in Europe, part of Africa and South America and PVY^C strain has been reported from Australia, India and Europe. PVY^C originally known as *Potato Virus C*, was the first to be recognized and identified in 1930s. Another PVY strain, PVY^{NTN}, previously found in Europe, has lately been confirmed in North America. PVY^Z and PVY^E have also been reported but are of less significance to potato production. The main sources of PVY inoculum are infected seed tubers. The virus can be transmitted by sap inoculation, grafting and aphids in a non-persistent manner. Aphids feeding on plants emerging from infected tubers acquire PVY within a few seconds and also transmit the virus to healthy plants within seconds. Thus, aphids probing on potato plants are potential vectors of PVY.

PVY belongs to the genus *Potyvirus* within the family *Potyviridae*. It has flexuous, 740x11 nm long filaments and single stranded RNA, occuring in low concentration in potato and serologically related to PVA. Symptom development by PVY^O depends on the potato cultivars, time of infection and location. Generally, it causes a mosaic symptom on the leaves, whereas in some cultivars it causes vein burning on the underside of the leaves and even necrotic "ringlets" on the leaves (Figure 1a). Symptoms due to PVY^N and ^{NTN} are similar to that described for PVY^O. However, one important difference between PVY^O and PVY^{NTN} is the development of tuber symptoms. PVY^{NTN} isolates cause severe superficial tuber necrosis (potato tuber necrotic ring spot disease) and may also cause necrotic foliar symptoms.

Table 1: Important characteristics and genome structure of potato viruses (Khurana, 2004).

Virus (acronym)	Virus genus/group	Family	Morphology/number of distinct particle size	Particle diameter	Vectors	Mode of transmission, spread	Geographical distribution
Potato leafroll virus (PLRV)	Polerovirus Group IV (+)ssRNA	Luteoviridae	Isometric/01	24	Aphid[P]	TPS	Worldwide
Potato virus X (PVX)	Potexvirus Group IV (+)ssRNA	Flexiviridae	Filamentous/01	13	-	Contact, TPS	Worldwide
Potato virus Y (PVY)	Potyvirus Group IV (+)ssRNA	Potyviridae	Filamentous/01	11	Aphid[NP]	TPS, mechanical	Worldwide
Potato virus A (PVA)	Potyvirus Group IV (+)ssRNA	Potyviridae	Filamentous/01	-	Aphid[NP]	Mechanical	Worldwide
Potato virus M (PVM)	Carlavirus Group IV (+)ssRNA	Flexiviridae	Filamentous/01	12	Aphid[NP]	Contact	Worldwide
Potato virus S (PVS)	Carlavirus Group IV (+)ssRNA	Flexiviridae	Filamentous/01	12	Aphid[NP]	Contact	Worldwide
Tomato leaf curl NewDelhi virus-potato (ToLCNDV-potato)	Begomovirus Group II (ssDNA)	Geminiviridae	Geminate particles	21-24 nm	Whitefly	-	India
Tomato spotted wilt virus (TSWV)	Tospovirus Group IV (-)ssRNA	Bunyaviridae	Enveloped particle/01	70-110	Thrips[P]	Mechanical	Hot climate, Worldwide
Potato aucuba mosaic virus (PAMV)	Potexvirus Group IV(+)ssRNA	Flexiviridae	Filamentous/01	11	Aphid[#C]	TPS, Contact	Worldwide
Alfalfa mosaic virus (AMV)*	Alfamovirus Group IV (+)ssRNA	Bromoviridae	Bacilliform/04-05	19	Aphid[NP]	TPS, Pollen	Worldwide (Uncommon)
Andean potato latent virus (APLV)*	Tymovirus Group IV (+)ssRNA	Tymoviridae	Isometric/01	28-30	Flea Beetle	TPS, Pollen	S-America (Uncommon)

(Contd.)

Virus name	Genus / Group	Family	Shape / Segments	Size	Vector	Transmission	Distribution
Andean potato mottle virus (APMV)*	Comovirus, Group IV (+)ssRNA	Comoviridae	Isometric/01	28	Beetle	Contact	S-America
Arracacha virus B - Oca strain (AVB-O)*	Nepovirus, Group IV (+)ssRNA	Sequiviridae	Isometric/01	26	Unknown	TPS, Pollen	Peru, Bolvia
Cucumber mosaic virus (CMV)*	Cucumovirus, Group IV (+)ssRNA	Bromoviridae	Isometric/01	30	Aphid[NP]	Sap, TPS	Worldwide (Uncommon)
Potato black ringspot virus (PBRSV)*	Nepovirus, Group IV (+)ssRNA	Comoviridae	Isometric/01	26	Nematode[SP]	Soil borne, TPS, Pollen	Peru
Potato deforming mosaic virus (PDMV)*	Begomovirus, Group II (ssDNA)	Geminiviridae	Segmented/02	18	Whitefly[SP]	TPS	Brazil
Potato latent virus (PotLV)*	Carlavirus, Group IV (+)ssRNA	Betaflexiviridae	Filamentous/01	-	Aphid[NP]	Contact	N-America
Tobacco rattle virus (TRV)*	Tobravirus, Group IV (+)ssRNA	Virgaviridae	Rod or tubular/02	22	Nematode[P]	Mechanically, TPS	Worldwide
Tobacco streak virus (TSV)*	Ilarvirus, Group IV (+)ssRNA	Bromoviridae	Quasi-isometric/01	22-35	Thrips	Pollen, TPS, mechanical	S-America
Potato yellow dwarf virus (PYDV)*	Nucleorhabdovirus, Group V ((-)ssRNA)	Rhabdoviridae	Bacilliform	75	Leafhopper[P]	Mechanical	N-America
Potato yellow mosaic virus (PYMV)*	Begomovirus, Group II (ssDNA)	Geminiviridae	Segmented/02	18-20	Whitefly[SP]	-	Caribbean region
Potato mop top virus (PMTV)*	Pomovirus, Group IV (+)ssRNA	Virgaviridae	Rod or tubular/02	18-20	Fungus[P]	Mechanical	W-Europe and S-America
Potato yellow vein virus (PYVV)*	Crinivirus, Group IV (+)ssRNA	Closteroviridae	Filamentous	-	Whitefly[P]	Infected tuber	S-America
Potato yellowing virus (PYV)*	Alfamovirus, Group IV ((+)ssRNA	Bromoviridae	Bacilliform	21	Aphid[SP]	TPS	S-America
Potato virus T (PVT) *	Trichovirus, Group IV (+)ssRNA	Flexiviridae	Filamentous/01	12	-	Contact, TPS, Pollen	S-America
Potato virus U (PVU) *	Nepovirus, Group IV (+)ss RNA	Comoviridae	Isometric/01	28	Nematode	Contact, TPS	Peru

(Contd.)

Potato virus V (PVV) *	Potyvirus Group IV (+) ss RNA	Potyviridae	Filamentous/01	12-13	Aphid^{NP}	TPS	N-Europe, S-America
Solanum apical leaf curling virus (SALCV)*	Begomovirus Group II (ssDNA)	Geminiviridae	Segmented/03	18	Whitefly^{SP}	TPS	Peru
Tobacco mosaic virus (TMV)*	Tobamovirus Group IV (+)ssRNA	Virgaviridae	Rod or tubular/01	18	-	Contact, Infected soil	Worldwide
Tobacco necrosis virus (TNV)*	Necrovirus Group IV (+)ssRNA	Tombusviridae	Isometric/01	26	Fungus^{P}	Soil borne spores, mechanical	Europe N-America
Tomato black ring virus (TmBRV)*	Nepovirus Group IV (+)ssRNA	Comoviridae	Isometric/02	5-6	Nematode^{P}	Pollen, TPS	Europe
Tomato mosaic virus (ToMV)*	Tobamovirus Group IV (+)ssRNA	Virgaviridae	Rod or tubular/01	18	-	TPS, Pollen, Contact	Hungary
Potato spindle tuber viroid (PSTVd)	Pospiviroid Circular (+)ssRNA	Pospiviroidae	Circular ss RNA only		Aphid^{CI}	TPS, Pollen, Contact	United states Canada, South Africa, Russia

TPS = True potato seed; P/NP = Persistently/ Non-persistently transmitted; SP =Semi-persistently transmitted; HC =Helper component involved for transmission,

CI = Coinfection of PLRV essential for aphid transmission of viroid; * = Viruses that are of quarantine importance in India or not reported in potato in India.

Potato leafroll virus (PLRV)

Potato leafroll virus (PLRV) is the second most important virus of potato globally. It causes severe yield loss (up to 90 per cent), and in some cultivars, a quality reduction due to internal damage to tubers (net necrosis). One estimate has suggested that the virus is responsible for 20 million tonnes production loss globally. PLRV occurs in extremely low concentration and is confined to the phloem cells. The virus persists in the aphid throughout its life cycle. Aphids transmit the virus in a persistent circulative non propagative manner. All instars (stages) of the aphid can transmit the virus, but the nymph stage is more efficient than the adult. The virus can be spread to long distances by winged aphids. It accumulates in tubers and, if planted, the virus is transmitted to progeny tubers as well as into the foliage.

PLRV belongs to the genus *Polerovirus* within the family *Luteoviridae*. It has isometric virions of 24 nm diameter and single stranded RNA. Symptoms of PLRV through primary infection causes chlorosis of young leaves with an erect habit of the infected plant while secondary symptoms include stunting of shoots, older leaves rolling upward and turning chlorotic, leathery and brittle. In some species, particularly stunting and yellowing without leaf rolling is observed of this virus. Inward rolling of lower leaflets, extending ultimately to the upper leaves, is typical (Figure 1b). The leaves become dry and brittle, and if touched the plant makes a characteristic rustling noise. Leaves are chlorotic and often show purple discoloration. PLRV infected plants usually produce normal-shaped, but small tubers.

Potato virus X (PVX)

Potato virus X (PVX) is found worldwide wherever potatoes are grown. It is economically important as its incidence is high (usually 15-20 per cent) despite lower potential for yield reduction. Combined infection with other viruses, particularly PVA and PVY, causes severe yield losses. PVX is transmitted in nature mainly mechanically. It is highly contagious, and once attached to a surface such as clothing, the virus can remain infective for many hours provided the surface remains wet. Consequently, a virus picked up from an infected plant can be transmitted to many other plants when moving through a crop. The virus accumulates in tubers, and the process of cutting seed tubers can spread the virus from one tuber to another. PVX is not transmitted by true seed or by aphids.

PVX belongs to the genus *Potexvirus* within the family *Flexiviridae*. It has filamentous virions of 515x13 nm and single stranded RNA. This virus causes mild or no symptoms in most potato varieties, but when PVY is present, synergy between these two viruses causes severe mosaic symptoms in potatoes. Visually, symptoms range from absent through a faint or fleeting mottle to a severe necrotic streak (Figure 1c). The light conditions can affect symptom detection, with low-light conditions making them more apparent. When symptoms are expressed, there is a pattern of light and dark green on leaflets; the lighter, small, irregular blotches being between the veins. Only occasionally does leaf distortion, rugosity, necrotic spotting or stunting occur.

Potato virus S (PVS)

Potato virus S (PVS) occurs worldwide especially in East European countries and is the most frequently found virus in potato. Globally it is known to causes up to 20 per cent tuber yield reduction. An incidence of 8.5 to 99.5 per cent has been recorded in the potato crop at different locations in India (Singh *et al.*, 1994). Reduction in tuber yield is very low but may go upto 10-20 per cent or more in combine action with PVX. It has been found to be responsible for breaking late blight resistance in some potato cultivars. Two strain groups of PVS have been recognized viz., PVS^O (Ordinary) and PVS^A (Andean) based on whether it causes non systemic or systemic infection in *Chenopodium* spp. PVS^O occurs worldwide while PVS^A is found only in the Andean region. PVS is most contagious and spreads easily through seed cutting and even plant-to-plant contact. It also spreads in a non-persistent manner by aphids, particularly by *Myzus persicae*.

PVS belongs to the genus *Carlavirus* within the family *Flexiviridae*. It has slightly flexuous filamentous particles 660x12 nm and single stranded RNA. It is normally latent and infected plants of most of the potato varieties look almost healthy but occasional/ transient leaf symptoms of faint rugosity, vein deepening and leaf bronzing can be seen (Figure 1d).

Potato virus M (PVM)

Potato virus M (PVM) occurs worldwide wherever potatoes are grown. It normally causes only minor reduction in yield but the losses may go up to 15-45 per cent as observed in eastern Europe and Russia where some cultivars may be 100 per cent infected (de Bokx and van der Want, 1987). It also has a very narrow host range infecting mainly potato and only a few species of *Solanaceae*. PVM is sap transmissible but it naturally spreads through infected tubers and aphids in a non persistent manner.

PVM belongs to the genus *Carlavirus* in the family *Flexviridae*. Serologically, it is related to PVS, slightly curved filaments of 650x12nm and single stranded RNA. Despite worldwide distribution, it has utmost significance in eastern European varieties but has also become important in India. PVM generally involves only transient mosaic, crinkling and waviness, rolling of margins of leaflets with leaves tending to roll and stunting of shoots (Figure 1e). Symptoms mainly occur in plants infected at very young stage. Severity is influenced by virus isolate and potato cultivar.

Potato virus A (PVA)

Potato virus A (PVA) also has a worldwide distribution. Yield reduction due to PVA may go up to 30-40 per cent but the losses are higher in combination with PVX and/or PVY. PVA is transmitted in non-persistent manner. As with PVY, the virus can be acquired rapidly from an infected plant (<1 minute) and transmitted equally rapidly by the aphids. PVA occurs less frequently than PVY and it causes severe disease in combination with the PVX & PVY.

PVA has flexuous 730x11 nm long filaments and it belongs to genus *Potyvirus*, family *Potyviridae* with single stranded positive sense RNA. It is serologically closer to PVY. It naturally infects only potato worldwide and even experimentally; it infects only a few Solanaceaeous species. PVA causes mild mosaic symptoms not dissimilar to those caused by PVX. Differences can be difficult to detect visually, but PVA mottles may appear on the veins, and infected leaves look shiny (Figure 1f). Infected plants may have a more open habit. Although visually similar to PVX, this virus is related to PVY.

Groundnut bud necrosis virus (GBNV)

Groundnut bud necrosis virus (GBNV) causes stem necrosis disease in potato (PSND). It has a wide natural host range including crops like tomato (*Lycopersicon esculentum*), tobacco (*Nicotiana* spp.), peanut (*Arachis hypogaea*), soyabean (*Glycine max*) and cotton (*Gossypium* spp.). Distribution of the GBNV is worldwide. It is known to naturally infect potatoes in Argentina, Australia, Brazil and central India (Khurana *et al.*, 1997). It is rather common in early crop of potatoes in Central/ Western plains and plateaux of India. It's incidence was recorded up to 90 per cent in some parts of Madhya Pradesh and Rajasthan and up to 50 per cent in Pantnagar (Pundhir *et al.*, 2012). PSND is important only in localized areas where both the vector (thrips) and virus sources occur. Several thrips species belonging to the genera *Thrips* and *Frankliniella* act as the virus vectors in a persistent manner but they can acquire the virus only as nymphs, with 4-9 days of latency and about 1hr for transmission. Viruliferous thrips move to early crop of potatoes from other preceding crops and cause upto 29 per cent losses. Late planting (after end of October) in the northern plans of India is helpful in reducing the disease incidence by avoiding crop exposure to the vector. It was also achieved through use of systemic insecticide either as tuber dressing or/and foliar sprays.

GBNV belongs to the genus *Tospovirus* in the family *Bunyaviridae*. It has large roughly spherical, enveloped particles, ranging from 70-110 nm in diameter and single stranded RNA. Infected plants show extensive necrosis with formation of concentric rings or spots on leaves and stems (Figure 1g). Shoots that are not killed have stunted/rosetted appearance and chlorotic necrotic ring spots on leaves. Tubers on such plants may be few, sometimes small and rarely deshaped yet without the virus.

Tomato leaf curl New Delhi virus-potato (ToLCNDV)

A geminivirus causing potato apical leaf curl has only recently been recorded in India. It is characterized initially by curling of apical leaves and later by yellow mosaic, leaf distortion and dwarfing (Figure 1h). Usharani *et al.* (2003) confirmed that this virus is a strain of *Tomato leaf curl New Delhi virus* (ToLCNDV) belonging to the genus *Begomovirus* within the family *Geminiviridae*. The virus has 93–95 per cent sequence identity with ToLCNDV isolates and <75 per cent sequence identity with other *Tomato leaf curl virus* isolates and *Potato yellow mosaic virus*. The incidence of this virus correlates positively with the whitefly population and the whitefly infestation period of potato crops. About 40-100 per cent of infections were recorded in Indo-Gangetic plains at Hisar with heavy yield losses in susceptible varieties. Significant decrease in size and number of tubers lead to marketable yield loss which may be as high as 50 per cent in early planted susceptible cultivars.

Potato spindle tuber viroid (PSTVd)

Potato spindle tuber viroid (PSTVd), the type species of *Pospiviroidae* consists of small, single-stranded circular RNAs. It is commonly 359 nucleotides in length. Under field conditions, mild strains with indistinct symptoms outnumber severe strains by a ratio of 10:1 and cause yield losses of 15–25 per cent, whereas severe strains with distinct symptoms causes 65 per cent yield loss (Figure 1i). PSTVd is probably limited to untested wild potato collections and potato breeding lines in different countries. The viroid is highly contagious and readily transmitted to plants by contaminated cultivating and seed-cutting tools. It is transmitted through pollen and true potato seed; therefore, breeding and release of new cultivars can be one of the sources of its introduction to fields. Aerial symptoms develop in warmer conditions but are masked in cooler ones. Primary symptoms are seldom evident on potato plants. Stem and blossom pedicels are slender, longer than normal, and remain erect. Leaflets are slightly reduced with fluted margins, tend to curve inward and overlap the terminal leaflet. Angles between stems and petioles are more acute than normal. Leaves near the ground are noticeably shorter and erect, contrasting with healthy leaves, which rest on the ground. Severe strains cause enhanced symptoms, twisting of leaflets, and rugosity of leaf surfaces.

Fig. 1: Virus symptomes on leaves/tuber a.PVY, b. PLRV, c. PVX, d. PVS, e. PVM, f. PVA, g. GBNV, h. ToLCNDV, i.PSTVd.

EPIDEMIOLOGY

The mode of transmission of most of the potato viruses is through the infected seed tubers. The intensity of the virus and viroid diseases depends on virus/viroid strain, their mode of transmission, host variety involved and environmental factors. The intensity of the disease varies from time to time and location to location during the crop season. The goal of the virus/viroid disease management is to maintain the intensity of the disease to a low level. This is most common in the case of contagious viruses (PVX and PVS) and viroid (PSTVd) which easily spread through contaminated hands, clothes, implements, farm machinery, seed trays, etc. In the case of severe mosaic (PVY) and leafroll (PLRV), the aphids, mainly *M. persicae* and *Aphis gossypii*, act as vectors and spread them both far and wide. In the case of *Potato aucuba mosaic virus* (PAMV), both mechanical/contact and aphid transmission are important yet the latter occurs only either if the virus source has combined infection of PVY, PVA and PAMV or if the aphids have had acquisition on plants infected with some potyvirus. Some strains of PVS may also spread non-persistently through aphids (*M. persicae*). PVS being contagious and latent, in almost all cultivars under diverse environments, it occurs as most frequent natural contaminant along with PVX, and/ or PVY and PVM. PSTVd is readily transmitted through pollen and true potato seed. There are essentially two approaches to manage potato viruses and viroid diseases, similar to other plant diseases. The first approach is to decrease the sources of infection (reservoirs) and secondly to minimize the rate of spread.

DIAGNOSTICS

Effective management of potato viral diseases depends essentially on their rapid detection, accurate identification and sieving them out through indexing. diagnostics can be employed for varying purpose which may include:

- To determine the presence and quantity of the virus in potato crop in order to take plant protection measures.

- To determine the extent of viral incidence and consequent yield loss.

- To certify seed potato planting materials for plant quarantine and certification programs.

- To assess the effectiveness of application of cultural, physical, chemical, or biological methods of containing the viruses.

- To assess viral infection in plant materials in breeding programs.

- To detect and identify new viruses rapidly to prevent further spread.

- To study taxonomic and evolutionary relationships of viral pathogens.

- To resolve the components of complex diseases incited by two or more viruses.

- To study pathogenesis and gene functions.

Diagnostic techniques for viruses fall into two broad categories: biological properties related to the interaction of the virus with its host and/or vector (e.g. symptomatology and transmission tests) and intrinsic properties of the virus itself (coat protein and nucleic acid). These diagnostic methods provides greater flexibility, increased sensitivity, and specificity for rapid diagnosis of virus diseases in disease surveys, epidemiological studies, plant quarantine, seed certification, and breeding programs.

Biological indexing, symptomatology and physical properties

Initially biological assays were developed which are still in use, since they are simple, require minimal knowledge of the pathogen and are polyvalent. Furthermore, biological indexing is still the only method available to detect uncharacterized but graft transmissible agents. Its sensitivity is considered to be very high due to the viral multiplication in the host plant used as indicator.

The symptoms on the potato plants commonly are used to characterize a disease having viral aetiology and for rouging of diseased plants in an attempt to control the disease. Visual inspection is relatively easy when symptoms are clearly characteristic of a specific disease. However, many factors such as virus strain, host plant cultivar/variety, time of infection, and the environment can influence the symptoms exhibited. Some viruses may induce no apparent symptoms or cause symptomless infection. In addition, different viruses can produce similar symptoms or different strains of a virus cause distinct symptoms in the same host. Usually, it is necessary that visual inspection for symptoms in the field is done in conjunction with other confirmatory tests to ensure accurate diagnosis of virus infection. Experimental host plants, under standardized conditions, will exhibit consistent and characteristic disease symptoms when infected with a particular virus. Several herbaceous plants are susceptible to a large number of viruses and can be used to diagnose potato virus infections. Physical properties of a virus (e.g., thermal inactivation point, dilution end point, and longevity *in vitro*), taken to be a measure of infectivity of the virus in sap extracts, were previously used to identify plant viruses. However, these properties are unreliable and are no longer recommended for virus diagnosis.

Electron microscopy

Electron microscopy (EM) provides very useful information on the morphology of the virus particles and is commonly used to examine viruses in crude extracts from infected plants. Examination of negatively stained virions by electron microscopy reveals flexuous rod-shaped particles with no obvious terminal structures in PVY and PVA. However, when examined using atomic force microscopy and immunogold labeling microscopy, these viruses were found to contain a protruding tip at one end of the virus particles, which is presumably associated with the 5'-end of viral RNA. Filamentous and rod-shaped viruses such as potyviruses, potexviruses, and tobamoviruses can more readily be differentiated in negatively stained leaf dip preparations than isometric viruses and other viruses. Though, EM is highly sensitive, however it is not suitable for routine testing of a large number of samples.

Unknown viruses are in many cases most effectively detected and identified by EM. Transmission electron microscopy is the most useful tool to diagnose virus through particles observation and its influence on potato cell organelles, which are closely related with the type of potato-host response to viral infection. The localisation of specific antigens with antisera and epitope analysis provides the opportunity to resolve the spatial details of infection with respect to the production of virus-specific "products" and the nature of the host response. A rapid hypersensitive response during which highly localised increased accumulation of electron-dense deposits of calcium pyroantimonate were detected in PVY^{NTN} and also PVY^{NWi} inoculated plants by electron microscopy.

Serology based techniques

Serology involves use of specific antibodies to detect their respective antigens in test specimen. Antibodies are composed of immunoglobulin (Ig) proteins produced in the body of an animal in response to the presence of antigens. Each antibody is specific to a particular antigen and will bind to it. Serological or immunological methods are generally specific, sensitive enough and have been used successfully for a number of years for the detection of potato viruses. However, there are some limitations to the use of antibodies in potato virus diagnosis. Firstly, the nature of the cross reactions between heterologous antibody-antigen complexes are not well understood. Secondly, diagnosis is based on only part of the organism's structure such as the coat protein of a virus which represents only a small proportion of the information about the virus. Thirdly, serology is only useful when the antiserum has been prepared or when an antigen is available for producing an antiserum. Although antibodies production may take several weeks, they are suitable for both laboratory and field conditions and can identify strains within species. However, different diagnostic protocols based on antibody-antigen are routinely being used for detection of major potato viruses and they have been briefly discussed below.

Immunosorbent electron microscopy

The Immuno sorbent electron microscopy (ISEM) for detection and identification of potato viruses by combining electron microscopy and serology are highly sensitive. PLRV is a small, phloem-restricted virus occurring in very low concentration and poses problem in detection with conventional electron microscopic detection. Optimum parameters were determined for the reliable and sensitive immune electron microscopic diagnosis of PLRV along with other important potato viruses. PLRV was best detected when the virus and its antibodies interacted in liquid phase followed by trapping on the grid coated with protein A and homologous antibodies.

Precipitation, gel diffusion and agglutination tests

Precipitin tests (either in liquid medium or in agar/agarose) rely on the formation of visible precipitate when adequate quantities of virus and specific antibodies are in contact with each other. These tests are routinely used by some investigators, but agglutination and double diffusion tests are more commonly used. In double diffusion tests, the antibodies and antigen diffuse through a gel matrix and a visible precipitin line is formed where the two diffusing reactants meet in the gel. The double diffusion method can be used to distinguish related, but distinct, strains of a virus or even different but serologically related viruses. In an agglutination test, the antibody is coated on the surface of an inert carrier particle (e.g., red blood cells), and a positive antigen–antibody reaction results in clumping/agglutination of the carrier particles which can be observed by the naked eye or under a microscope. Agglutination tests are more sensitive than other precipitin tests and can be carried out with lower concentrations of reactants than are necessary for precipitation tests.

Enzyme-linked Immunosorbent Assay (ELISA)

The ELISA has significantly increased the ability to detect and study potato viruses, and is currently the most widely used method for the detection of potato viruses due to its simplicity, adaptability, rapidity, sensitivity and accuracy. The double antibody sandwich (DAS-ELISA) test on a solid phase (usually plastic) is most commonly used. Virus is first selectively trapped by a specific antibody adsorbed on a solid surface, a specific enzyme-labeled antibody (conjugate) is added to the immobilized virus, and the reaction is measured visually or spectrophotometrically, after adding a suitable enzyme substrate.

Dot blot assay

In immune blots or dot-blots assays, antibodies or virus particles bound to nitrocellulose membrane filters are used. Dot blot ELISA tends to be rapid, easy to perform and conservative of reagents and often more sensitive than ELISA carried out in a microtiter plate. Immunoblot assays use the same reagents used in microtiter plate ELISAs, except that the substrate produces an insoluble product which precipitates onto the membrane. Positive reactions can be determined visually. Assays in which antibodies or antigens are bound to nitrocellulose or nylon membranes have been used to detect PVS, PYX , PVY and PLRV.

Tissue blotting, tissue squashes and southern blotting

In this technique blots are made by pressing the freshly cut tissue surface gently but firmly on a nitrocellulose membrane. Antigens in tissue blots are detected by enzyme labeled probes and this technique has been used to detect PVX and PVY from tubers in the field. Southern blotting combines transfer of electrophoretically separated DNA fragments to a filter membrane and subsequent fragment detection by probe hybridization. In this method total DNA is isolated from infected plant and transferred on membrane followed by its hybridization with radiolabeled probe which is generally made from partial or full length viral nucleic acid. Southern blotting method has been used for quantitative determination of many begomoviruses like ToLCNDV.

Lateral flow immune assay kits (LFIA)

It is based on the interaction between the target virus and immunoreagents (antibodies and their conjugates with colored colloidal particles) applied on the membrane carriers (test strips). When the strip is dipped into the sample being analysed, the sample liquid flows through the membrane and triggers immunochemical interactions resulting in visible coloration in test and reference lines. LFIA kits for the detection of five viruses viz., PVX, PVA, PVS, PVM and PVY either individually or in combination of two viruses have been developed.

Nucleic acid based techniques

Nucleic acid based techniques of potato virus diagnosis are well described and the sensitivity of the methods exceeds that of serology and have the advantage of targeting the genome of the virus. Some of the disadvantages of nucleic acid based techniques are the equipment, facilities, expertise and the time required to determine the optimum conditions for a particular application of the technique. Further, while several serological tests can be conducted in the field for rapid diagnoses, nucleic acid based methods are still carried out in laboratories. However, several nucleic acid based methods are very robust, accurate and higly sensitive in detection of potato viruses and have been discussed below in brief.

Conventional PCR or RT-PCR

RT-PCR protocols have been standardized for detection of major potato viruses that can detect very low level of virus inoculum. Several variations of RT-PCR like Immuno capture PCR (IC-PCR), Print-capture PCR (PC-PCR), nested PCR, multiplex RT-PCR and multiplex nested RT-PCR have been standardized for detection of potato viruses to differentiate strainal variation of a particular virus. Multiplex RT-PCR is a time- and reagent-saving amplification technique in which multiple primer sets are used to amplify multiple specific targets simultaneously in the same sample (Kumar et al., 2014). IC-RT-PCR, DB-RT-PCR and PC-RT-PCR assays were successfully used for the detection of PVY virions and PC-PCR for the detection of ToLCNDV-potato.

Real-time PCR

Real time RT-PCR protocols have been developed for potato viruses in several laboratories. TaqMan® duplex RT-PCR have been used for the detection of TRV, PMTV, PLRV and PVY. Agindotan et al. (2007) reported an assay where four common potato-infecting viruses, *Potato leafroll virus*, *Potato virus A*, *Potato virus X* and *Potato virus Y*, were detected simultaneously from total

RNA and saps of dormant potato tubers in a quadruplex real-time RT-PCR. Multiplex real-time RT-PCR has also been used to detect single and mixed infections of PLRV and PVY in seed potatoes using molecular beacons.

Rolling circle amplification (RCA)

RCA is an isothermal method for reliable diagnosis of geminiviruses and presumably all viruses with small single-stranded circular DNA genomes. This technique has been found useful for characterizing viral DNA components of several geminiviruses from experimental and natural host plant sources. RCA amplified viral DNA can be characterized by restriction fragment length polymorphism (RFLP) analysis and directly sequenced up to 900 bases in a single run, circumventing cloning and plasmid purification. RCA is better, easier and cheaper than polymerase chain reaction (PCR) or antibody-based detection of geminiviruses. RCA in combination with PCR assay increases the sensitivity and specificity of the assay.

Nucleic acid spot hybridization

Nucleic acid hybridization techniques have been standardized for detection of many potato viruses. Nucleic acid hybridization of DNA or RNA probes has the advantage of being able to detect the nucleic acid of the virus in both forms, single-stranded and double-stranded. cRNA probes can be labelled with isotopes or nonradioactive probes. cRNA probes are preferable to cDNA probes for detecting RNA viruses. An RNA extraction from infected tissue is blotted onto a membrane and the probe hybridized to it and detected. Both radioactive probes (^{32}P labelled) and non radioactive probes (fluorescent probes) were used for the detection of PSTVd.

Microarrays and macroarrays

Microarrays are high-density arrays with spot sizes smaller than 150 microns. A typical microarray slide can contain up to 30,000 spots. Macroarrays are generally membrane-based with spot sizes of greater than 300 microns. Arrays printed with probes corresponding to a large number of virus species can be utilized to simultaneously detect all those viruses within the tissue of an infected host. A convenient, cost-effective macroarray and microtube hybridization (MTH) system in which cDNA probes immobilized on nylon membrane, target viruses were amplified and labelled with biotin and then hybridization was carried in hybridization oven and colorimetrically detected using nitro blue tetrazolium (NBT)/ bromo-4-chloro-3-indolyl phosphate (BCIP) reagent. A microchip using short synthetic single-stranded oligomers (40 nt) instead of PCR products as capture probes for detection of PVA, PVS, PVM, PVX, PVY and PLRV, in both single and mixed infections.

Next Generation Sequencing (Pyrosequencing)

De novo sequencing of viruses using deep sequencing is a new technique that has successfully identified known and unknown viruses from long or short reads. Next Generation Sequencing (NGS) technology is able to sequence viruses from samples without the need for laborious and costly purification, cloning and screening techniques. NGS technologies can be used as a diagnostic tool to identify a potato virus in an unbiased fashion when no prior knowledge of the aetiology of the virus is available. This technique has been exploited for identification and determination of complete genome of a novel strain of PVS using GS FLX 454 Life Sciences (Roche). Small RNA deep sequencing technique has also been utilized to identify different potato viruses and found to have rich genetic diversity.

MANAGEMENT OF POTATO VIRUSES

Management of potato virus and viroid diseases is a matter of vital importance and concern to the potat growers and the scientists. Till date, there is no direct method available to control the virus diseases, consequently, current measures rely on indirect tactics to manage them. The possible strategies for potato virus disease management include (i) eradicating the source of infection to prevent the virus from reaching the potato crop, (ii) minimizing the spread of the virus by controlling its vector, (iii) utilizing virus-free potato seed material, and (iv) incorporating host-plant resistance to the virus. The choice of management strategies depends on the nature of the particular virus and viroid, but the decision to use the control measures depends on the assessment of the economic risks involved. The recommended control measure should be sufficiently effective to reduce the disease to an acceptable level and should be simple, environmentally safe, and inexpensive to apply. The entire range of methods aimed at the control of plant virus and viroid diseases like quarantine, eradication, sanitation, production of virus free seed potatoes, avoiding vectors, control of vectors, forecasting, breeding for resistance, transgenic approach and integrated management are briefly discussed here.

Exclusion of viruses through quarantine

Exclusion means the practice of keeping out any material or objects that are contaminated with pathogens and preventing them from entering the production system for which quarantine rules and regulations are to be implemented. Exclusion can take many forms, for example, prohibiting the movement of plants from an infested country into another which is free of a particular pest or disease. In every country the entry and further spread of virus and viroid diseases is restricted by the use of virus-free planting material, in a limited quantities received from the importing/ exporting countries. In India the legislation of P.Q. Order 2003 prohibited imports of 14 crops from various regions/countries. At quarantine stations, rapid, accurate and sensitive diagnostic tests have to be used for testing the exporting and importing seed or vegetative plant material for the presence of virus and virus-like diseases besides other pests and diseases. Large number of potato viruses e.g., APLV, APMV,TRV, PMTV, PVV, PVT, PYVV etc which do not occur in India need to be checked for their entry into the country.

Sanitation

PVX, PVS and PSTVd are readily transmitted through contaminated hands, clothes of workers, farm machineries/ tools. Use of 3 per cent trisodium phosphate or sodium hypochloride solutions (1:50 dilutions) can inactivate these viruses in hands, tools and other surfaces. In the field, general practices of sanitation like destroying cull piles to prevent sprouting, and roguing and removing infected plants from the field reduces PLRV and PVY inoculum.

Production of virus free plants through eradication

Meristem tip culture

In many instances, a new clone with good agronomical and high yielding characters can be rescued by the method of meristem-tip culture. Excising a small (0.2–0.5 mm) piece of tissue from the meristematic area and culturing on a nutrient medium can result in a pathogen-free plantlet for regeneration. Virus free plants, regenerated from meristem tips, are genetically stable and yield true-to-type plants. The number of virus particles was directly proportional to the size of the meristem. Combination of meristem tip culture along with other techniques like thermotheraphy, cryotheraphy increases the probability of virus elimination.

Thermotheraphy

Different viruses have a varying response in their sensitivity to heat, viz., PVY and PVA are easily eliminated at 36/39 °C, while PSTVd is eliminated after chilling at 5-6 °C for 8-12 weeks. The plant age, treatment duration and season(s) strongly influence the survival of infected plants upon heat exposure. Normally, well established plants should be treated. Heat treatment invokes temporary damage or abnormalities of colour and shape of foliage and plant growth. These changes may not be important since the plants start turn normal a few weeks after the heat treatment is over. Thermotherapy prior to meristem culture helps in elimination of the viruses otherwise difficult to eliminate. Temperatures fluctuating between 35 and 43°C are most favourable for plant survival for several months than 38°C constant. Preconditioning of plants at 27-35 °C for few days to one week, prior to treatment at 38°C is helpful. Treatment of plants for 16h during day at 36 °C and 8h during night at 29°C for 20-24 weeks is ideal for eradication of PVX and PVS without detrimental affect on the health of the plants. The temperature and length of treatment vary with the heat tolerance of the cultivar. Most potato cultivars can withstand 37°C up to a few weeks. The potato viruses in order of increasing difficulty of their eradication are PLRV, PVA, PVY, PAMV, PVX, PVM, PVS and PSTVd. In fact, PVS and PSTVd are the most difficult to eradicate whereas PVA and PVY are often eliminated by meristem culture alone without prior heat treatment. It has been observed that PVA and PVY were eliminated from 85-90 per cent of the meristem cultures while PVX and PVS, being stable, were eliminated upto 10 per cent. Heat treatment of sprouts on tubers before meristem culture gave rise to 16 out of 18 PVS free plantlets. Combination of meristem tip culture and thermotherapy was very effective in eradicating PVX. The virus elimination reported in potato through meristem tip culture followed by thermotherapy and obtained highest percentage (43.79) of virus free plants at 35±1°C with highest survival response (24.55) at 27±1°C. A severe strain of PSTVd from a potato clone by growing viroid infected plant at 5-8 °C successfully eradicated for six months followed by meristem tip culture. Prolonged cold treatment, however, severely damages a large number of meristem tips.

Chemotherapy

Chemotherapy is the application of antiviral agents either given to the infected plant before bud excision or incorporated into the culture medium. The requirements of a useful antiviral chemical include abilities to inhibit multiplication, spread, or symptom induction of the virus; be selective enough not to harm the host; have broad-spectrum activity against a number of virus diseases; move systemically in the host; and not have harmful effects on the environment. Some of the compounds were used in *in vitro* studies for virus elimination from meristems, for example, Ribavirin (virazole) has inhibited virus multiplication of CMV, PVY, PVS and PVM in potato. In tobacco, 2-thiouracil was proved to be effective against PVY. PVS is eliminated when the mother plants were first treated with anti-metabolites but incorporation of riboside in the medium greatly helps in eradication of PVX, PVY, PVM, PLRV. The synthetic riboside, Ribavirin is very effective when used *in vitro* as it resulted in elimination of one or more potato viruses. It is best to use the heat treatment of *in vitro* plants and/or supplement the medium with the antiviral agent(s). Simultaneous thermotherapy at approximately 37°C and chemotherapy had higher efficiency for the elimination of PVY (83.3 per cent) with Ribavirin (RBV) about 70 per cent with 5-Azacytidine (AZA), and about 50 per cent with 3-Deazauridine (DZD).

Chemotherapy for three weeks on MS medium containing 20mg/ml ribavirin followed by thermotherapy of the same plants on same medium at 37±1 for two weeks prior to mericloning eliminated PVX and PVS.

Electrotherapy

Electrotherapy is a simple method of virus eradication without the need to use any special or expensive equipment. In this technique an electric current is applied to plant tissues in order to disrupt or degrade viral nucleoprotein and thus eliminating its virulence. The use of electric current for the production of virus-free plants has been reported for the elimination of PVX from potato plants in which an electric current of 15 mA for 5 min led to 60–100 per cent elimination in various cultivars. Electrotherapy has also been used successfully for elimination of PVY, PVA, PVS and PLRV. The electric current of 15 mA for 10 minutes produced the highest degree of virus elimination for PLRV, PVY and PVS (26 to 100 per cent). The electric current of 35 mA for 20 minutes was the most effective electrotherapy treatment for eliminating PVY and PVA with regeneration of 54 to 70 per cent.

Cryotherapy

The technique of cryotherapy of shoot tips has also been used to eliminate virus and virus-like pathogens from the vegetatively propagated plants. In cryotherapy, infected shoot tips were exposed briefly to liquid nitrogen (-196 $^{\circ}$C). The cryotherapy results in a high frequency of virus-free regenerates. Thermotherapy followed by cryotherapy of shoot tips can be used to enhance virus eradication. Uneven distribution of viruses and obligate vasculature-limited microbes in shoot tips allows elimination of the infected cells by injuring them with the cryo-treatment and regeneration of healthy shoots from the surviving pathogen-free meristematic cells. Cryotherapy of shoot tips is easy to implement. It allows treatment of large numbers of samples and results in a high frequency of pathogen-free regenerants. Difficulties related to excision and regeneration of small meristems is largely circumvented. To date, several pathogens in potato like CMV, PLRV, PVY have been eradicated using cryotherapy. Using cryogenic protocols i.e. encapsulation-dehydration, encapsulation-vitrification and droplet and obtained PLRV and PVY free plants at 83-86 per cent and 91-95 per cent frequencies, respectively which were higher than those by meristem culture and thermotherapy.

Avoidance and control of vectors

Avoidance of vectors

Several aphid species are known to spread or transmit potato viruses like PVY, PLRV, PVA, and PVM from diseased to healthy potato tissue in non persistent-persistent manner. Thus, the seed crop in North Western Indian plains is raised only in aphid free periods or locations in the designated seed producing areas through the Seed Plot Technique. The seed crop is being grown during aphid free period (October-December/January) including use of healthy seed, application of systemic insecticides, field inspection for roguing all infected/ offtype plants and dehaulming the crop as soon as the aphids cross the critical limit of 20 aphids/100 compound leaves. Cucumber and Squash plants can be used as bait plant to manage TYLCV. However, Cucumber is a much better host for whiteflies and once they land on this host they do not leave it as long as the plants remain in a good condition for colonization.

Control of vectors by insecticides

Chemical control of aphids should begin just prior to the expected time of decline in bioefficacy of soil or seed treatment insecticides. Insecticides are often effective against the spread of persistently aphid-transmitted viruses like PLRV but not against the spread of non-persistently aphid-transmitted viruses. Sometimes, use of conventional insecticides even increases the incidence of the virus within the crop, presumably due to increased probing's by the agitated vector. Non-persistently aphid-transmitted viruses are transmitted quickly even by the short duration probings of less than a minute by the vector. Consequently, only the newer classes of insecticides, called synthetic pyrethroids, which can quickly knock down the vector, hold some promise. The pyrethroid application gave maximum control of the aphid vectors *Macrosiphum euphorbiae* and *M. persicae* and halved the incidence of PVY. The chloronicotinyl, imidacloprid, pyridine azomethine and pymetrozine were highly effective in reducing transmission of PLRV from infected to healthy potato plants by *M. persicae*. The synthetic pyrethroid, esfenvalerate, was effective in reducing inoculation of PLRV by virus-infected aphids into potatoes due to its repellent effect, but not virus acquisition by aphids from infected plants. Insecticides are less effective in controlling PVY infection, because the PVY is non-persistent and borne on the aphid's stylet, and may be transmitted before the aphid is killed.

Control of vectors by botanicals

Efforts have been made in different countries to test the efficacy of certain plant products (botanicals), to induce systemic resistance for managing the incidence of plant virus diseases. The botanical insecticides composed of essential oils may be an alternative to the more persistent synthetic pesticides for the management of vectors responsible for disease spread. A proprietary emusifiable concentrate containing 25 per cent essential oil extract of *Chenopodium ambrosoides* (EOCA) as the active ingredient at 0.5 per cent caused mortality (43.6 per cent) and insecticide soap (55.2 per cent) and were effective against *Myzus persicae*. The extract of *C. ambrosoides* at the concentration of 0.5 per cent gave excellent control of thrips, *Frankliniella schultezi* (95.7 per cent) than insecticidal soap (83.6 per cent), neem oil (17.7 per cent) and water (10.8 per cent). EOCA proved to be more effective than commercial products in controlling the major virus vectors such as *M. persicae* and *F. schultezi*, than neem extract, insecticidal soap endosulfan and abamectin.

Control of vectors by oils

Intensive researches of virus disease management have also led to the discovery of oils for the control of virus diseases. The finding that vectors carrying viruses seldom transmit after probing a leaf surface which is lightly coated with oil offers a novel way of minimizing the disease spread. The motivation for using these non-toxic materials at recommended concentrations are great because they are less likely to cause the environmental pollution than the chemical pesticides, have excellent spreading and sticking properties, are not subject to resistance development, and are economical to use. The transmission of PVY by *M. persicae* was impeded by coating either the source plants or the test plants with mineral oil (liquid paraffin). However, the use of mineral oils can be effective only if seed potato production is located under low infection pressure conditions.

Forecasting for viruses/ insect vectors

There are several factors viz., host, viruses, insect vectors and the environment which are usually involved in deciding the outcome of interaction. Infact, the environment viz. temperature, directly affects the vector behaviour/multiplication and its translocation within the host. However, great variations occur in virus transmission efficiency of different aphid species in nature. Forecasting for the aphid borne viral diseases has been tried in case of PVY and PLRV, based on the relationship between aphid migration and spread of PVY which has been studied by exposing bait plants to the aphid vectors in the fields. Number of alate aphids and their vector efficiency, the time of aphid migration in relation to plant growth, and also the availability of virus source have been used for developing forecasting methods. Simulation models have also been used to describe the epidemiology of non-persistently transmitted viruses taking care of aphid behaviour. Results indicated that the alates (and not apterae) were mainly responsible for spread of PVY. In another simulation model, relationship between important variables and parameters like plant disease-vector dynamics at the level of individual field was carried out. Some of the most important variables are alates as virus vectors, PVY infected plants as virus sources, the susceptibility of the crop based on the planting date, date of haulm killing and PVY infected daughter tubers. The model output successfully predicts the extent to which the proportion of progeny tubers infected with PVY will increase during late summer. The present model differs from the earlier one in adding/ changing the following parameters/ variables: Newly PVY infected plants, totally infected potato plants, spread of PVY, latent period, cultivar susceptibility, date of haulm killing, roguing of PVY infected plants and possible risk of virus spread from off the field virus sources. The model output and the real data showed high degree of agreement. Vector efficiency is determined by vector behaviour and its mobility hence models were tried to predict the relationship between vector activity and the incidence of PVY. REF values assigned to different aphid species vary and depend on vector efficiency, relative abundance, time of migration, age of potato crop, etc. Thus, *Brachycaudas helichrysi* had a higher REF than *M. persicae* at Harpenden during 1984 due to higher efficiency and larger number; *Rhopalosiphum padi* was more important in Sweden and the Netherlands due to early migration in larger numbers. Occurrence of young PVY infected source plants early in the season, when alates are moving about, lead to greater risk for PVY infection in the progeny tubers than if alates migrate late in the season because of mature plant resistance. Aphid migration can also be predicted by the use of suction traps at 12 m above ground level. A close correlation between suction trap counts and virus spread was observed in Sweden when the incidence of virus source plants, potato variety and crop age (plant resistance) were also considered. Attempts have also been made on similar lines to predict the incidence of PLRV. A link between the number of aphids trapped in suction trap, and the winter temperature, spring migration of aphids and spread of PLRV in the field was observed. Winter temperature during February and dates of first catches of an holocyclic specimen of *M. persicae* had a positive correlation, hence it was used to forecast the spring migration of *M. persicae* as well as to advise farmers whether or not to use the pesticide for aphid control while planting seed crop. Delayed planting (beyond 25th October) and application of systemic insecticide helped in preventing the crop from thrips-borne stem necrosis in Central/ Western India.

Use of resistant cultivars

Use of resistant cultivars is relatively inexpensive to deploy and has no adverse environmental consequences. The farmer will not have to incur additional production cost. The biggest obstacle in virus breeding programs is to identify and incorporate on a large scale, multiple resistance factors in elite genetic materials. The *Ryadg* gene/locus shows the great potential for use in breeding because it confers resistance to all currently known PVY strains viz., PVY^O ,PVY^C, PVY^N, PVY^Z, and PVY^E in addition to *Potato virus X*. Ry_{sto} from *S. stoloniferum* providing resistance to PVY^O have been shown to be effective against all strains of PVY and also to PVA and PVV. Resistance to PLRV controlled by a single gene from *S. etuberosum*, designated Rlr_{etb} has been crossed into a potato cultivar and resulted in decrease in virus levels. Besides *Solanum acaule* possesses a dominant gene (Rx) conferring extreme resistance, almost approaching immunity. Once the resistance conferred by Rx is elicited by coat protein of the infecting virus, it can suppress replication of a completely unrelated virus(es) hence may take care of important potato viruses. Resistance to aphid vectors should be useful as it could help to reduce virus spread. A type of resistance to aphids has been found in the wild Bolivian potato species *Solanum berthaultii*. In addition, resistance to aphids based on reduced feeding or colonization has been observed, but this may not be sufficient to prevent infection by potyviruses.

Transgenic approach

The transgenic approach would be more appropriate in situations where sufficient levels of resistance to the virus are not available in the related germplasm or the resistance is difficult to transfer by normal crossing techniques because of either reproductive isolation or linkage with other undesirable traits.

Pathogen derived resistance

In 1988, Hemenway transformed potato with the coat protein gene (cp) of the PVX, which was one of the first attempts to obtain pathogen-derived resistance to major potato virus. Transformation of plants with viral CP genes was soon found to work against many other viruses as well. However, in many cases the highest levels of resistance were associated with low levels or no detectable production of the CP and could even be attained using non-translatable or antisense CP gene constructs. The evidence for RNA-mediated resistance led to the discovery that transgenic resistance could result from 'homology-based gene silencing' an inducible cellular RNA surveillance mechanism targeted against viruses containing sequences homologous to the transgene.

Transgenes included the complete or partial sequences of CP, movement protein gene and nuclear inclusion protein gene (NIb). Moreover, multiple genes derived from different viruses have been successfully introduced into potato simultaneously for multi-virus resistance. Transgenic plants expressing dsRNA derived from the 3' terminal part of the coat protein gene of PVY which is highly conserved among different resistant strains of PVY viz., PVY^N, PVY^O

and PVY^{NTN}. A chimeric expression vector containing three partial gene sequences derived from the ORF2 gene of PVX used for PVY and Coat protein gene of PLRV to develop transgenic potato plants with broad spectrum resistance against viruses and found that the transgenic plants are immune against all three viruses.

Non-pathogen derived resistance

Many genes other than potato have been used to induce restance against potato viruses. Transgenic potato clones expressing a gene encoding for ribosome inactivating protein (RIP) have high degree of virus resistance isolated from pokeweed. Similarly, a mammalian oligonucleotide synthetase gene into potato has been reported to confer extreme resistance to PVX in field grown plants. Mutated *S. tuberosum* eIF4E gene was expressed in transgenic potato plants to confer resistance to PVY. Ribozymes could be used effectively to control replication of viroid RNA in the nucleus. They have also showed that transgenic potato plants expressing a ribozyme against PSTVd minus-strand RNA were resistant to the viroid.

Artificial miRNAs

MicroRNAs (miRNAs) processed from nuclear-encoded transcripts control expression of target transcripts by directing cleavage or translational inhibition. Artificial miRNAs (amiRNAs) are designed to mimick the intact secondary structure of endogenous miRNA precursors and processed *in vivo* to target genes of interest. The strategy of expressing amiRNAs was first adopted to knock out/down endogenous genes for functional analysis. The technology is widely used in engineering antiviral plants and animals. Compared to conventional RNAi strategies, amiRNAs have many advantages: (1) No off-targets and increased biosafety of transgenic crops due to the short sequence of amiRNAs when compared to a long viral cDNA fragment (2) use of tissue- or cell-specific promoters helps for tissue- or cell-specific knock out/downs of genes of interest (3) amiRNAs are especially useful in targeting a class of conserved genes with high sequence similarities, like tandem arrayed genes, because a short conserved sequence is more easily found in these genes. The amiRNA targeting sequences that encode the silencing suppressor HC-Pro of PVY and the TGBp1/p25 (p25) of PVX were designed and used to transform tobacco. The amiRNAs efficiently inhibited HC-Pro and p25 gene expression and conferred highly specific resistance against PVY or PVX infection in transgenic *Nicotiana tabacum*; this resistance was also maintained under conditions of increased viral pressure.

Resistance by means of plantibodies

Another approach is the expression of antibodies, commonly used in animals to recognize pathogens. Although the immune system linked to these proteins in animals is not present in plants, affinities of selected antibodies can be high enough to disrupt essential functions of a viral protein in plants. Though technically complex, the generation of single-chain variable fragment (scFv) antibodies, the development of the phage display approach and the generation of synthetic scFv libraries have greatly improved the applicability of this strategy. Expression of a synthetic gene encoding a single chain Fv fragment of an antibody directed against the nuclear inclusion a (NIa) protein of potato virus Y (PVY) in transgenic plants conferred resistance to PVY[o] by inhibiting a crucial step in the virus multiplication, such as polyprotein cleavage. Transgenic potato (*Solanum tuberosum*) plants expressing scFvP1-1 (raised against the C terminus of P1 protein) showed high levels of resistance following PLRV inoculation by viruliferous aphids.

INTEGRATED MANAGEMENT OF POTATO VIRUSES

No single approach as outlined above will yield desirable result. A combination of them or most of them will be the only lasting solotion. Schedule of integrated control of potato viruses include: inspection of seed production areas and rejection of fields with mosaic incidence higher than the prescribed level, killing of vines of seed crop at the recommended date or earlier and not to allow re-growth of the vines, destroy volunteer potato plants and weeds in and around the seed crop, monitor the population of vectors and application of insecticides to keep the aphid vectors below the critical level, use of properly disinfected tools, maintaining proper isolation of the seed crop from virus sources, use of the best quality certified seed tubers for planting, avoiding use of cut tubers as seed for seed crop, planting seed crop at a specified period to avoid exposure of the crop to the vectors, minimising chances of virus spread through farm machinery and stop irrigation 10-15 days before harvest to allow skin curing.

REFERENCES

Agindotan BO, Shiel PJ and Berger PH (2007) Simultaneous detection of potato viruses, PLRV, PVA, PVX and PVY from dormant potato tubers by taqman real time RT-PCR. *Journal of Virological Methods* **142**: 1-9

De Bokx JA and Van Der Want JPH (1987) Viruses of Potatoes and Seed-potato Production. PUDOC, Wageningen, Netherlands: 259p

Khurana SMP (2004) Potato Viruses and Their Management. In, Diseases of Fruits and Vegetables: Diagnosis and Management. SAMH Naqvi (ed), Kluwer Academic, Dordrecht, Boston and London: 389-440

Khurana SMP, Pandey SK, Singh RB and Bhale U (1997) Spread and control of the potato stem necrosis. *Indian Journal of Virology* **13**:23-28

Kumar R, Jeevalatha A, Raigond B, Kumar R, Sharma S and Singh BP (2014) A multiplex reverse transcription PCR protocol for simultaneous detection of four potato viruses in potato plants and dormant tubers. *International Journal of Innovative Horticulture* **3** (1): 22-29

Pundhir VS, Akram M, Ansar M and Rajshekhara H (2012) Occurrence of stem necrosis disease in potato caused by *Groundnut bud necrosis virus* in Uttarakhand. *Potato Journal* **39**(1): 81-83

Singh RP (1988) Control of viroid and contact transmitted virus diseases. In, Potato Pest Management in Canada. Boiteau G, Singh RP and Parry RH (eds), Canada Agriculture, NB: 309-25

Singh S, Kumar S and Khurana SMP (1994) Incidence and relative concentration of common potato viruses in five cultivars. *Indian Journal of Virology* **10**: 44-50

Usharani KS, Surendranath B, Khurana SMP, Garg ID and Malathi VG (2003) Potato leaf curl – a new disease of potato in northern India caused by a strain of Tomato leaf curl New Delhi virus. *New Disease Reports* **8**:2

Vreugdenhil D, Bradshaw J, Gebhardt C, Govers F, Mackerron DKL, Taylor MA and Ross H A (2007) Potato biology and biotechnology: Advances and Perspectives. Elsevier, Oxford, Amsterdam: 856p, ISBN-13: 978-0-444-51018-1

17

Important Potato Pests and Their Management

Mohd Abas Shah[1], Aarti Bairwa[2], Kailash C. Naga[2], Subhash, S.[2], Raghavendra K.V.[3], Priyank H. Mhatre[4], Anuj Bhatnagar[3], Kamlesh Malik[3] Venkatasalam, E. P.[4] and Sanjeev Sharma[2]

[1]ICAR-Central Potato Research Station, Jalandhar-144003, Punjab, India
[2]ICAR-Central Potato Research Institute, Shimla-171001
Himachal Pradesh India
[3]ICAR-Central Potato Research Institute Campus, Meerut-250110
Uttar Pradesh, India
[4]ICAR-Central Potato Research Station, Muthorai- 643 004, Tamil Nadu, India

INTRODUCTION

Insect pests cause variable and complex problems for potato farmers. Insect pests account for 16% of the crop losses of potato worldwide, and reductions in tuber yield and quality can be between 30% and 70% for various insect pests. India has a great diversity of insect pests that attack potato. These pests can damage potato plants by feeding on leaves, reducing the photosynthetic area and efficiency by attacking stems, weakening plants and inhibiting nutrient transport, and by attacking the potato tubers destined for consumption or for use as seed. In India, approximately 60 billion rupees (US$1.2 billion) worth of potato tubers are lost annually due to pest damage, which accounts for 10–20% of total production (Chandel *et al.*, 2013). The potato pests are grouped into soil pests, foliage feeders, sap feeders, and storage pests. In seed production, aphids and whiteflies are of greatest concern. In ware production, the key pests may be insects which attack tubers, such as tuber moth, whitegrubs, and cutworms. In some situations, foliage feeders such as noctuid moths and coccinellid beetles are also important.The Potato cyst nematode (PCN) (*Globodera* spp.) and root knot nematodes (RKN) (*Meloidogyne* spp.) are amongst the most economically important nematode pests of potato. Cyst nematodes have amazing adaptation for long term survival in the soil, even in the absence of a suitable host which makes them challenging to the farmers, scientists and policy makers. They are subjected to stringent quarantine and/or regulatory procedures, wherever they occur and presents a serious threat to domestic and international commerce in potatoes. RKNs are prevalent all over the world and can cause significant crop loss in both warm and cool climatic conditions, depending on their species. In this chapter, the current information on biology, ecology and management of the major potato pests is discussed in the Indian context.

SUCKING PESTS

Aphids (Aphididae: Hemiptera)

Aphids are the most important pests of potato worldwide. Aphids are sap feeding insects but the major damage inflicted by aphids in potato crops is by transmission of numerous potato viruses limiting disease free seed production with a progressive decline in yield. The losses in yield by the aphid transmitted viruses range from 40 to 85%. Common potato viruses transmitted by aphids include *Potato virus Y* (PVY), *Potato leaf roll virus* (PLRV), *Potato virus A* (PVA), *Potato virus M* (PVM) and *Potato virus S* (PVS). More than 22 species of aphids are recorded worldwide that colonise potato plants. Earlier, five major species infesting potato under Indian conditions were known viz., *Myzus persicae* (peach-potatoaphid or green peach aphid), *Aphis gossypii* (melon aphid or cotton aphid), *A. fabae* (black bean aphid), *Rhopalosiphoninus latysiphon* (bulb and potato Aphid) and *Rhopalosiphum rufiabdominalis* (rice root aphid) with notes on biology, life cycle, migration and management, in addition to two minor species *Rhopalosiphum nymphaeae* (water lily aphid) and *Tetraneura nigriabdominalis* (rice root aphid). Later, Bhatnagar *et al.* (2017) compiled information on 13 species of aphids recorded on potato crops in India viz., *M. persicae* (peach-potato aphid), *A. gossypii* (cotton aphid), *A. fabae*, *A. spiraecola* (spiraea aphid; green citrus aphid), *A. nerii* (oleander aphid), *A. craciivora* (cowpea aphid or groundnut aphid or black legume aphid), *Macrosiphum euphorbiae* (potato aphid), *Brevicoryne brassicae* (cabbage aphid or mealy cabbage aphid), *Aulacorthum solani* (glasshouse-potato aphid), *Lipaphis erysimi* (mustard aphid or turnip aphid), *Hydaphis coriandari* (coriander aphid), *Rhopalosiphum rufiabdominalis* and *Rhopalosiphum maidis* (corn leaf aphid).

Apart from the colonising aphid species, hundreds of non-colonizing species visit potato crops briefly and contribute to virus prevalence. Non-colonizing species do not reproduce on potato, but may transiently visit potato plants. Such species can occur in very large numbers, making their effect on virus spread large despite their lower virus transmission efficiency. Hundreds of non-colonising aphid species have been reported from potato fields and tested for virus transmission ability. Around 65 species of aphids are known to transmit PVY and 13 species are on record that transmit PLRV.

Biology and life cycle of major aphid species

Aphids reproduce asexually on potato crops. The nymphs undergo four moults to become adults. Both winged and wingless adults are produced at different points of time or together with variable proportions. The winged aphids are mainly responsible for the spread of the viruses. Individual species show a lot of variation with respect to life cycle patterns. The life cycles of most important aphids infesting potato crops are described as follows.

Myzus persicae (Sulzer)

The peach potato or green peach aphid, *Myzus persicae* is the most economically important aphid crop pest worldwide (Figure 1). It is a notorious polyphagous pest infesting nearly 250 plant species belonging to 77 genera in India, inflicting heavy losses to a variety of crops and is also an important vector of many plant virus diseases. It is universally distributed present in all ecological conditions prevailing in the country.

It is a notable example of a heteroecious (host-alternating) aphid species. As the day length drops below a critical level in the autumn, apterous holocyclic viviparae produce gynoparae and males on secondary (herbaceous) hosts which migrate to the primary host, peach, *Prunus persica* L. (Rosaceae). The gynoparae then give birth to oviparae that lay the overwintering eggs after mating with males. This is the typical life cycle in temperate regions however, other reproductive strategies also exist. Anholocyclic clones are unable to produce any sexual morph and overwinter as parthenogenetic females on weeds or winter crops. Other genotypes are able to invest in both reproductive and overwintering strategies.

In tropics where some varieties of peach are grown and the mean monthly temperature never falls below 20°C except at high altitudes, the production of sexual morph production may be directly inhibited by high temperature all the year round whereas parthenogenesis may continue to be uninterrupted. Apparently distinct anholocyclic biotypes of *M. persicae* such as that on tobacco may have originated in the tropics in this way. At the same time, long-range displacements into and out of the tropics of *M. persicae* from other latitudes is possible although not properly known. In subtropical zone temperature is not low enough to permit the production of sexual morphs by October north of the equator, and by April south of the equator, then it will be too late for migration to *Prunus* and maturation of the oviparae before leaf fall. Therefore an induced holocycle in these regions is likely to be abortive. During the winter months temperature ceases to be inhibitory to sexual morph determination, and in winter or spring gynoparae and males may migrate to primary hosts. Mating and oviposition on peach trees in February and March have been recorded at localities in Egypt, Pakistan and India. As far as is known, any eggs laid in spring do not hatch. Where an abortive holocycle persists in climatic conditions which strongly favour anholocycly, this implies immigration of holocyclic genotypes from other regions. This situation warrants further investigation. Winters are so mild in this subtropical zone that they present no obstacle to continuous parthenogenesis, and anholocycly predominates.

The winged morphs of *M. persicae* usually appear on potato crop in the field from November onwards in most parts of Indo-Gangetic plains and does not migrate to the primary host plant for egg laying. Its population increases up to the end of February and early March. However, by the end of March many alatae are formed and migrate to mid and then to higher hill where the climate is mild and suitable and a number of secondary host plants are available. The aphids keep on multiplying till November on high hills and thereafter, its return migration starts from hills to plains and *vice versa*. It is, thus, clear that *M. persicae* is present on the secondary host plants throughout the year either in the plains or hills. The life cycle continues through migration of alate virginoparae out of the subtropics to mild climate areas and then back as the temperature becomes suitable.

Aphis spp.

Among the common species are *A. gossypii, A. nasturtii, A. fabae, A. spiraecola, A. nerii* and *A. craccivora* (Figure 2). Most of these are highly polyphagous and exhibit wide variation in life cycles. Adults of *A. gossypii, A. fabae* and *A. craccivora*, both apterae and alatae reproduce parthenogenetically throughout the year in India. However, sexual morphs have also been reported. Similarly, *A. spiraecola* usually reproduces anholocyclically but is holocyclic with *Spiraea* as primary host in the temperate areas. *A. nasturtii* life cycle is heteroecious holocyclic worldwide with asexual phase on *Rhamnus cathartica*. Apparently, this species also continues to reproduce asexually throughout the year in the Indo-Gangetic plains. The immediate implication of such life cycles is that the aphids are always active in the plains limited only by the prevailing temperature. Due to their broad host ranges, the availability of food is apparently not a limiting factor.

The other groups of aphids infesting potato include the grain aphids which migrate from wheat crops or other grasses growing along with potato crops. These have been inflicted with virus transmission in various studies. The remaining species have been recorded in small numbers but can be of major importance in particular years or at particular locations. Detailed treatments can be found in Bhatnagar *et al.* (2017) and Shah *et al.* (2018).

Nature and symptoms of damage

Both nymphs and adults suck the sap from phloem of potato stems, leaves and roots using their stylets during which virus acquisition and inoculation occur. The non-persistant viruses are transmitted during the brief stylet probes while as the persistent viruses require longer acquisition period ca. 20 minutes. Therefore, the latter are generally transmitted by colonising aphids only. Direct damage due to feeding injury of aphids is almost non-significant in potato while as the transmitted viruses lead to a wide variety of symptoms such as mosaics, yellows, mottle, roll, crinkle, rugose and rosette of potato leaves.

Management

1. For effective management of aphids in potato crops, continuous monitoring of the aphid populations is necessary. Aphid populations can be monitored using yellow sticky traps, yellow water pan traps or manual scouting by leaf count method. The first two methods help to monitor the flight activity of winged aphids while as the last method provides information on both winged and wingless aphids as well as nymphs. The leaf count method consists of counts from 100 compound leaves from upper, middle and lower levels, i.e. taken from 33 plants in a zig-zag manner at random and one leaf from 34th plant.

2. Infestation of the predominant aphid vector, peach-potato aphids can be avoided/kept low by adjusting the planting dates in the Indo-Gangetic plains, where about 90 per cent of seed crop is grown, i.e. upto 15th October in North-Western plains, upto 25th October in central plains and upto 5[th]November in the North-eastern plains. The haulm cutting of seed crop should be done as soon as the aphid number crosses the critical level i.e. 20 aphids/100 leaves. The manipulation of planting dates and haulm cutting in response to increasing aphid numbers to avoid peach-potato aphid period is referred to as the Seed plot technique.

3. Spray of Imidacloprid 17.8% S.L @ 0.03% at 75% crop emergence is recommended and same may be repeated at an interval of 12-15 days depending on the stage of crop and level of infestation. The following sprays should preferably be alternated with Dimethoate 30 EC @ 0.03 per cent, Flonicamid 50 WG @ 0.03% and mineral oil @1-3%.

4. Application of mineral oils @1-3% successfully protects potato plants from aphids and aphid transmitted potato viruses, in particular the non-persistent viruses.

5. Application of Phorate 10G @ 15kg/ha in furrows at planting or earthing up keeps the aphid vectors under check up to 45-60 days provided there is enough soil moisture.

Whiteflies (Aleurodidae: Hemiptera)

The cotton whitefly, *Bemisia tabaci* (Gennadius), being a pest of tropical and subtropical areas is a major component in the potato seed production complex in India (Chandel *et al.*, 2010) (Figure 3). Although it does not inflict any discernible direct damage to potato, it transmits the *Tomato leaf curl New Delhi virus* (ToLCNDV) in potato leading to Potato Apical Leaf Curl Virus (PALCV) Disease. In potato, a variant of ToLCNDV named ToLCNDV [potato], is now one of the most important diseases of potato in India. The disease incidence is higher particularly in the Indo-Gangetic plains (40–100% infection) which leads to heavy yield losses in susceptible varieties. Besides *B. tabaci*, the greenhouse whitefly, *Trialeurodes vaporariorum* is also known to infest potato under glass house condition and in temperate areas. The greenhouse whitefly is not of much concern and is manageable through general IPM practices in potato crops.

Seasonal occurrence and distribution

The whitefly, *B. tabaci* exhibits different patterns of population dynamics across the Indo-Gangetic plains on potato crops. The whiteflies numbers are generally highest in the beginning of the crop season (October and November). The infestation pattern is determined by the prevailing temperature at the location and the cropping sequence adopted by farmers. Locations where the daily minimum temperature falls below 10-12°C are characterised by very low or undetectable population of whitefly adults for 4-6 weeks on potato crops during core winter months (December and January) e.g. Hisar, Modipuram, Jalandhar and Gwalior. Yellow sticky trap catch data indicate very high flight activity in the early part of the growth period, probably due to high rate of immigration from preceding crops and weeds, followed by a sharp decline as temperature falls and no or minimal activity thereafter. The initial incidence of whiteflies is higher at locations where preceding crops sustain high whitefly populations e.g. cotton and solanaceous vegetables. The data on the number of eggs laid and nymphs per plant show a similar trend and very few or none of the immatures are present on the potato plants roughly after the midseason. In contrast, locations where the daily minimum temperature remains well above the critical temperature, crops sustain whitefly population throughout the growing season. At locations where the adjoining crops support high growth rate of whiteflies give rise to a linear increase in whitefly population on landscape level which is reflected on all such crops including potato e.g. Kalyani and Deesa. In contrast, whitefly population doesn't show sustained growth at locations where potato is not surrounded by crops that support high whitefly population growth rate for whiteflies. At such locations e.g. Chhindwara, Raipur and Bhubaneswar, the whitefly population increases briefly but tends to fall back soon. The degree of increase and the subsequent fall is determined by the relative acreage under whitefly supporting crops such as tomato, brinjal, cucurbits and that under whitefly non-supporting crops e.g. onion and garlic. Overall, potato crop does not appear to sustain a high whitefly population on its own and thus potato crop is a poor bridge for whitefly to sustain the cold winters. The extent of immigration from adjoining crops will depend on their relative acreage, among other factors (Shah *et al.*, 2019).

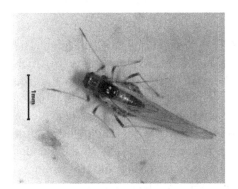

Fig. 1: Alate (left) and apterous (right) viviparous females of *Myzus persicae*

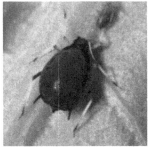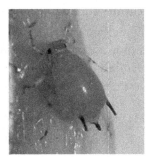

Fig. 2: Apterous viviparous females of *Aphis gossypii* (left), *A. fabae* group (mid.), and *A. spiraecola* (right)

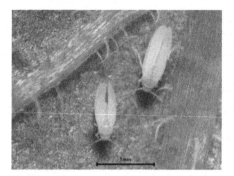 

Fig. 3: Adults of cotton whitefly, *Bemisia tabaci* on potato leaves

Being a vector of PALCV, tolerance limit for whitefly on potato crops is extremely low which is often difficult to maintain. The whiteflies have a broad host range with more than 600 host plants including many crops, vegetables, fibre crops, spices, ornamentals plants, and many weed plant species.

Biology and life cycle

Whiteflies have six life stages *viz.*, egg, four nymphal instars, and the adult. The female lays around 150-300 eggs singly on the underside of the leaf which hatch in 4-7 days on common hosts of the pest. The first instar known as the crawler, has legs and is the only mobile nymphal stage that moves to look for feeding sites. The other instars are sessile and they complete their life cycle on the same leaf. The nymphs become adults in 10-14 days. The healthy adult lives for 10-20 days. The total life cycle is completed in 20-30 days under favourable weather conditions. As potato is cultivated during cold temperatures, the duration of immature growth is considerably extended.

Nature and symptoms of damage

Whitefly causes severe damage to potato plants indirectly by transmitting the important viruses in the genus *Begomovirus* belonging to the family Geminiviridae in subtropics. Both nymphs and adults suck plant sap from the underside of the leaf during which virus acquisition and transmission takes place.

Management

1. Potato cv. Kufri Bahar is tolerant to whitefly and can be planted in whitefly prone areas.

2. Maintain field sanitation by removing and destroying the weeds, alternate hosts and crop residues of vectors and viruses.

3. Place yellow sticky traps (15 x 30 cm²) just above the canopy height @ 60 traps per hectare at equidistance from each other for mass trapping.

4. Seed treatment with Imidacloprid 17.8 SL at 0.04% for 10 minutes and its foliar application at 0.03% at 75% crop emergence followed by Thiamethoxam 25WG @ 0.05% after 15 days is recommended. The sprays can be repeated as per requirement. Various new chemistry molecules like Spiromesifen, IGRs, and knockdown insecticides are also being used across the locations.

Leafhoppers (Hemiptera: Cicadellidae)

Leafhoppers are polyphagous pests with worldwide distribution causing significant economic losses to many crops including potato. In India, the leaf hoppers are distributed in all potato growing regions. The potato leaf hopper (*Empoasca devastans* Distant) is known to be most important species, and has long been recognized as a major pest of potato. Potato also witnesses other leafhopper species in various potato growing regions which include *Amrasca biguttula biguttula* (Ishida), *E. solanifolia* Pruthi, *E. fabae* Harris, *E. kerri motti* Pruthi, *Alebroides nigroscutulatus* Distant, *Seriana equata* Singh and *E. punjabensis* Pruthi. The leaf hopper (*E. devastans* Distant) is a polyphagous pest feeding on various plant species like okra, groundnut, jute, soybean, cotton, niger, sunflower, eggplant, including potato. The pest is reported in Karnataka, Gujarat, Maharashtra, Punjab, Haryana and Utter Pradesh where it causes serious menace to potato production.

Biology and life cycle

Adults *Empoasca devastans* are pale green marked with a row of white spots on the anterior margins of the pronotum (Figure 4). The female lays on an average of 200-300 eggs into veins and petiole of leaves. It takes about ten days to hatch which give rise to light green, translucent, wingless nymphs which feed by remaining under surface of leaves. It takes about 15 days to develop adults into adult. The adults survive for 30-60 days depending on the weather conditions. The leafhopper can complete two generations on potato in a year.

Nature and symptoms of damage

Both nymph and adult are responsible for enormous direct losses by sucking cell sap from undersides of leaves. During feeding, leafhoppers inject watery saliva containing enzymes that reduce plant photosynthate movement leading to yellowing, browning, cupping and curling of leaves. The hopper damage can be easily identified by v-shaped burn on leaf tips, starting from the tip of leaves. The margin of the leaves get broken and crumble into pieces when crushed. The severely infested field gives brunt appearance in the patches commonly referred as "hopper burn". The leaf hoppers cause indirect losses by spreading potato viruses like potato yellow dwarf virus and beet curly top virus. It is also act as a vector of purple top of potato, which is caused by aster yellows mycoplasma- like organisms. The hopper can breed throughout the year but are most active during October to March, coinciding with potato growing season but more sever during early growth phase. The hopper build up are adversely affected by heavy monsoon rains at hilly regions.

Management

1. Cultural practices like regular inspection, removal of weeds near field vicinity which act as a source of re-infestation, following proper plant spacing, judicious use of nitrogenous fertilizers and balance use of plant nutrients and delay planting reduce the hopper incidence.

2. In early planted potato crops, leaf hopper incidence can be minimized by spraying insecticides like Dimethoate 30EC @ 2 ml/ litre. Soil incorporation of Phorate 10G @ 10-15 kg/ ha at the time of field preparation gives promising results against leaf hoppers. The spray can be repeated with foliar application of Imidacloprid 17.8 SL @ 0.03%.

Mites (Acarina: Tarsonemida, Tetranychidae)

Two mite species are known to infest potato crop, the broad mite, *Polyphagotarsonemus* latus Banks and red spider mite, *Tetranychus neocaledonicus* André. Among these the yellow mite or broad mite is a major pest of potato which has worldwide distribution and is a serious threat to potato production during *Kharif* season in Maharashtra and Karnataka in India. It also affects the potato production in Punjab and western Utter Pradesh when planted early during *Rabi* season. The mite incidence has also been recorded around Gwalior (MP) and Kangra valley (HP). Apart from potato the mite in known to feed on various agriculture and horticulture crops like tomato brinjal, tea, chilli, cotton, beans, mango, apple, citrus, coffee, grapes, guava, papaya and pear etc.

Biology and life cycle

The mites are tiny in size; the female measures about 0.2 mm with swollen profile. It has light yellow to amber body colour with a light and indistinct median stripe that fork near posterior end of the body. Males look similar to female except size and lake median strip. The *P. latus* have four life stages egg, larvae, nymph and adult. The adult female lays about 30 to 76 eggs on underside of leaves which hatch in two to three days. The newly emerged larvae are slow moving and do not disperse much after two to three days the larvae turn to quiescent larvae (nymph) stage. The quiescent male pick quiescent female up and carry them to the new foliage. When females emerge from the quiescent stage is generally known as yellow mite.

Nature and symptoms of damage

Mites are usually found on the upper part of the plant. They feed on apical shoots and on lower surface of young leaves by sucking the cell sap. During feeding mite inject toxic saliva that produces typical symptoms like bronzing, curling and discoloration of leaves (Figure 4).

Management

1. Early planted potato is more prone to mite infestation; delay in planting date results in lower mite damage.

2. Dry weather conditions also favour the mite build up; therefore maintaining sufficient moisture through irrigation reduces the mite attack.

3. The mite damage can be avoided by application of wettable sulfur @ 3 gms/ liter. Application of Spiromesifen 22.9 SC (40–75 ml/100 L) gives effective control of the mite infestation; repeat the spray after seven days if necessary.

1. Predatory mite, *Amblyseius barkeri* Hughes (Acarina: Phytoseiidae) is known to keep the population of broad mite under check, hence emphasis may be given towards conservation of the predatory mite.

Thrips (Thysanoptera: Thripdae)

Thrips are one of the important pests of potato responsible for direct as well as indirect losses by transmitting tospo viruses. *Thrips palmi* Karny, *Scirtothrips dorsalis* Hood, *Caliothrips collaris* (Bagnall) and *Haplothrips sp.* (Thysanoptera: Phlaeothripidae) are the thrips species associated with potato. Among these,*T. palmi* is one of the predominant species distributed across the potato growing states of India, viz., Chhattisgarh, Himachal Pradesh, Delhi, Haryana, Punjab, Jammu and Kashmir, Karnataka, Madhya Pradesh, Maharastra, Rajasthan, Orisha, Tamil Nadu, Utter Pradesh and West Bengal.

Biology and life cycle

Adults of *Thrips palmi* are pale yellowish to whitish in colour with numerous dark setae on the body and have pale slender fringed wings. Adult measures about 0.8 to 1.0 mm in length, with females slightly larger than males. Development duration varies from 12 to 20 days depends on weather conditions. Females may lay about 200 eggs in leaf tissues, in a slit cut made by female. The eggs take on an average 4.3 to 16 days to hatch. Larvae resemble adults except for the wing pads and smaller size. The larvae take 5-15 days to develop to inactive and non-feeding prepupal stages. The pre-pupae and pupae take about 3-12 days to convert into fully grown adult depending on prevailing temperature.The thrips build up is favoured by high temperatures (30–35°C) and low humidity during September–October and these conditions coincide with the early growth phase of potato production in plains.

Nature and symptoms of damage

Both adult and nymph feed gregariously on leaves, firstly along the midribs and veins and also attack stems particularly at or near the growing tip. They are found in large numbers amongst the flower petals and developing ovaries and the feeding leaves many scars and deformities. Both adults and nymphs are responsible of crop damage by scraping epidermal tissues near the leaves tips and suck the oozing sap resulting in silvering of leaves.With continuous feeding the tips of leaves wither, curl up and die. Thrips also act as a vector of potato stem necrosis disease which brings about 15-30% yield loss in potato in northern Gujarat, parts of Madhya Pradesh and Rajasthan.

Management

1. Late planting of potato witnesses lower thrips incidence, hence early planting should be avoided.

2. Kufri Sutlej, Kufri Badshah, and Kufri Jawahar varieties have been found promising against thrips incidence.

3. The thrips population must be monitored regularly as soon as thrips or its damage symptoms appears, application of Imidacloprid 17.8 SL (0.05%) and Thiamethoxam25WG (0.025%) provides good control. If the thrips population still persist, then the spray can be alternated with Dimethoate 30EC @ 0.03 %.

4. Conservation of Anthocorid bugs, *Orius maxidentex* and *Carayonocoris indicus* may play a significant role in controlling the thrips population naturally.

SOIL AND TUBER PESTS

Cutworms (Noctuidae: Lepidoptera)

Cutworms are polyphagous and destructive insect pests and are cosmopolitan in distribution. In India, cutworm problem is severe in northern region. Common species include *Agrotis segetum* Denis & Schiffer muller and *Agrotis ipsilon*. *A. segetum* is commonly found in hills whereas *A. ipsilon* is more common in plains. The peak activity of cutworms reported to occurs during May-June in Shimla hills, in August in Peninsular India and in March-April in Bihar and Punjab. In Bihar the tuber infestation upto 12.7 % and in Himachal Pradesh 9-16 % has been reported.

Biology and life cycle

Female moth lays eggs either singly or in clusters on vegetation, moist ground or in cracks in the soil. The average fecundity varies from 600 to 800 eggs. Egg period lasts for about 10-28 days. Young caterpillars are smooth, stout, cylindrical, blackish brown dorsally and greyish green laterally. Fully grown larvae measure 4 to 5 cm in length. Newly hatched larvae feed on the leaves and later on the stems. Older larvae feed at the base of the plants or on roots or stems underground. They are nocturnal in habit and hide during day time under the soil or plant debris. Larvae construct burrows or tunnels in the soil about 2.5-5 cm deep near the host plant and at night they move up to the soil surface for feeding. Pupation takes place in an earthen cell in the soil. Adults are dark grey, black or brown coloured medium sized moths with markings on the front wings. They are active at night. The life cycle can be completed in 6 weeks under warm conditions.

Other species: *Agrotis flammatra*; *A. spinnifera*; *A. interacta*

Nature and Symptoms of Damage

- Caterpillars are the only damaging stage of pest. Young larvae cause damage by feeding on leaves, stems/seedlings at ground level.

- Irregular holes are found in the infested tubers.

- Larvae feed during night time on young shoots or underground tubers and hide in soil during day time near to the stem.

- After tuber formation, they start feeding on tubers and roots results in irregular holes of different size, resulting in reduced tuber yield and market value.

Management

1. Flooding of potato fields before planting helps in killing of cutworms in the soil.

2. Deep summer ploughing before planting to expose various immature stages of pest to natural enemies and high temperature

3. Use of well decomposed FYM (Farm yard manure)

4. Installation of light traps for mass collection and destruction of adult moths

5. Hand collection and destruction of early instar gregarious larvae

6. Garlic as intercrop with potato minimizes the damage in potato crops at Shimla

7. Soil application of Chlorpyriphos 20 EC @ 2 ml per lit of water at 2 % plant damage.

White grubs (Scarabaeidae: Coleoptera)

White grubs, the immature stages of June beetles are highly polyphagous and cosmopolitan in distribution. A large number of species are known to damage potato crops however, *Holotrichia longipennis* (Blanchard) and *Brahmina coracea* (Hope) are the predominant species damaging potato in the hills.*Holotrichia serrata* is prevalent in Karnataka, Maharashtra, Andra Pradesh, Tamil Nadu, Kerala, South Rajasthan, the tarai belt of Uttrakhand, and south Bihar.

Biology and life cycle

In summer, soon after first shower, overwintering beetles emerge from the ground at dusk, feeds on the leaves of trees and mate during night. At dawn they return to the ground, where the females lay 15-20 eggs in earthen cells several centimeters below the surface. Most of the beetle lay eggs in grassy surface. Eggs hatch 3 to 4 weeks later. The young grubs feed on the plant roots throughout the summer, in the monsoon; they burrow below the soil to a depth of 1.5 meters and hibernate. Grub is creamy white in colour with dark brown head and it becomes curved (C shape) when disturbed. Larvae pupate inside the soil. The adult beetles start coming out of soil at dusk soon after pre monsoon showers generally in May end or early June and settle on nearby tress namely *Acacia* spp., roses, peach, plum, pear, apricot, apple (in hills) and neem tress (in plains) for mating and feeding. The beetles after feeding on host foliage return to burrows in the soil early in the morning for egg laying.

Other species: *Holotrichia seticollis* Moser, *Anomola dimidiate* Hope and *Melolontha indica*

Nature and symptoms of damage

*   Grubs are the damaging stage. Second and third instar grub which feeds on the underground parts of plant such as roots and rootlets, making shallow and circular holes in the tubers (Figure 5).

*   Economic damage results from deep holes in the tubers underground.

- Severe damage occurs in potato field near pastures or grazing fields.

Management

1. Two to three deep ploughings immediately after harvest or before potato planting to expose the resting stages to birds or other natural enemies.

2. Potato should not be planted directly in pasture or grass fields.

3. Flooding of fields wherever it is possible and adopt crop rotation.

4. Application of fully decomposed FYM to the field.

5. Removal of weed plants in the vicinity of crop field. The host trees (*Rubinia, Polygonum*, Kaithe and temperate fruit crops) of adult beetles should be pruned.

6. Setting up of bonfire or installation of light traps during May-June at night for mass collection of beetles and killing them in kerosene/chemical treated water.

7. Seed potatoes should be planted at 8-10 cm depth than normal depth of 6 cm.

8. Application of entomogenous fungus, *Metarrhizium anisopliae* @ 5 gm per lit of water.

9. Application of Phorate 10G @ 10-15 kg/ha near plant base at the time of earthing up or drenching of ridges with Chlorphyriphos 20 EC @ 2.5 ml per liter of water when adult beetles appears.

10. Spray host trees with Chlorphyriphos 20 EC @ 2 ml per liter of water immediately after first monsoon showers.

Potato tuber moth (Gelechiidae: Lepidoptera)

Potato tuber moth (PTM), *Phthorimaea operculella* (Zeller) is a major pest of potato throughout the world. PTM is principally a storage pest but can damage the crops in field as well. In India, the damage has been reported from Maharastra, Bihar, Madhya Pradesh, Himachal Pradesh (Kangra), Tamil Nadu, North Eastern hill states, plateau region and Karnataka. The range of infestation could be 30 to 70 % in stored potato.

Fig. 4: Cotton leafhopper (left) and potato leaf damaged by mites (right)

Fig. 5: White grubs (left) and red ant (right) damaging potato tubers

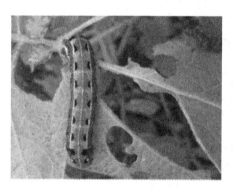 

Fig. 6: Tobacco caterpillar (left) and gram pod borer (right) damaging potato leaves

Biology and Life cycle

Adult moth lays eggs either singly or in groups on the underside of leaves or tubers. Eggs are white in colour. Upon hatching, larvae measure 1 mm long. The larva feeding on leaves is purple to green in colour while on tuber whitish purple. Maximum population growth of PTM occurs at temperature range of 20–25 °C. Life cycle of PTM is completed in 21-30 days at 27-35°C. Upper and lower threshold limit of temperatures for PTM are 40°C and 5°C. The male and female moths are brownish grey in colour and wings are folded in the form of roof like structure. In addition to temperature, precipitation also influences development and abundancre of *P. operculella.* The damage is severe under low rainfall and high temperature conditions.

Nature and symptoms of damage

- Larvae are the damaging stage of the pest. Upon hatching, young larvae mine into the leaves. Larvae also bore into stems of the plant causing wilting of plants.

- The larvae enter into the tubers and feed on them causing mines.

- The activity of larvae in tubers placed in heaps, results in production of heat which causes rotting of the produce significantly. In country stores, 18-83 % tuber damage due to PTM has been reported in the NE hill states.

Management

1. Seed tubers should be planted at a depth of 10 cm which reduces a damage to great extent as compared to traditional planting depth of 6 cm

2. Earthing up of field should be done 6-7 weeks after planting so that the tubers buried 25 cm below soil surface to avoid tuber exposure to egg laying and infestation.

3. Installation of PTM sex pheromone traps @ 20 traps/ha for mass trapping of male moths and monitoring of field populations.

4. Spray Deltamethrin 2.8EC (1350ml) or Cypermethrin 25EC (450ml) in 750 litres of water/ha of water on 30 days old crop.

5. Attract and kill oil formulation which contains the pest's sex pheromone and contact insecticide as active ingredients are applied at a density of 2,500 droplets (100 µl) per ha in the field (1 droplet/4m^2) reduce the male population and hence leaf and tuber infestation.

6. Use of 2-2.5 cm thick layers of dried *Lantana* or *Eucalyptus* leaves below and on the top of the potato heaps helps to reduce the PTM infestation.

7. Chlorpropham (isopropyl-N-(3-chlorophenyl) carbamate) commonly known as CIPC, the sprout suppressant, is reported to be effective against PTM damage in country stores when applied @ 30 ppm.

8. *Bacillus thuringiensis* ssp. *kurstaki* (Btk)-talcum formulation is used to protect tubers in storage against PTM. Potato tubers should be treated (powdered) directly after harvest and placed in storage. Granulosis virus based PhopGV-biopesticide formulated in talcum can also be used to protect stored potatoes.

Wireworms: (Coleoptera: Elateridae)

Wireworms are the larvae of various click beetles and are occasional pests of potato crop. Common species include *Drasterius* spp. *Agronichis* spp., and *Lacon* spp. When the potato crop is planted in field taken out of pasture or grass or when potatoes follow cereal crops, wireworms seems to be a problem.

Biology and life cycle

The eggs are laid in grassland or grassy stubble in May and June. Incubation period lasts for about 1 month, soon after hatching young wire worms initially feed on the organic matter in the soil. Older worms feed on the roots of many crops and weeds and bore into stem and other plant organs, including potato tubers. The mature larvae pupate in the earthen cell in soil and adult emergence starts following spring. Major damage to potato crop is caused by the larvae in their second and third years of development. The wireworms take 4-5 years to complete their life cycle and spend their entire time feeding in the soil. The presence of wireworms can be checked using buried baits. Pieces of carrot can be buried about 7.5 cm deep at 10-20 marked sites throughout the field. After 2-3 days, the carrot pieces are retrieved and checked for wireworms. If more than 1.32 wireworms per square meter are found, the field should either be treated before planting potatoes, or not be used for potato production. However, this action threshold may vary from one region to another.

Nature and symptoms of damage

- Grubs are the damaging stage. They bore into the tubers making cylindrical holes and pave a way for secondary infection, hence reducing the quality of the crop. Major damage occurs from tuber initiation till harvest.

- Tubers may contain wireworms when they are lifted from the field, but stored potatoes rarely contain them.

Management

1. When wireworm population in soil is more, avoid growing of potatoes and go for planting of legumes as rotational crop and keep the fields weed free.

2. Insecticides incorporated into ridges before planting can reduce tuber damage, but are unlikely to give complete control where wireworm levels are high.

3. Phorate 10 G @ 10 kg/ha mixed with sand applied in furrow is recommended for the control of wireworms in potatoes in India. It can also be side-dressed after potato shoots begin to emerge.

Mole Cricket (Gryllotalpidae: Orthoptera)

Mole cricket is a minor pest of potato crops in some parts of India. In West Bengal, it is sporadically sever. They are soil-dwelling insects which become pests of potatoes when they feed on tubers. They may also feed on the base of potato stems and kill the plants. From West Bengal, 5 to 6 percent plant damage, along with 10 to 15 percent tuber damage due to mole cricket infestation.

Biology and life cycle

Mole cricket has fossorial legs for digging. Adults are strong fliers but they are very rarely seen because they live in soil and are active during night. Adults produce a chirping sound which attracts other mole crickets. Adult female lays eggs during the rainy season at around 100 to 150 mm deep in the soil in earthen chambers prepared by her. A single female frames 3 to 4 chambers in her lifetime and deposits up to 25 to 30 eggs in each chamber. The immature nymphs are similar to adults but lack wings. Nymphs develop into adults underground in connected tunnels.

Nature and symptoms of damage

Nymphs feed on the roots of potato and other cultivated and wild plants. They also tunnel into newly planted seed tubers and also the developing tubers of potato. Both nymphs and adults come out of the soil during the night and devour the leaves of plants.

Management

1. Their presence and risk of damage can be detected with attractant baits. The bait consists of rice/wheat bran mixed with 120ml vegetable oil can be used for monitoring of mole cricket.

2. Soil application of Phorate 10G or Chlorpyrifos 10G @ 10kg/ha will not only manage mole crickets but also many other soil borne root and tuber feeders and sucking insect pests of potato.

Termites (Isoptera: Termitidae)

Various species of termites, like *Microtermes obesi* (Holmgren), *Odontotermes obesus* (Rambur), and *Eromotermes* spp. have been reported to damage potato. Rain-fed potato crops are more prone to termite attack than the irrigated ones. Large scale damage occurs in red sandy/loamy soils and in drought conditions. However, black soils and frequently irrigated areas are free from termite damage.

Biology and life cycle

Termites are colony forming insects. The queen termite (fertile female) lays up to 3000 eggs per day and she will lay tens of millions of eggs during her life. The eggs hatch and the nymphs develop into different caste. Caste determination is post-embryonic and each nymph upon hatching can become a worker, soldier or reproductive forms.

Nature and symptoms of damage

The worker caste damage roots and make deep holes in potato tubers. The tubers become hollow and filled with soil. The leaves of affected plants turn yellow, wilt and will ultimately dry up. If infested plants are pulled out, many feeding holes/punctures on the roots and potato tubers are found.

Management

1. Termites often attack sick or water stressed plants than healthy plants. Hence maintain conditions for healthy plant growth to prevent termite damage.

2. Systematic digging up of termite mounds and ploughing fields, bunds and water channels to destroy termite nests to expose them to predators like ants, birds, *etc.* and destruction of queen is a permanent measure for the management of termites.

3. The use of fresh or un-decomposed farmyard manure should be avoided. The damage can also be reduced by irrigating the fields regularly.

4. Digout/harvest the potato tubers at right time, as termites often attack the crop left in the field after maturity. If there is any risk of termite infestation found earlier, avoid leaving the crop in the field after harvest.

5. Soil drenching with Chlorpyriphos 30 EC @ 10 ml/L in standing crop is effective.

Red Ants (Hymenoptera: Formicidae)

Red ants (*Dorylus orientalis*; *D. labiatus*) are important soil pests of potato in many parts of the country e.g. West Bengal, Uttar Pradesh, Bihar and North-Eastern states. In Bihar, this pest is reported causing 70-90% damage to potatoes and 35-40% damage is reported from West Bengal. The pest appears during December and remains active until April, causing more than 10% of the damage in irrigated potato crops. High temperatures with dry weather favours population build-up of red ants.

Biology and life cycle

Alike termites, the red ants are also social insects that attack plants from below ground. The ants that are commonly seen in the field are workers. Mostly the worker caste will attack the plants and damages the potato tubers. Red ants are known to damage crops like potato, cauliflower, cabbage, groundnut, sugarcane, and coconut seedlings.

Nature and symptoms of damage

They cause damage on underground potato tubers and damaged tubers exhibited minute holes (2-3 mm diameter) on tubers and clean out the soft peels (Figure 5). In case of severe attack plants wilt in direct sunlight and will eventually dry up.

Management

1. At the time of first earthing-up, apply mustard cake to the soil @ 150 kg/ha, which repels red ants and protects the potato plants and tubers.

2. Soil drenching of infested potato fields with Chlorpyrifos 20 EC @ 0.06% keeps the red ant damage under check.

FOLIAGE FEEDERS OR DEFOLIATING PESTS

Hadda Beetles (Coleoptera: Coccinellidae)

Common species of Hadda beetles or epilachna beetles include *Epilachna dodecastigma* and *Epilachna vigintioctopunctata* (Fabricius). These are distributed throughout the country and infest a wide range of crops including potato, tomato, brinjal, many other *Solanum* species and occasionally cucurbits. The 12-spotted *E. dodecastigma* is more prevalent in the higher hilly regions of India whereas the 18-spotted *E. vigintioctopunctata* is restricted to plains and mid hills. During hot summer days, there is a considerable decline in the pest population where as in the winter; beetles undergo hibernation inside the soil or beneath the dried leaves around the field. They can be a major problem on spring crops of potato.

Biology and life cycle

The adults are active fliers and lay 500-750 eggs in clusters on the lower surface of leaves. The eggs hatch in about 3-4 days. The grubs measures about 6 mm in length, yellowish with six rows of branched spines on their dorsal surface. The larval period lasts for about 8-10 days on potato and pupation takes place on leaves with pupal period of about 3-6 days. The total life cycle is completed in about 21-36 days. Adults overwinter under grass and weeds. The pest completes 7-8 generations from March to October.

Nature and symptoms of damage

- Both adults and grubs cause the damage. They feed voraciously on the leaves by scraping chlorophyll resulting in skeletanization of leaves.

- The leaves present a lace-like appearance, turn brown, dry up and may fall off.

- In case of severe infestation, crop gets completely defoliated before tuber maturation.

Flea beetle (Coleoptera: Chrysomelidae)

Flea beetles are so named because of their enlarged hind legs with very stout femora with which they jump vigorously when disturbed. They are cosmopolitan in distribution with a host range of potato, tomato, turnip and brinjal. Common species infesting potatoes is *Psyllodes plana* Maulik.

Biology and life cycle

The adult flea beetles overwinter under leaves, grass and debris around field borders. Beetles terminate their diapause in April. They feed on weeds before migrating onto potato crop. The eggs are laid in the soil at the base of plants and are creamy white in colour. Individual female lays on an average 50-80 eggs. From the eggs cylindrical, brown-headed, white larvae hatch after a week. The grubs undergo three moults and when full grown, measures about 5 mm in length. The grubs burrow into the soil and feed for 2-3 weeks on the rootlets within 3-8 cm of the surface. Pupation takes place in the soil in an earthen cell and pupa measures about 0.5 mm long. The adults vary in colour from shiny black to black and generally there are one to two generations per year.

Nature and symptoms of damage

- Both adults and grubs cause the damage. They attack plants soon after their emergence from soil and damage continues until the crop is harvested. Grubs commonly feed on roots, often riddling them with tunnels or eating of rootlets.

- The adult flea beetles chew small round holes in the leaves, often starting on the ventral side of leaf causing typical symptom called "Shot hole"

- On small plants, this "buckshot" damage results in plant death.

- On larger plants, these feeding sites paves way for the secondary infection by pathogens causing blights and wilts.

Management of Hadda beetles and Flea beetles

1. Deep summer ploughing to kill the overwintering population.

2. Wherever the population is abundant, mechanical collection and destruction of grubs and various stages of the pests is recommended.

3. Thorough irrigation of infested crop minimizes increase in population of hadda beetles.

4. Since flea beetle adults feed on weeds in early spring and late fall, keeping downweeds around the fields holding these pests in check.

5. Under severe infestation, spraying of Dichlorovas 76 EC @ 1 ml or Chlorpyriphos 20 EC @ 2 ml per liter water is suggested. Application should be initiated as soon as the beetles or their eggs are found on the plants.

Caterpillars (Lepidoptera)

Polyphagous caterpillars such tobacco caterpillar, *Spodoptera litura* (Fabricius) and gram pod borer, *Helicoverpa armigera*(Hubner), Cabbage semilooper, *Plusia orichalcea* (Fab.), Oriental armyworm, *Mythimna separata* (Walker), Bihar hairy caterpillar, *Spilosoma obliqua* (Walker), occasionally damage potato crops as well (Figure 6). This is widely distributed all over the world. In the Indo-Gangetic plains, the caterpillars usually infest crops from mid-November and mid-December. If entered into protected structures, they can severe damage to potato plants.

Biology and life cycle

The moths are carried over from the preceding or adjoining crops and lay eggs on potato crops. The eggs hatch and start feeding on the foliage. The populations usually do not increase beyond one generation due to low temperatures.

Nature and symptoms of damage

Freshly hatched tiny larvae feed gregariously by scraping the leaves but mature larvae feed solitarily. More mature caterpillars segregate and feed voraciously on leaves, often completely defoliatin the plants. During severe infestation, an entire crop may be defoliated overnight.

Management

1. Alternate weed hosts of these insects, on which the first generation of caterpillars may develop, should be destroyed. The visible egg masses, as well as leaves with gregarious young larvae, should be collected and destroyed mechanically.

2. Summer ploughing could be used to expose and kill pupae in the soil.

3. Need based sprays of Chlorpyriphos 20 EC @ 0.2% or Indoxacarb 15.8 EC @ 0.06% or Flubendiamide 20 WDG @ 0.06% are able to check the population growth of caterpillars on potato crops.

POTATO NEMATODES

The Potato cyst nematode (PCN) (*Globodera* spp.) and root knot nematodes (RKN) (*Meloidogyne* spp.) are amongst the most economically important nematode pests of potato. PCN viz., *Globodera rostochiensis* (Woll) and *G. pallida* (Stone) also popularly called the Golden nematodes, are one of the economically important pests hindering the sustainable production of potato in many countries worldwide including India. Amazing adaptation for long term survival in the soil, even in the absence of a suitable host present them challenging to the farmers, scientists and policy makers. They are subjected to stringent quarantine and/or regulatory procedures, wherever they occur and present a serious threat to domestic and international commerce in potatoes. RKNs are prevalent all over the world and can cause significant crop loss in both warm and cool climatic conditions, depending on their species.

Potato cyst nematodes (*Globodera* species)

The Andes Mountains of South America, the original home for potato is also the place of origin for PCN. Presently *G. rostochiensis* have been reported from 78 countries and *G. pallida* from 61 countries covering five continents viz., Africa, America, Asia, Europe and Oceania. In India, potato cyst nematode was first reported in 1961 from The Nilgiris (Tamil Nadu). Accordingly, Tamil Nadu Government imposed domestic quarantine during 1971 to ensure strict checking of seed potato for marketing from Nilgiris (Prasad, 1993), and recently from Himachal Pradesh. In addition, Government of India restricted the movement of potato seeds from some areas of Himachal Pradesh, Jammu & Kashmir and Uttarakhand hills(Gazette Notification S.O.No 5642 (E) dated 2nd November, 2018).

Globally, an average yield loss of nine per cent, which is about 43 million tonnes, has been estimated due to this nematode alone. The economic threshold level for crop loss due to PCN is usually around 20 eggs per g of soil, which may vary with the environmental interactions and host tolerance.

Other than potato, PCNs are known to attack, tomato and eggplant. Other hosts of PCN mainly includes the species from Solanaceae family viz., *Datura, Hyoscyamus, Lycopersicon, Physalis, Physoclaina, Salpiglossis* and *Saracha* and *Oxalis tuberosa Molina*. In India, the differential host reactions of PCN populations from The Nilgiris and Kodaikanal hills revealed that the pathotypes Ro1 of *G. rostochiensis* and Pa2 of *G. pallida* are the most prevalent forms. The other prevalent pathotypes are Ro2 and Ro5 of the former and Pa1 and Pa3 of the later.

Biology and life cycle
The hatching of cysts is stimulated by the chemical substances called hatching factors present in the potato root diffusates (PRD) of the host plant roots. Maximum activity of PRD is reached three weeks after plant emergence and most actively from the root tips. The second stage juvenile (J_2) coming out of the cysts moves actively in soil and invade the roots by rupturing with its stylet. It enters through the epidermal cell walls and eventually settles with its head towards the stele and feeds on cells in pericycle, cortex or endodermis by forming a feeding tube. This induces enlargement of root cells and breakdown of their walls to form a large 'syncytium' or 'transfer cell' with dense granular cytoplasm that provides nourishment for nematode development. The sex of the nematode is determined during J_3 stage. The females become sedentary, swollen, remain attached to the roots and posterior part of the body comes out by rupturing the root cells. Males retain their thread shape and come out of the roots to locate and mate the females. After fertilization males die in soil whereas females keep on feeding to lay eggs. The immature females of *G. rostochiensis* are golden yellow in colour while that of *G. pallida* are white or cream in colour. After the female dies, the body wall turn brown and thickens to form a hard brown cyst that is resistant to adverse weather conditions. Cysts can easily dislodged in soil at harvest and each cyst contains about 200-500 eggs. The life cycle is completed in 35-40 days and generally, one generation is completed in one crop season. However, there are evidences of second generation of *G. rostochiensis* being completed because of its shorter dormancy (45 to 60 days) and long duration varieties. *G. pallida* has a long dormancy of 60 to 75 days therefore it can complete only one generation per season. Only 60 to 80% of juveniles can be stimulated to hatch in the presence of root diffusates and it never reaches 100%. This is a part of the survival strategy of PCN and some juveniles will remain dormant for several years before hatching (Prasad, 2006).

The PCN normally spreads by the movement of infested soil containing cysts and larvae, movement of seed potatoes from infested fields to the clean fields, irrigation and rain water, raising of seedling in infested area and planting to clean area, movement of compost from infested area, agricultural implements, through shoes of the workers, hoofs of cattle and through the use of old gunny bags in which the potatoes from infested plots were packed/stored previously.

Nature and symptoms of damage

The disease caused by this nematode is referred as 'potato sickness'. When the infestation is sufficiently heavy and localized, small patches of poorly growing yellowish plants appear in the field, wilting may occur during hot parts of the day (Figure 7). More evenly distributed infestations may cause a sudden failure of crops in whole fields. Heavily attacked plants remain severely stunted with dull and unhealthy looking foliage. As the season advances, the lower leaves turn yellow and brown and wither, leaving only the young leaves at the top, the entire plant now presenting a somewhat 'tufted head' appearance. The browning and withering of the foliage gradually extends and ultimately causes the premature death of the plant. The root system is poorly developed and the yield and size of the tubers are reduced considerably depending upon the degree of infestation. Close examination of the roots of infected plants reveal the presence of small pinhead sized, white (*G. pallida*) or yellow (*G. rostochiensis*) female nematodes sticking to the roots (Figure 8).

Fig. 7: Symptoms of field infected with potato cyst nematode

Fig. 8: Infection of *Globodera species*; yellow females of *G. rostochiensis* (left) and white of *G. pallida* (right)

Management

As no single method of control is fully effective in giving desirable level of nematode suppression, an integrated nematode management package incorporating judicious blend of various management options such as host resistance, chemical, biological and cultural methods is being advocated to bring down the PCN population to levels that permit profitable cultivation of potato.

1. Crop rotation with non-solanaceous crops is widely recommended for management of PCN. Crop rotation of three to four years involving potato, french beans and peas and radish, cabbage, cauliflower, turnip, garlic, carrot, green manure crop like lupin etc. brings down the cyst population. Among different non-solanaceous crops, radish and garlic reduce the PCN population.

2. Intercropping with French Beans (3:2), mustard (1:1) with Carbofuran, and radish (2:1) decrease the PCN population with enhanced yields (Priyank et al., 2019).

2. Trap cropping with susceptible potato cutivars such as Kufri Jyoti reduce the PCN population but the trap crops should be destroyed before the completion of PCN life cycle. On other hand S. sisymbriifolium was found to be one of the most promising resistant trap crops for both the species of PCN (Priyank et al., 2019).

3. PCN resistant cultivars such as Kufri Swarna, Kufri Neelima, Kufri Sahyadri can be adopted for good yields (Aarti et al., 2019).

4. Application of Carobofuran 3G @ 2 kg a.i./ha at the time of planting and soil fumigation with Dazomet (Basamid 90G) @ 40-50 g/m^2 with soil covered with polythene sheet is known to significantly reduce the PCN population. Treatment of PCN infested tubers with 2% sodium hypochlorite solution (4% available chlorine) up to twelve cycles of repeated use for the duration of 30 min. (per soak) is reported to cause 100 % cyst disintegration (Aarti et al., 2019).

5. Application of biological control agents viz., *Pseudomonas fluorescens* and *Paecilomyces lilacinus*, organic amendments like neem cake (5 t/ ha) blended with *Trichoderma viride* (5 kg/ha), and incorporation of radish leaves @ 1 kg/m^2 and covering with polyethylene sheet are known to suppress PCN population in soil.

6. **Integrated management**: Inter-cropping of potato with mustard in 1:1 plant ratio combined with Carbofuran 3G @1.0 kg a.i./ha application reduced PCN infestation and enhanced potato yield. Application of *P. fluorescens* (2.5 kg/ha) + neem cake (1 t/ha) + mustard intercrop

(between potato rows) + Carbofuran 3G @1.0 kg a.i./ha increased the tuber yield and decreased the PCN population. Effective PCN reduction (47% in 2 years) could be achieved by rotating susceptible potato variety with the resistant one and application of Carbofuran @ 2.0 kg a.i./ha for each crop of potato. For eradication of PCN by soil solarisation (4 weeks) followed by application of neem cake (5 t/ha) in combination with *T. viride* (5 kg/ha) recorded decrease in PCN population.

Root knot nematode (RKN)

Some RKN species have been reported from all potato growing countries of the world and the most important ones are *M. chitwoodi* (Golden), *M. fallax* (Karssen), *M. incognita* (Kofoid and White), *M. javanica* (Treub), *M. arenaria* (Neal) and *M. hapla* (Chitwood). In India, the dominant RKN species affecting potato both in hills and plains has been *M. incognita* while *M. javanica* infestation is in mid hills and plains of northern India. *M. hapla* is confined to hilly tracts of Uttar Pradesh, Himachal Pradesh, Jammu and Kashmir and Tamil Nadu. However, *M. arenaria* is reported form the plains of Uttar Pradesh (Prasad, 1993).

Root-knot nematode found to infect more than 2000 plants worldwide that includes many important cereals, pulses, vegetables, fruits and ornamental plants. This makes them very difficult to manage because they can always survive and reproduce on other host crops including weeds.

Biology and life cycle

The vermiform second stage juveniles hatch out from the egg masses penetrate the young potato roots and starts feeding. This leads to the formation of giant cells, which result from the re-differentiation of vascular root cells. Whereas, hyperplasia of surrounding root cells leads to the formation of the gall which is an easily recognized symptom of plant infection by RKNs. The second stage juveniles (J_2) undergo moulting and pass through J3 and J4 stages and finally become adult females or males. The adult females are sedentary and pear shaped while males are migratory and vermiform or thread shaped. Males move out of the root to locate and mate the females. The females lay about 300 to 400 eggs in gelatinous matrix, usually adhering to the root galls. These eggs readily hatch and infect fresh roots. At the time of tuber formation, the juveniles usually enter the tubers since the root system would have started decaying. Life cycle of RKN is completed in 25-30 days during summer and 65-100 days in winter. In the hilly regions, two generations are completed by the time of tuberization and hence tuber infestation is invariably observed even under low initial inoculum levels, whereas, in plains, tuber infestation could be low because of the shorter crop duration and preference of the fresh roots by the newly hatched juveniles.

Nature and symptoms of damage

The above ground symptoms due to RKN infection include stunting and yellowing of plants with chlorotic leaves due to the hindrance in water and nutrient uptake in roots. In roots, the characteristic swellings called 'galls' are formed. The galls on the root are small and often unnoticed but the warty 'pimple-like' blemishes on the tubers due to nematode infection which reduces the commercial value and keeping quality of potato tubers (Figure 9). Brown spots are evident in the flesh of the cut tubers due to the presence of nematodes. An initial inoculum of 2 juveniles per cc of soil results in an overall yield reduction of about 40% with 100% tuber infestation (Prasad, 1993).

Fig. 9: Potato tubers infested with root-knot nematode (Meloidogyne spp.)

Management

1. Avoiding seed tubers from infected areas and planting of nematode-free tubers.

2. Deep ploughing during summer months exposes the infective juvenile stages of the nematode to direct sunlight, thereby killing them.

3. Adopting good sanitation and keeping the field free of weeds which otherwise would serve as alternate hosts for root-knot nematode.

4. Crop rotation with non-host crops like maize or wheat helps in minimizing the nematode damage.

5. Early planting of spring crop in first week of January and late planting of autumn crop in second and third week of October reduce nematode infection in potato due to the lower temperature prevailing during crop period.

6. Growing antagonistic crop such as marigold, *Tagetes patula* in alternate rows with potato reduce nematode population in soil. Similarly, growing *Cassia* species and incorporation of saw dust @ 26 q/ha also helped in reducing tuber infestations.

7. Application of Carbofuran @3 kg a.i./ha reduces nematode infestation. The efficiency of pesticides increased with split application, one at planting and the second at the time of earthing up.

8. A pathogenic bacteria *Pasteuria penetrans* can be used for managing root-knot nematode damaging potatoes (Prasad, 2008).

Integrated management: RKN in North India is managed using nematode-free seed tubers, crop rotation with maize or wheat and application of Carbofuran 3G @1-2 kg a.i. /ha at the time of potato planting. A two-year adoption of INM for root-knot gives efficient and economical production system (Prasad, 2008).

Conclusion and Future Outlook

The potato is plagued by a number of serious insect pests that can completely destroy the crop if left uncontrolled. At the present time, their management relies predominantly on synthetic insecticides and poses a serious threat to the environment in several parts of the world. Depending on the variety used, targeting the different market segments (e.g., seed, fresh market, processing),insect management strategies may vary due to crop value and specific quality needs. A better understanding of the structure and function of a potato ecosystem is necessary for developing truly sustainable crop protection solutions. Judicious use of the available techniques, new and old is the only way out. Genetic engineering, gene knockdown and other such techniques offer exciting options which need to be explored. Others have been around for a while now e.g., biological and bio-rational control. Integrating these techniques into an economically viable and environmentally friendly system that can be constantly adjusted to the changing conditions is a daunting task. It is also an open-ended process, as developing a universal set of approaches that will work forever in every potato field in the world is impossible.

Breeding resistant varieties from diverse sources of needs to be explored for resistance to wide spectrum of PCN pathotypes and RKN species. This also necessitates identification of molecular markers for resistance to both the species of *Globodera* and all the species of RKN. The possibilities should be explored to inhibit the activity of genes responsible for production of hatching factors in root diffusates and also identify, characterize and inhibit the activity of genes involved in important aspects of the life cycle of PCN and RKN through transgenes. Though eradication of PCN is very difficult if once established in an area, more recently, Western Australia has been declared free of potato cyst nematodes, after a battle of about 24 years, opening up big opportunities for its $45 million potato industry. This indicates the possibility of the eradication of this nematode from India also with the strong background of science based bio security policies, strict regulatory and sanitation measures and management operations.

REFERENCES

Aarti B, Sanjeev S, Venkatasalam EP, Mhatre PH and Naga KC (2019) Management of potato cyst nematode.http://www.krishisewa.com/articles/disease-management/1007-management-of-cyst-nematode-in-potato.html.

Bhatnagar A, Jandrajupalli S, Venkateswarlu V, Malik K, Shah MA and Singh BP (2017) Mapping of aphid species associated with potato in India using morphological and molecular taxonomic approaches. *Potato Journal* **44**(2): 126-134

Chandel RS, Banyal DK, Singh BP, Malik K and Lakra BS (2010) Integrated management of whitefly, *Bemisia tabaci* (Gennadius) and potato apical leaf curl virus in India. *Potato Research* **53**(2): 129-139

Chandel RS, Chandla VK, Verma KS and Pathania M (2013) Insect pests of potato in India: biology and management. In, Insect Pests of Potato Global Perspectives on Biology and Management. Giordanengo P, Vincent C and Alyokhin A (eds.). Academic Press, Waltham (MA): 227-268

Prasad KKS (1993) Nematodes– Distribution, biology and management. In: Advances in Horticulture, Vol. 7, Potato, Chadha KL and Grewal JS (eds). Malhotra Publishing House, New Delhi: 635-647

Prasad KKS (2006) Potato cyst nematodes and their management in the Nilgiris (India). Central Potato Research Institute, Shimla, India. Technical bulletin-77: 20p

Prasad KKS (2008) Management of potato nematodes: an overview. *Journal of Horticultural Science.* **3**(2): 89-106

Priyank HM, Divya KL, Venkatasalam EP, Aarti B, Sudha R and Berliner J (2019) Potato cyst nematode: A hidden enemy of potato cultivation in Nilgiris. *Bhartiya Krishi Anushandhan Patrika* **34**(1): 50-53

Shah MA, Jandrajupalli S, Venkateshwarlu V, Malik K, Bhatnagar A and Sharma S (2018) Population ecology of aphid pests infesting potato. In, Sustainable Agriculture Reviews, vol. 28. S. Gaba et al. (eds.), Springer, Cham: 153-181

Shah MA, Malik K, Bhatnagar A, Katare S, Sharma S and Chakrabarti SK (2019) Effect of temperature and cropping sequence on the infestation pattern of *Bemisia tabaci* in potato. *Indian Journal of Agricultural Sciences* **89**(11): 1802-1807

18

Potato Seed Production: Present Scenario and Future Planning

Rajesh K Singh[1], Tanuja Buckseth[1], Ashwani K Sharma[2], Jagesh K Tiwari[1] and SK Chakarbarti[1]

[1]*ICAR- Central Potato Research Institute, Shimla-171001 Himachal Pradesh, India*
[2]*ICAR-Central Potato Research Institute Regional Station Kufri, Shimla-171 012, Himachal Pradesh, India*

INTRODUCTION

India is the second largest producer of potato in the world contributing about 12.52% of production from 11.29 % area under potato with an average yield of 22.31 t/ha. Potato is grown in almost all the states in India under diverse agro-climatic conditions. About 90% of the total potato area is located in the sub-tropical plains, 6% in the hills and 4% in the plateau region of the peninsular India. Indo-Gangetic plains account for about 76% of the potato area and about 87% of the potato production in the country. Availability of assured quality seed is the most important production constraint in all the potato growing regions of the country.

Seed is the basic vital input in agriculture and its timely availability in adequate quantity as per demand decides the strength growth of agricultural economy in the country. Seed has played a very important role in India's green revolution and shall continue to be the important component in the days to come. There is a saying "as you sow, so shall you reap", which traditionally relates to the quality of seed determining the production, has been the wisdom of our forefathers. The traditional farmers' practices are to save their own seeds generally selected from healthy plants from fields which are free from diseases pests, harvest them separately, clean, cure and then store in containers and inspect regularly for any kind of damage; or they get quality seed from other seed producing farmers on barter basis, indicating thereby the importance of seed in their production system.

There is limited accessibility of quality planting material in case of vegetatively propagated plants in India. Generally, tubers are used as seed in potato overwhelming various problems like low multiplication rates, increased production cost various seed borne diseases befitting the risk of degeneration over the years. Such tubers when planted resulting in low yeild and diseased plants in future generations. This was coined as 'degeneration' or selenity by the Dutch workers in 1972. Such selenity acts as a major constraint in producing clean potatoes. Potato production is constrained by various factors that include availability of suitable varieties, agro- techniques, quality seed and storage infrastructure, especially in tropics and sub-tropics. But most important being the availability of quality seed as it has a direct bearing on crop yield. Potato seed is bulky and the rate of multiplication is slow (5-6 times), this poses a unique challenge for seed production and seed supplying agencies. Since the crop is multiplied vegetatively using tuber as seed, it gets degenerated very fast necessitating replacement of the seed every year (ideally) or after every two years. Further the seed is either unavailable or out of the reach of farmers owing to high price. The seed related issues are further aggravated because of restricted accessibility of reasonable seed producing regions quality seeds can alone add to the increased yield by 15 - 20%.

HISTORY OF POTATO CULTIVATION IN INDIA

Spread in India

The precise introduction of the potato in India is not known. Probably it was brought in either by the Portuguese who first opened the trade routes to India, or later by the Britishers. However, the stocks introduced in all likelihood were of those then grown in Europe, and not directly from South America. The earliest reference of the potato occurs in account of the voyage of Edward Terry, 1655 (Loomba and Burton, 2007) who was chaplain to Sir Thomas Roe, British Ambassador to the court of the Mughal Emperor Jahangir from 1615-1619. Terry, in his description of Indian solid and its produce, wrote "In the northernmost part of this empire they have a variety of pears and apples; everywhere good roots as carrot, potatoes, and others like them......are grown'. Terry's account thus places the potato as a crop in the northernmost parts of India, probable in the hills, earlier than 1615. Similarly, Fryer's travel records, (1672-81) mention the potato as a well-established garden crop in Surat and Karnataka in 1675. However, Habib (1963) in his book The Agrarian System of Mughal India (1556-1707) wrote that ordinary (now called Irish) potatoes were not among the vegetables grown. According to him, mention of 'potatoes' referred to in Terry's 1655 (Loomba and Burton, 2007) and Fryer's 1672-81 (Hutchinson, 1974) travel accounts actually meant varieties of yams which were grown and formed an article of popular diet in northern and southern India during that period.

By late 18th or early 19th century the potato was an important established vegetable crop in the hills and plains of India. Early introductions resembled the Andigena potatoes and were adapted to short winter days having long dormancy and capable of withstanding higher temperatures under country stores. They were grown by various names in local dialects depicting some character, viz. Phulwa-flowering in the plains; Gola-round potatoes, Satha-maturing in 60 days, etc. and came to be known as *desi* varieties.

Between 1924 and end of World War II, the State Agricultural Departments and other agencies introduced a large number of European potato varieties with a view to selecting those suitable for local conditions. These efforts, however, proved of little value mainly due to poor adaptation of temperate long day adapted European varieties in the sub-tropical short days available in plains of India. Only few foreign varieties approached the yield levels of local varieties under commercial culture, however, none proved to be a very good yielder. Fast degeneration of seed stocks was another important factor in non-establishment of European varieties in the sub-tropics. Lack of adequate cold storages for storing seed potatoes over long periods of hot summer also posed problems. Whereas, the local old varieties kept well in country stores, the imported varieties could not survive in these conditions. The introductions from Europe thus made no impact on potato culture in sub-tropical plains of India. However, a few introductions such as Magnum Bonum, Up-to-Date, Royal Kidney, Great Scot, Craig's Defiance, etc. survived in the hills. In India till 1950, 16 each of *desi* and European varieties were identified mostly under cultivation.

Origin and Domestication

Originally potato was found 8000 years ago as a wild species by the inhabitants of Andes in South America at an elevation of 3800 m above sea level. The tubers are rich source of energy and large amount of stored carbohydrates made potatoes the food of poor. In India, the potato for the very first time was introduced amid the 17th century by the Portuguese merchants. It has been clearly mentioned in Terrys record of a banquet in Ajmer that it was given to Sir Thomas Roe by Asaph Chan in 1615. It had successfully established in gardens of Karnataka and Surat in 1675. Potato cultivation was introduced in the hills of Dehradun by Major Young, an armed force officer. Whereas, it was introduced in the Shimla hills during 1828, where the householders used to cultivate potatoes in small strips followed by its introduction in Nilgiri hills by 1830. With the course of time trade between the hills and plains developed. It had turned into a profitable source of income by 1839. Potato was likewise traded to Tibet in the early 19th century through India.

Import of Foreign Varieties

Prior to freedom and during the colonial period, seed potatoes of indigenous varieties were alone utilized for potato cultivation. Whereas the seeds of high yielding potato varieties were first imported from the Netherlands in 1960 and continued thereafter. Import of European varieties was customary for commercial potato production. In the year 1946-1947, almost 1072 maunds of seed potatoes of three distinctive varieties namely Arran Consue, Up-To-Date and Kerrs Pink were distributed to local farmers. However, deterioration of seed stock soon took place due to absence of any seed potato production program. Foreign varieties were imported each year until the point that the world war-II set in, wherefrom, the state needed to scan other avenues for importing seed potatoes. After the stoppage of foreign imports in 1939, varieties available within the nation were tested by Dr. B. N. Uppal and concluded that the assortment of Up-to-date (Numbri) was suitable for cultivation in Bombay. From that point, India has been producing its own seed.

Seed Potato in India

For seed production in India, sincere efforts were made to maintain healthy standards of potato seed stocks with in 1933 in UP (Uttar Pradesh) by Agriculture Department. They attempted to increase the pathogen free seed supplies of Dunbar Cavalier and Majestic cultivars imported from England. This aided in anchoring around 100 quintals of Majestic and 250 quintals of the other assortment i.e. Dunbar Cavalier at Almora and Nainital hills of Garhwal district. To conquer the issue of dormancy, a second round of multiplication was done in the mid hills of Bhawali. But this resulted in degeneration of stocks due to occurrence of bacterial brown rot disease in seed tubers and thus the scheme failed (Pushkarnath, 1976).

In 1935, a thought was envisioned to establish an undeniable research foundation in India. This provoked the opening of a Potato Breeding Institute in Shimla and two seed creation farms at Bhowali (Kumar slopes, UP) and Kufri (Shimla Hills) as a piece of Indian Agricultural Research Institute, Delhi. Taking a look at the issue of seed degeneration, a research plan under the Indian (Imperial) Council of Agricultural Research was developed in 1941. It surveyed the various diseases affecting potatoes. Potato cyst nematode, popularly known as golden nematode is a 'major constraint to potato production throughout the world and this quarantine pest is restricted to Nilgiri hills of South India. It was first detected in1961 at Ootacamund and brisk execution of isolate estimates confined its further spread to other potato developing areas.

Hills-Ideal for Seed Potato Production

A project with the name 'Production of partially disease free potatoes' was set up. The fundamental approach was to clean up all the seed crops present in fields, which could be obviously distinguished for viral infection. It brought about threefold increase in the yield of Darjelling Red Round variety. But it was soon realized that potato seed generation on a huge scale requires production and development of India's own varieties. As the climatic conditions in India were entirely different from North America and Europe from where the potatoes were imported in. The experience additionally brought the acceptance that an association which is specialized in science and excels in the craft of seed production was required for commercial production of potato.

An overview was portrait by Davies in 1934 on the different reasons for seed potato degeneracy, and it was the aphids particularly *Myzus persicae* which was observed to be the fundamental driver. Additionally, with the advancements winning around then, hills were observed to be the perfect location for seed potato production which was pathogen free. According to the guidelines provided by him for the location and maintenance of seed potatoes in hilly areas, the principal seed production station for potatoes was established in the high hills of Kufri (Shimla hills, Himachal Pradesh). It satisfied the states of lower temperature in summers (underneath 20 °C) with the inability of aphid migrants to alight the potato crops and their colonization. It was additionally risky to grow potato crops for seed purpose when aphid population transcends above 20 on 100 leaves. Till 1935, the seed potato was being imported from various European countries on yearly basis, but during Second World War, European countries put a blanket ban on the export of potato seed to India.

Establishment of CPRI

European varieties were imported to India and endeavors were made to acclimatize them to the agro-climatic conditions prevailing in India. But soon deterioration of seed stock took place due to absence of any seed potato production program. With this, it was realized that India can't rely on exotic varieties alone and therefore, needs to have its own program for research and development of potato. In 1945, a scheme for developing Central Potato Research Institute was advanced by Sir Herbert Steward, then Agricultural Advisor of Government of India. The plan was implemented by Dr. S. Ramanujam in 1946 and it was the first occasion when that scientific orientation was given to the seed potato production. In 1946-1947, 1072 mounds of seed potatoes of three varieties- *Arran consue, up-to-date and Kerrs pink* were distributed to local farmers. However, it took three years for the entire scheme to attain a concentrated shape in the form of CPRI, Patna with its first Director

Dr. S Ramanujam. The three units functioning independently were (i) Potato Breeding Station, Shimla (ii) Seed Certification Station, Kufri (iii) Potato Multiplication Station, Bhowali were now converged into CPRI. The administration of Himachal Pradesh developed a research scheme for producing disease free seed tubers. This venture effectively delivered clones of 'Up-to-Date" which was imported from Northern Ireland. As hills were the only source of pathogen free seed, in 1956 the headquarters were shifted to Shimla. With the course of time, trade was successfully established between the hills and plains.

In between, 1956 to 1983, a few local research stations were built up in various potato developing zones of the country to address local issues of potato development. The organization has seven territorial research stations situated at Jalandhar (Punjab), Patna (Bihar), Modipuram (Uttar Pradesh), Kufri (Himachal Pradesh), Gwalior (Madhya Pradesh), Shillong (Meghalaya) and Ootacamund (Tamil Nadu). Potato seed creation was confined to high slopes till mid 1960's.

Overdependence on Hill Seeds

Viruses are transmitted by vectors known as aphids, consequently potato seed started to be produced in the high hills which almost free from aphid during the potato cropping period. This increased the strain on the indigenous seed multiplication systems being practised in the hills. Subsequently, after 22 years in 1966 another research station was setup in Fagu (Kufri, H.P.), as all hilly areas have not been found suitable for production and maintenance of healthy seed potatoes which can meet the standards. The issue of rapid degeneration had made the plains fully reliant on the hills in order to satisfy their prerequisite for seed. Thus, the entire pressure was laid onto the hills for quality seed potato production which offered ascend to an extensive no. of issues, for example:

(i)   Limited areas having elevation of 750 m or above which are technologically suitable for seed production.

(ii) Seeds produced in hills cannot be utilized in plains for sowing in October owing to dormancy.

(iii) Limited means of transporting the freshly harvested tuber but dormant seed tubers over long distances resulting in delay in planting.

(iv) Crop conditions due to irrigation practices: In the Indian hills viz., Kashmir valley, Kinnaur and Lahul Spiti potato planting takes place during the summer seasons followed by harvesting towards the end of monsoons. However, there was no uniform distribution of rainfall which is critical for the growth of plant. Therefore, in case of scanty rainfall, it's very difficult to provide required irrigation due to highly uneven topography, exceptionally porous nature and non availability of ground water. Another factor which adds is limited area for cultivation and the seed not being of the right physiological age to be immediately used in plains.

Also, a high rate of degeneration makes the seed break down after a couple of multiplications. Since the eastern, north-eastern, Deccan and South western parts of the nation, having hotter climatic conditions, and were not suitable for customary quality seed production. To overcome these issues, it was important to fabricate another framework for seed production as well as certification in plains which could limit the reliance on slopes. In 1971, an undertaking under the name All India Coordinated Potato Improvement Project (AICPIP) with its headquarters in CPRI, Shimla was set up.

Development of Seed Plot Technique for Plains

Till mid-1960's the seed potato cultivation was confined to high hills only as it had low aphid population during cropping season. Potato was cultivated as a spring crop in the Indo-Gangetic plains which was totally exposed to aphids bringing about degenerated seeds. Different analyses were led by CPRI for seed potato creation in various agro-climatic areas. A noteworthy achievement underway of value potato seed was culminated through the introduction of '' Seed Plot Technique' in 1965. This made India, the only country in South Asia with its own breeding program for potato. This aided in beating seed decline because of aphids as the seed crops could be grown during very little or no aphid population in the plains of Northern India. Due to the advancement of Seed Plot Technique, the significant production of disease free seed moved from the slopes to the fields in plains. Despite the fact that the potato was presented as a late spring crop in temperate hills, and it at last turned into a noteworthy winter crop for the plains. The vast stretch of fertile belt reaches out from Punjab to Assam, the seed produced through seed plot method gave 30-40% higher yields free from many soil and tuber borne diseases. It was adequate for producing enough breeder seed for the nation every year.

Seed Plot Technique benefitted Indian agriculturists as well as aided in sparing expansive measure of cash which was generally spent on the import of foreign seeds. The primary guideline of this technique relied on developing healthy seed potato utilizing low intensity of aphid infestation period in between October and starting of January incorporated with coordinated pest and disease management. Though, roguing and dehaulming of the seed crop needed to be completed by the last week of December or mid of January before the aphids infestation reach to the economic threshold level. This strategy opened a window in sub-tropical plains for production of quality seed. The seed production could be taken up in bigger zones under sub-tropical plains including Uttar Pradesh, Bihar, Punjab, Haryana and parts of Madhya Pradesh. Farmers could undoubtedly receive seed of right physiological age at the exact time of planting, with no issue of dormancy.

The selection of seed plot system prompted enormous development in cold stores for storing the produce which in the following season could be used for autumn planting. As a result of the advancement of seed plot method, potato tuber moth (PTM) got eliminated and seed rottage diminished. Thus, seed plot technique met the seed prerequisite as well as prompted higher productivity with minimum losses.

Virus Indexing

Exploiting the seed plot framework, a proficient consistent strategy for outfitting the quality of healthy seed stocks was envisioned after 1970 through clonal choice, tuber indexing and stage-wise field multiplication of healthy tubers. Indexed tubers concept for production of disease free potato seed took shape once diagnostic methods were developed for major potato viruses. In this method, tubers are selected from storage/ growing crop that are uniform in type, free from detectable diseases, and within an acceptable size range. Each tuber is coded for identification, and an eye, preferably from the stolon end, is removed and planted in the greenhouse. Resultant plants are evaluated for vigour and viral diseases. The tubers which are free from all the viruses are used for production of disease free seed by multiplying them in the field for four (Stage 1, 2, 3, 4) generations. This has been made feasible utilizing ELISA (Enzyme Linked Immune Sorbent Assay) with its first use in 1984, by CPRI. In this method, tubers are first screened against different infections like PVX, PVA, PVY PVS, PLRV, PVM, PVY^N and PVY^C using ELISA. Tuber indexing is a system of testing the picked tubers to be free from any disease by growing up their eye plugs in net houses. Another essential technique grasped for virus detection and elimination is Immunospecific Electron Microscopy which is in use since 1987. Scotland pioneered this concept which has since been used world over for potato seed production, including India. The new diagnostics developed over the years have improved the efficiency of the system.

SEED POTATO PRODUCTION SYSTEMS

Conventional system

The conventional method of seed potato production is to repeatedly propagate a sample which has been proved to be free of pathogens in a system called clonal multiplication. The clonal multiplication system has been practiced effectively in the Netherlands, South Africa, Kenya and India. This potato seed production system is laborious, expensive and time consuming. Because of a low multiplication rate (usually 1:6), it takes many years to produce significant quantities of seed to meet the demand of potato industry. Another constraint is the degeneration of seed stocks from one generation to another due to accumulation of bacteria, fungi, viruses and viroids. Modern seed programmes tend to minimize the use of clonal systems or use a combination of clonal and rapid multiplication systems. The production of potato seed under conventional system has not been effective in avoiding or reducing the buildup of pathogens and has consequently led to reduced quality potato seed and low crop yields. In this system, seed potato growers select better quality tubers for seed and discard those of poor quality. The diseased and healthy plants are identified and separated and the healthy tubers are used for the next season's production. If the mother potato plant becomes infected with a disease during the growing season, each of the new offspring is likely to be infected as well. During the growing season, growers peep seed fields visually for signs of disease and remove infected plants through the process of rouging. However, visual inspection, particularly for primary infection, is unreliable, time consuming and requires a well experienced eye. Inevitably, the quality of seed potato produced in subsequent generation's declines substantially. The conventional method of propagation is one of the slowest methods of seed multiplication. This method has also shown to be done at certain time of the year specific particularly in tropical and sub-tropical regions where potato is a winter crop. In addition, the method requires a seed producer to have enough land if he has to enter into commercial seed production. This, however, is associated with high labour cost in managing big fields.

Thus, the only way-out to overcome the above said limitations is augmentation of seed potato production through hi-tech seed production system to improve the quality and also to reduce the field exposure.

Hi-tech system

The virus-indexed and pathogen-free mericlones are subjected to a rapid tissue culture plant propagation method to generate abundant clean (pathogen-free) cultures. Several facets of the generic science of plant tissue culture have both current and potential future applications. These include micropropagation, i.e. clonal propagation through axillary shoot proliferation using explants (isolated plant tissues, e.g. leaf sections or stem internodes used in *in vitro* culture) containing pre-existing meristems, de novo shoot production following the induction of adventitious meristems by the application of plant growth regulators and somatic embryogenesis, which is the process of embryo initiation and development from vegetative or non-gametic cells. All these processes hold significant applications in potato biotechnology. The widespread introduction of plant tissue culture to seed potato production during the 1970s enabled the rapid multiplication of disease free seed material and resulted in more productive seed stock. Micropropagation now underpins many seed potato production systems and specifically provides the nuclear stock material, in the form of microplants (plants derived through *in vitro* axillary bud proliferation) or micro-tubers (*in vitro* produced tubers) for their subsequent use in a chain of potato seed production programmes.

Micropropagation involving *in vitro* asexual multiplication allows quick and round-the-year production of disease-free good quality pre-basic seed and thus it is a way out to supplement the ever high requirement of quality seed. The use of micropropagation in pre-basic seed production has resulted into mass production of potato plants in a very short period of time. Seed potato production through micropropagation is mostly based on the production of *in vitro* plantlets or microtubers, and on the subsequent production of minitubers as first *ex vitro* generation (Figure 1).

Fig. 1: Integration of tissue culture in potato seed production (Naik and Buckseth, 2018). (a) Rapid micropropagation of virus-free mericlones on semi-solid medium. (b) Multiplication of micropropagated plants in liquid medium for production of microtubers. (c) Potato microtubers developed under dark in MS medium supplemented with 10 mg l^{-1} BA plus 80 g l^{-1} sucrose. (d) Harvested microtubers. (e) Pathogen-free high-density potato crop raised either from in vitro plants or microtubers in net house. (f) Harvested minitubers from high-density net house crop. (g) Healthy field crop raised from minitubers harvested in previous season.

Minitubers can be produced from plantlets planted at high densities in the greenhouse in beds or in containers using different substrate mixtures, or even in hydroponic culture. Lommen (1995) presented alternative production techniques for minitubers using very high plant densities and non-destructive, repeated harvesting of minitubers by lifting plants carefully from the soil mixture and replanting them after harvest. These techniques allowed minitubers of ideal size to be produced, the number of tubers could be increased considerably, while total yield was reduced. Many seed programmes prefer to use minitubers, defined as the small tubers, (usually 5–25 mm), that can be produced throughout the year under semi *in vitro* conditions in glasshouse and screen-houses using *in vitro* propagated plantlets, planted at high density. Hydroponic and aeroponic techniques have been evaluated and proved suitable for minituber production in greenhouse; hydroponic culture in inert aerated substrate (such as wood, perlite, vermiculite), in which film culture, i.e. hydroponic systems in which roots grow directly in either a pure circulating nutrient solution or in a circulating nutrient solution system with very little substrate.

Recently hydroponic/aeroponic systems have been developed for production of minitubers from *in vitro* plants. In addition to reducing the cost of production, these systems facilitate round the year production and adoption of phytosanitary standards. Production of minitubers using aeroponics is an alternate soil-less culture technology. This technology consists in enclosing the root system in a dark chamber and spraying a nutrient solution on roots with a mist device. The modified device consists of aeroponic chamber, pump, spraying tube, timer and nutrient solution reservoir. Production and utilization of minitubers using aeroponics have been reviewed by Buckseth *et al.* (2016).

The multiplication rate of potatoes is very low compared to other crops, from between four to six times under optimal conditions. For this reason, a large portion of crop area is devoted to the production of seed tubers and it takes a considerable time to build up a sufficient amount of commercial tubers. With every field multiplication, the build-up and transfer of pathogens can increase, leading to seed degeneration. Therefore it is essential to investigate methods of increasing the number of minitubers (G0) produced from disease free in-vitro plantlets. Therefore, aeroponic technique offers many interesting opportunities for developing enhanced production systems, mainly for mini-tubers. Although requires some technical knowledge to design, establish and run, the benefits offered are sufficient for such systems which have been widely adopted by seed production companies worldwide (Figure 2).

Fig. 2: Aeroponics in potato (Naik and Buckseth, 2018). **(a)** Diagrammatic presentation of aeroponic system. **(b)** Minitubers developed in aeroponic chamber. **(c)** Harvested minitubers. **(d)** Minituber crop in net house, and **(e)** Minituber crop in the field.

Aeroponics seems to have lot of scope for augmenting the existing seed production in the country. Being a relatively new technique for seed potato production in India, optimization of several factors in seed production through aeroponics need to be addressed (Buckseth *et al.* 2016). The minitubers ("Nucleus Seed" or G-0 stage), thus, produced are cold stored and used for field planting in next season. The pathogen-free nuclear seed is multiplied 2 times (G-1 and G-2 stages) under strict sanitary and phytosanitary conditions on research farms to produce "Basic Seed" (produce of G-2 stage). This basic seed is further multiplied by registered growers and other seed producing agencies for 3 more years (G-3, G-4 and G-5 stages) to produce "Certified Seed" (produce of G-5 stage) as per minimum seed certification standards (Indian Minimum Seed Standards, DAC 2013). In all these multiplications, limited generation system wherein the planting of each seed class is limited as per the eligibility by compliance with established disease tolerance levels and the number of field multiplications in particular country is followed. During potato production, the plant is constantly exposed to sources of contamination. The probability of a seed tuber or seed lot, becoming infected with pathogens progressively, and increases every year. To minimize this, seed certification agencies have enacted regulations to basically restrict or limit the number of years the seed lot can be eligible for the seed certification process.

Role of ICAR-CPRI Shimla

The conventional seed production technology based on 'seed plot technique' is successfully being used in India since last five decades for quality potato seed production. However, the quantity of good quality seed being produced by using conventional method can hardly meet 20-25% requirement. Keeping that in view, ICAR-CPRI, Shimla has standardized a number of hi-tech seed production systems based on tissue culture and micropropagation technologies. Adoption of these systems of seed production will improve the quality of breeder seed, enhance seed multiplication rate and reduce field exposure of seed crop by at least 2 years. The systems were thoroughly tested at seed production farm of ICAR-CPRI before passing them on to farmers and other stakeholders. Adoption of hi-tech seed production systems developed by the institute has led to opening of more than 20 tissue culture production units throughout the country. Several Government/Private seed producing organizations procure virus-free *in vitro* mother cultures of important notified and released potato varieties every year from ICAR-CPRI, Shimla for further multiplication in their hi-tech seed production programmes. The latest hi-tech seed production system standardized by the institute is based on the concept of soil-less, aeroponic technology. The aeroponic system of seed production has the potential to further revolutionize potato seed sector after about 50 years of introduction of "seed plot technique" by the institute. The aeroponic system has been perfected in the year 2011 and so far it has been commercialized to 14 stakeholders from different states like Uttar Pradesh, West Bengal, Punjab and Haryana. Each firm is licensed to produce 10 lakh minitubers by aeroponic system. Even if each firm is operating at half of its potential, about 6.5 million minitubers are currently being produced by them. Besides, large number of private companies like Technico, Pepsico, Mahindra-HZPC etc. have developed their own system of minituber production.

Potato tissue culture raised mini-tubers (PTCMT) have also been included in the "Indian Minimum Seed Certification Standards". The minitubers (hi-tech seed) so derived are treated as equivalent to Breeder seed. The minitubers can be multiplied in three subsequent generations, i.e. foundation 1, foundation 2, and certified by following the prescribed standard. Site requirements, crop inspection and quality standards for foundation and certified seed production have been prescribed in the "Indian Minimum Seed Certification Standards". At present ICAR-CPRI produces ~ 3,187 metric tonnes of nucleus and breeder seed of 25 popular potato varieties; out of which 70% is through conventional system whereas, 30% through hi-tech systems. As there is limited scope to increase quantity of breeder seed production at ICAR-CPRI farms due to limitation of farm land, possibilities are being explored with the help of SAUs/KVKs/Pvt. farmers to identify the new areas of seed production, multiplication of breeder seed into FS-I, FS-II and Certified Seed under MoU and to produce seed through hi-tech systems with the help of entrepreneurs/private companies.

WAY FORWARD

In mitigating the problem of shortage of good quality seeds, strategies to after multiply the seed tubers such as tissue culture in conjunction with aeroponic systems have been tried. These technologies need to be given serious thought and should be promoted in most developing countries so as to increase potato yields. In areas having high disease pressure, the new system of seed potato production based on micro-propagation and aeroponics has the advantage of better health status of seed stocks due to the reduced number of field multiplications over the conventional (clonal multiplication) system. In terms of the need for a greater efficiency of seed potato production and for a reduced energy input, research on soil-free techniques will continue to be the subject of focus in both established and developing potato-producing areas, in the near and distant future. Advances in engineering technology will also assist in the development of more automated and controlled seed propagation systems. However, there are also options for simplifying the seed potato production systems for adaptation to low-technology situations, which has greater scope and relevance towards the increasing trends of potato production in developing countries.

REFRENCES

Buckseth Tanuja, A K Sharma, K K Pandey, B P Singh and R Muthuraj. 2016. Methods of pre-basic seed potato production with special reference to aeroponics—A review. *Scientia Horticulturae* **204**: 79–87.

Habib I.1963. The agrarian system of Mughal India, 1556-1707. Asia Pub. House.453p.

Hutchinson, Joseph Sir. 1974. Evolutionary studies in world crops: diversity and change in the Indian subcontinent. Cambridge University Press. 183 p.

Khurana S M P, Minhas J S and Pandey S K. 2003. The Potato: production and utilization in Sub-tropics. Mehta Publishers, New Delhi. 445p.

Lommen, W.J.M., 1995. Basic studies on the production and performance of potato minitubers. Doctorate thesis, Wageningen Agricultural University, Wageningen, The Netherlands, 181 pp.

Loomba A., Burton J. (2007) Edward Terry (1590–1655). In: Loomba A., Burton J. (eds) Race in Early Modern England. Palgrave Macmillan, New York. pp. 250-254.

Naik Prakash S and Tanuja Buckseth. 2018. Recent advances in virus elimination and tissue culture for quality potato seed production. In: Biotechnologies of Crop Improvement, Volume 1 Satbir Singh Gosal and Shabir Hussain Wani (eds.) Springer Nature Switzerland AG. 131-158p.

Naik P S and Karihaloo J L. 2007. Micropropagation for Production of Quality Potato Seed in Asia- Pacific. Asia-Pacific Consortium on Agricultural Biotechnology, New Delhi, India: 54p

Naik P S, Chakrabarti S K, Sarkar D, Birman R K. 2000. Potato biotechnology: Indian perspective. Potato, global research & development. Proceedings of the Global Conference on Potato, New Delhi, 6–11 December (1999) In: Khurana S M P, Shekhawat G S, Singh B P, Pandey S K (eds.) Potato, Global Research and Development, 1, pp. 194–211.

Naik P S, S M Paul Khurana. 2003. Micropropagation in potato seed production: need to revise seed certification standards. J. Indian Potato Assoc.30: 123-132.

Nugaliyadde M M, De Silva H D M, Perera R, Ariyaratna D, Sangakkara U R. 2005. An aeroponic system for the production of pre-basic seed potato. Ann. Sri Lanka Dept. Agric.7: 199–288.

Pushkarnath. 1976. Potato in sub-tropics. Oriental Longman Ltd., 289p.

Ranalli P. 1997. Innovative methods in seed tuber multiplication programmes. Potato Res. 40: 439–453.

Singh RK, Buckseth T, Tiwari JK, Sharma AK, Singh V, Kumar D, Venkataslam EP, Singh RK, Sadawarti MJ, Challam C, Chakrabarti SK (2019). Seed potato (Solanum tuberosum) production systems in India: A chronological outlook. Indian Journal of Agricultural Sciences 89 (4): 578–87.

19

Potato Post-Harvest Management Strategies

Brajesh Singh[1], Pinky Raigond[1], Arvind Kumar Jaiswal[2] and Dharmendra Kumar[1]

[1]*ICAR-Central Potato Research Institute, Shimla-171 001, HP, India*
[2]*ICAR-Central Potato Research Station, Jalandhar-144 001, Punjab, India*

INTRODUCTION

Potato production in India has shown a steady increase in the past, presently making the country second largest potato producer in the world. As per the latest available data, during 2018-19, approximately 52.5 million metric tons of potatoes have been produced (NHB, 2019). Increase in production, often resulting in gluts at harvest, has led to several post-harvest problems like storage and proper utilization of the produce. About 90% of the potato crop in the country is harvested during January-February in the Indo-Gangetic plains comprising the states of Punjab, Haryana, UP, Bihar, West Bengal, MP and Gujarat where the harvest is followed by rising temperatures of hot and dry summer, and further by warm and humid rainy season. Since potato tubers contain about 80% water, under such circumstances, a semi-perishable commodity like potato, cannot be stored for more than 3-4 months without refrigeration because of very high losses due to shrinkage, sprouting and attack by microorganisms.

Therefore, potatoes are mostly stored in refrigerated cold stores maintained between 2-4°C and 90-95% relative humidity (RH). Low temperature storage has advantages of natural control of sprout growth, low evaporative losses and minimum risk of diseases and pests. These conditions are ideal for storage of seed potato but these cause cold-induced sweetening leading to excessive accumulation of sugars in most of the potato cultivars making them unfit for processing due to browning in chips. Also, such tubers are less preferred for consumption because of sweet taste. Hence, potatoes meant for processing are either stored under non-refrigerated conditions for short term or stored at

10-12°C with the use of some sprout suppressant like CIPC (isopropyl N-(3-chlorophenyl) carbamate) to minimize the accumulation of reducing sugars. The chips produced from such tubers are light in colour.

Potato processing on the other hand is essential to sustain the increasing potato production in the country and to provide proper remuneration to the cultivators. Potatoes can be processed into a variety of products. Potato chips, French fries and dehydrated chips are the most popular processed potato products in the organized and unorganized sectors. The present chapter deals with post-harvest management of potatoes in terms of dormancy and sprouting issues and storage technologies for seed, table and processing potatoes. Besides, areas suitable for growing processing potatoes in the country, potato processing scenario in the organized and unorganized sectors are also discussed.

POST-HARVEST LOSSES

The post harvest losses result in reduction of the quantity as well as quality of potatoes. Quantitative losses are apparent and attempts are made to reduce these losses whereas qualitative losses are not apparent and their importance is underestimated. Since qualitative losses can greatly reduce the value of potatoes, adequate attention should be paid to it. Post-harvest losses result from physiological as well as pathological reasons.

Physiological losses

Respiration and evaporative loss of water from tubers result in physiological losses in potatoes. The magnitude of these losses depends largely on the environmental conditions. Two important storage environmental factors that affect the storage behaviour are temperature and relative humidity. Physiological losses occur through natural respiration and evaporative loss of water through skin which is the result of magnitude of environmental conditions. Respiration rate varies with variety and response to storage temperature. In general, tuber respiration is relatively high at low storage temperatures, decreases as storage temperatures increase, and then increases again as storage temperatures are elevated (Burton et al., 1992). Weight loss is higher during the early part of the storage season due to higher tuber respiration rates, higher storage temperatures for wound healing, and higher transpiration. The relative humidity of the storage also affects transpirational weight loss and hence, it is recommended that relative humidity should be kept as 95% or above.

Optimal holding temperatures for potatoes in storage depend on the potato variety and the intended end use of the product. Physiological damage can occur from exposure to high or low temperatures both before and during storage. Physiological weight loss is the most relevant quality reducing process.

Respiration is a natural biochemical process under the physiological activity of the tuber tissue. It can be controlled by an appropriate temperature control to reach and keep low physiological activity of the tubers. Respiration consumes oxygen and as a result converts starches to sugar. The tuber cell, oxidize glucose into nutrient that is required by tuber to stay alive, produce water, carbon dioxide, heat and by-products. The heat produced in the process reduces the relative humidity of the air within the cells, also increases water holding capacity and continues moisture loss through evaporation from tuber skin. Respiration also increases when tubers are mishandled, transported over rough tracks or subject to bruises or cuts. For most varieties, temperatures above 15°C may cause dramatic increase in respiration (Talburt and Smith, 1987).

Evaporation on the other hand is a physical process which can be managed by controlling the storage environment in terms of temperatures, air humidity and air flow rate. It should be taken into consideration that temperature change in storage should be made in gradual way. This gradual temperature reduction results in little changes in the sugar content of tubers and also avoids deterioration of quality of processed product. Most of the tuber shrinkage that occurs during the first month of storage results from water lost before the completion of the wound healing process (Ezekiel *et al.*, 2003). Maintaining high relative humidity in potato storage prevents some of the early season tuber dehydration and helps controlling the total shrinkage during the season. Shrinkage in storage is directly proportional to the length of the storage season and inversely proportional to the relative humidity conditions maintained within the store (Figure 1). The recommended RH in stores is 95% or above for minimizing early storage tuber losses due to dehydration.

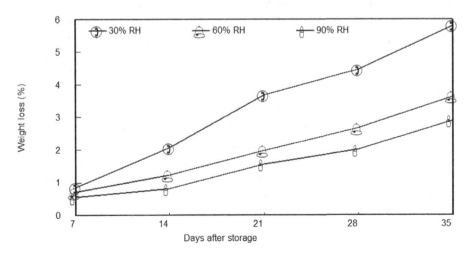

Fig. 1: Effect of different relative humidity levels on the weight loss in sprouting tubers of Kufri Chandramukhi (storage temperature 28-30°C) (Singh and Ezekiel, 2003).

Pathogenic losses

Losses caused by pathogens are higher than the losses due to physiological causes. Physical damage to tubers during harvesting and handling aggravates the attack of bacteria and fungi leading to quantitative losses. Generally, the more common storage diseases caused by fungi are late blight, dry rot and pink rot. The most severe bacterial disease that causes rotting is soft rot. When infection occurs in the field, rotting begins there itself and continues during storage. When infection occurs after harvesting, it is generally through mechanical injury as in the case of dry rot. High humidity and condensation of water on tuber surface can lead to infection by soft rot causing bacteria *Erwinia spp.* Qualitative losses are caused by diseases such as common scab, powdery scab, black scurf, silver scurf and wart, which affect the appearance of the tuber and thus reduce the market value of potatoes. Among the insect pests, tuber moth causes highest damage during storage and is common in potatoes stored under higher temperatures, as is the case with non-refrigerated storage. The larval damage results in direct weight loss and tuber moth infection greatly reduces market value of the tubers.

POTATO STORAGE MANAGEMENT

The quality of potatoes during storage depends largely on the quality of tubers being stored and therefore, its management starts from the production stage itself. Cultural practices affect the quality of tubers, therefore, under sub-optimal conditions which may include use of excess N fertilizers, excessive soil moisture and harvesting before the periderm firmness may lead to reduced storability. Adequate pest and disease management is also necessary for producing potatoes with good keeping quality. Hence, management of pre-storage factors that affect the keeping quality of potato is a prime step in good storage management. The potatoes need to be harvested in dry weather and therefore, irrigation has to be stopped about two weeks before dehaulming. Harvesting is done 10-15 days after haulm cutting for facilitating skin setting. Besides this, curing/ suberization (the process by which wounds are healed in potatoes under optimum conditions of 25°C temperature and 95% relative humidity) is also important. Potatoes harvested under wet soil conditions must be dried before storage because even a little moisture on the surface of the tubers may lead to infection and rotting of tubers during the storage. Only mature tubers with good skin setting are ideal for storage. All the issues which affect the storage of potatoes and the technologies/methods for their storage are being discussed in the subsequent heads.

Dormancy

Dormancy is a state in which tubers do not sprout even when placed under conditions ideal for sprout growth (18-20°C, 90% RH and darkness). Duration of dormancy is normally counted from the date of harvest. Based on the dormancy duration from the date of harvesting (it is assumed that harvesting is done 10-15 days after dehaulming) in the plains, Indian potato cultivars can be divided into three categories: Short dormancy: <6 weeks, *e.g.* Kufri Bahar; Medium dormancy: 6-8 weeks, *e.g.* Kufri Jyoti & Long dormancy: >8 weeks, *e.g.* Kufri Sindhuri (Ezekiel *et al.*, 2003). Even within a cultivar, dormancy duration can be affected by growth conditions and tuber weight. Dormancy duration is not related to the crop duration of a cultivar. In other words, it is not necessary that a late cultivar should have long dormancy.

Soil and environmental conditions during crop growth have a strong influence on the dormancy duration. Cold and wet weather is known to increase the dormancy duration while dry and warm weather reduces it. Potatoes grown under short-day conditions (plains) have shorter dormancy duration than those grown under long-day conditions (hills). For example, the dormancy duration in Kufri Jyoti, Kufri Chipsona-1 and Kufri Chipsona-2 was 55, 47 and 40 days, respectively, when grown in the northern plains under short day conditions and 126, 84 and 91 days, respectively, when grown in the north-western hills under long day conditions (Ezekiel *et al*, 2003). Season (year) to season variation in the duration of dormancy can also be substantial due to variation in the environmental conditions during crop growth.

Storage temperature has a strong influence on the dormancy duration. Higher storage temperature hastens dormancy release, while storage at a temperature of 4°C and below prolongs dormancy by preventing sprout growth (Figure 2).

Fig. 2: A dormant and a sprouted tuber.

Exposure to light during storage also affects dormancy duration but its effect is much less compared to temperature effect. RH in the storage atmosphere has very little effect on dormancy duration. Concentration of gases in the storage atmosphere also affects dormancy duration. Low concentration of oxygen (O_2) and high concentration of carbon-dioxide (CO_2) hastens the release of dormancy and stimulates sprout growth.

Sprouting and its management

Sprout growth begins at the end of dormancy period. Once sprouted, potatoes start loosing weight, their appearance is affected by shrivelling and they loose marketability for table and processing purposes. Besides, in seed potato sprouting before the required time is undesirable since the shrivelled tubers loose vigour. Sprouting is influenced by several factors and the major factors that influence sprout growth are cultivar, temperature, humidity, light, concentration of CO_2 and O_2 and size of tubers (Burton, 1989). The rate of sprout growth is cultivar dependent provided all the other conditions are similar. Considerable variation has been observed in the length and weight of sprouts. The pattern of sprout growth also varies with the cultivar.

Temperature has strong influence on sprout growth. Generally sprout growth does not take place when the storage temperature is less than 4°C. It is observed at 5°C and increases with increasing temperature. It is optimum between 15 and 20°C and decreases with increase in temperature above 20°C. Compared to the influence of temperature, humidity has only a slight effect on sprout growth. Light inhibits sprout growth and it has been shown that red light above 650 nm and blue light below 500 nm both contribute to the inhibition of sprout growth (Burton, 1989). Potato sprouts grown in the light develop chlorophyll and are shorter and sturdier than those grown in the dark.

For potatoes meant to be used for table and processing purposes, sprouting is an undesirable characteristic and therefore, it is important to avoid sprouting in these potatoes. When stored at 2-4°C sprout growth does not take place. Potatoes stored at this temperature are suitable for seed but not for ware and processing purposes. Potatoes meant for ware and processing are stored at higher temperatures generally at 10-12°C and at this storage temperature, sprout growth takes place which needs to be checked with some sprout inhibitor. Several sprout inhibitors have been used on potatoes world over. Some of the compounds used for checking sprout growth include several alcohols, acetaldehyde, ethylene *etc*. Naphthalene acetic acid (NAA) and its derivative methyl ester of alpha naphthalene acetic acid (MENA), 2-4 dichlorophenoxy acetic acid (2,4-D), 2,4,5-trichlorophenoxy acetic acid (2,4,5-T), 1,4-dimethyl naphthalene (DMN), *etc*. have also been reported to show sprout suppression effect. Although a number of chemical sprout inhibitors have been tried, only a few have gained commercial acceptance, which include Tetrachloro nitrobenzene, Maleic Hydrazide, 1,4-DMN, S-Carvone and Isopropyl-N- chlorophenyl carbamate (CIPC). CIPC is the most commonly used sprout suppressant and presently, it is the only chemical registered in India for commercial application on potatoes (Ezekiel *et al.*, 2003).

Chlorpropham (CIPC) is a mitotic inhibitor and potatoes to be treated with CIPC need to be mature and well cured. Sufficient time (2 to 3 weeks) has to be allowed after harvest for skin curing and wound healing. If CIPC is applied with unhealed wounds, rottage can be unacceptably high. During cold storage, the potatoes are packed in bags having bigger pore size, preferably in leno bags having pore size of 8 x 8 mm for allowing sufficient quantity of CIPC to enter the bags and get deposited on the tuber surface. Storage temperature is brought down gradually by lowering 1°C per day to 10-12°C with RH being maintained at 90-95% to minimize weight loss during storage. For treating cold stored potatoes, hot fogging concentrate (HN) formulation of CIPC is used. It can be applied in the form of fog using a fogging machine without disturbing the stored potatoes (Figure 3). The commercial preparation of the chemical contains 50% a.i. and 35-40 ml of this preparation is required for fogging one tonne of potatoes. However, the dose can vary depending upon the loss of fog during treatment due to leakage, the pore size of the bags used, the quantity of potatoes stored, *etc.* After the CIPC treatment, the store has to be kept airtight for 24-48 hours. Due to higher storage temperature, there is accumulation of CO_2 in the store, which might adversely affect the quality of potatoes. Therefore, arrangement for regular flushing out of accumulated CO_2 is also done to keep the CO_2 concentration lower than 0.1%. First fogging is done at the first sign of sprout growth and second fogging is done at about 45 days after the first fogging. The timing of first fogging is important. If it is done at the right time, CIPC is most effective in suppressing sprout growth.

 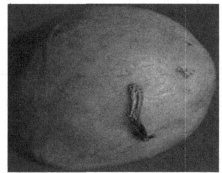

Fig. 3: Fogging machine with CIPC fog and a potato showing sprout burning.

Most of the potatoes (about 90%) in India are harvested in February-March in the Indo-Gangetic plains when the temperatures begin to rise and thus, they have to be stored during the hot summer months. The storage method depends on the required duration of storage and destined use of potatoes. It can be broadly divided into two categories *viz.*, refrigerated storage and non-refrigerated storage.

Refrigerated storage

Refrigerated storages are generally used for long-term storage of seed, table and processing potatoes. This is essentially done in commercial cold storages since potatoes harvested in February-March have to be stored until October or so for 6-8 months.

Storage at 2-4°C

There are more than 7000 cold storages in India and of the total capacity of cold stores, approximately 80% is used for storing potatoes alone. Low temperature storage is the most common method in the country (Figure 4). It is effective because at 4°C and below, sprout growth does not take place in most of the varieties. Further, the metabolic processes are slowed down therefore; losses caused by respiration are minimum. In India, cold stores were developed mainly for the storage of seed potatoes but under the present circumstances, even most of the ware crop is also being stored at low temperature, though the requirement for seed storage in only about 8%. Since the primary purpose of storage at 2-4°C is to check sprout growth, storage of seed potatoes at 2-4°C is ideal. But storage of ware potatoes at 2-4°C is not desirable since low temperatures induce accumulation of large amounts of sugars making the tubers sweet and therefore, less suitable for consumption. Besides, the cold stored potatoes are unfit for processing as high level of reducing sugars causes browning in potato chips, even the keeping quality of cold stored potatoes deteriorates quickly once they are removed from the cold store as tubers begin to sprout fast and marketing of cold stored potatoes is also a problem, due to their poor keeping quality. The high sugar content of potatoes stored at 2-4°C can be reduced to some extent by reconditioning at 20°C for 6 weeks. However, reconditioning cannot restore the sugars to the initial levels. Therefore, it is recommended that 2-4°C storage be used for seed potatoes only and the potatoes meant for table and processing should be stored at higher temperatures either under elevated temperatures of 10-12°C for long-term (6-8 months) or under non-refrigerated conditions for short-term (3-4 months). For seed storage, it is recommended that after proper skin curing, sorting and grading, the seed should be treated with a solution of 3% boric acid and after shade drying, be stored in cold stores.

Fig. 4: A traditional cold storage and stack of potatoes inside the store.

Elevated temperature storage at 10-12°C

Potatoes for ware and processing purposes are generally stored at 8-10°C in the developing world. This storage temperature does not allow excessive sugar accumulation, therefore, potatoes are not sweet in taste. Since sugar level is within the acceptable limit, the potatoes are also suitable for processing. But this temperature does not slow down sprout growth at the release of dormancy and therefore, some sprout suppressant has to be used for checking sprout growth. CIPC (isopropyl N (3-chlorophenyl) carbamate) is the most common sprout suppressant used on potatoes. It is a mitotic inhibitor and checks sprout growth by inhibiting cell division. This chemical was registered for use in India only in 1998, and at present about 1000 cold stores in the country are using this chemical on potatoes stored at 10-12°C (Figure 5). The details of the treatment and recommendations have already been described in previous section.

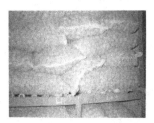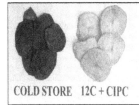

COLD STORE   12C + CIPC

Fig. 5: A commercial cold store using 10-12°C storage technology, potatoes stored in leno bags and comparison of chip colour from two refrigerated storages.

Non-refrigerated storage

The infrastructure for refrigerated storage in the country is unevenly distributed and sometimes it is beyond the reach of small and marginal farmers. In many states the storage facility is either lacking or is not adequate to accommodate the ware/ seed potatoes. Many farmers in these states use indigenous storage practices like pits, heaps, trenches and basements to hold some of their produce for short-term to fetch better prices during off-season. Though these on-farm storage methods are economical and practical, they are not efficient because of higher losses due to increased rotting (10-21%). Therefore, improvements were done in the traditional storage practices, particularly, potato heaps as a part of marketing strategy to increase remunerations from potato cultivation by holding the produce on-farm for short periods (3-4 months) and to bypass the immediate post-harvest period when the prices are the lowest (Ezekiel et al., 2003).

Improved heaps help to reduce the daily range of variation in temperatures while maintaining a high relative humidity (RH). While the day time variation in ambient temperature and relative humidity is considerable, the atmosphere inside the heap is quite stable. Minimum-maximum temperatures in the heap during storage (March to June) ranges between 13-30°C as compared to 8-44°C of the ambient and RH remains consistently high (60-95%) compared to wide variation and lower levels (27-87%) of the ambient as observed during experimentation. Weight loss and rottage are significantly reduced (<10%) in potatoes stored in heaps as compared to those stored at room temperature (Figure 6). Weight loss up to 10% is considered acceptable because no visible shrivelling takes place up to this level, the stored potatoes remain firm and desprouted potatoes fetch prices comparable to the cold stored potatoes, which are 40-50% higher than that at the time of harvest. Treatment of potatoes with sprout suppressants before heap storage has been found to reduce the losses further.

Fig. 6: A modified heap and potatoes after removal (treated with CIPC).

Physiological and pathological losses in potatoes are higher under non-refrigerated conditions, therefore, appropriate pre and post-harvest measures are to be strictly followed for successful storage in heaps as described previously. Storage in heaps may be completed by the end of February so that the lower temperatures prevailing during this period can be taken advantage of. Heaps may be protected from unseasonal rains either by erecting a thatched roof or covering with a water proof sheet. If a water proof sheet is used, it needs to be removed after the rain stops. Stored potatoes may be sent to the market before the temperatures inside the heaps reaches 30°C.

Potatoes stored in heaps are highly suitable for processing due to low reducing sugar content. These are also not sweet in taste and are more preferred as table potatoes. Therefore, the prices fetched by these potatoes in markets as table potatoes are also more than the cold stored potatoes.

POTATO PROCESSING

Potato is the fourth most important food crop and is a wholesome food. In India, potatoes have been utilized largely for consumption as fresh potatoes and the major part of potato harvest (approx. 68.5%) goes to domestic table consumption (Singh *et al.*, 2014). Whereas, in the developed countries, table potato utilization is merely 31%, rest being frozen French fries (30%), chips and shoestrings (12%) and dehydrated products (12%) (Rana, 2011). The processing of potatoes in the country was not in vogue till 90's and with the openings of organized processing by multinationals and indigenous players, potato processing industry has grown manifolds. During 2007-08 about 7% of potato production was used by processing industry and the sector is still increasing at a rapid rate. The pattern of Indian potato industry suggests that the demand for potatoes for processing purpose is expected to rise rapidly over next 40 years for French fries (11.6% ACGR) followed by potato flakes/ powder (7.6%) and potato chips (4.5%). The demand for processing quality potato is expected to rise to 25 million t during the year 2050 (Singh *et al.*, 2014).

Suitable areas for growing processing potatoes

Potatoes are grown in almost all agro-ecological zones in the country and different varieties are used in these zones depending on their productivity. Major quantity of potatoes is meant for consumption as table potatoes and is not suitable for processing. Dry matter and reducing sugars are the two parameters that are of utmost importance to the potato processing industries as described above. Although both of these traits are characteristic of a variety, but different environmental and agronomic conditions such as temperature during crop growth, day length, light intensity, soil type, availability of moisture, time of irrigation, rainfall, tuber maturity, *etc.* affect these traits. Out of all these parameters, temperature plays a major role in the dry matter and reducing sugars accumulation. Night temperature of 10°C or more during the last 30 days of crop growth improves the quality of the potatoes and produces potatoes with high dry matter and low reducing sugars (Marwaha *et al.*, 2003). Areas suitable for potato processing in different parts of the country have been identified based on the temperatures during the later phase of crop growth (Figure 7). Generally varieties grown in the eastern and the southern parts of the country contain high dry matter and low reducing sugars.

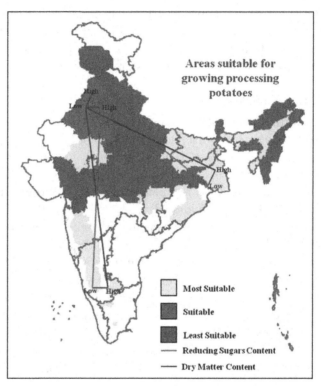

Fig. 7: Map showing the areas suitable for growing processing potatoes.

For preparing good quality potato chips, the dry matter of the tubers should be more than 20% and the reducing sugars content should be less than 150 mg/ 100g tuber fresh weight (Sukumaran and Verma, 1993). If the level of reducing sugars is more than this, the fried product becomes dark in colour and unacceptable. The dark colour is formed due to a reaction called Maillard reaction between reducing sugars and amino acids. Maillard reaction takes place during the high temperatures of frying. For ranking of colour in potato chips, colour cards have been developed by ICAR-Central Potato Research Institute (ICAR-CPRI), Shimla. These colour cards have scores of 1-10, higher the colour score more is the browning. A colour score up to 4 is acceptable while 5 and above is unacceptable (Figure 8). The Maillard reaction which produces dark colouration in potato products is also responsible for production of another un-advantageous trait 'Acrylamide' as a by product of the reaction between amino acid asparagines and reducing sugars. Since acrylamide has been categorized as a carcinogen, there is high concern for its concentration in the processed potato products. All the Indian varieties have been profiled for the formation of acrylamide in the chips and French fries and the results have clearly shown that the varieties known for good processing quality having low reducing sugars also produce less acrylamide in the products and hence, it is recommended that for processing of potatoes, only processing varieties *viz.*, Kufri Chipsona-1, 2, 3, 4, Kufri Himsona and Kufri Frysona should be used. The traits of these varieties have already been discussed under the chapter on 'Potato Breeding for Processing'. Growing of processing varieties in the identified suitable areas may ascertain the required quality for processing in potatoes.

Fig. 8: Colour cards for potato chips.

Common processed products

Potato chips

Potato chips (Figure 9) originally known as "Saratoga Chips" were first prepared by housewives in the Unites States of America during the middle of the nineteenth century. The production and popularity of potato chips has increased tremendously with the large scale manufacture of chips which began during the Second World War. The name "chips" has been used in America and "crisps" has been more popular in Europe. Potato chips have been considerably popularized by the potato chip manufacturing companies. With increasing popularity among all the sections of the society, potato chips comprise 85% of the Rs. 2500 crore salty snack business in India (Rana, 2011). The sector is still increasing with the production of potato chips from 0.38 million t in 2006-07 to

0.61 million t in 2010-11 and is expected to increase further to 3.55 million t by the year 2050.

Fig. 9: Potato chips and French fries.

French fries

French fries (Figure 9) are very popular in the developed countries including USA. In India McCain is the largest French fry Company. Approximately 0.6% of potato production was used for preparation of French fries in 2006-07. Although India has huge French fry manufacturing capacities, but at present its actual processing is quite low. The limited number of outlets serving the product is one of the major reasons of its low utility and high price. However, the demand for French fries is increasing gradually. Generally, frozen French fries (par-fries) are prepared and sold. In India 0.02 million t potatoes were used in French fry production during 2007-08 (Rana, 2011).

Besides chips and French fries, several products of potatoes in combination of other food ingredients are being manufactured and sold in the market in ready to eat forms. Some of these include Smiley's, Wedges, Nuggets, Bites, Tikkis, *etc*. Processed potato products like potato chips are no doubt popular but the cost of production of potato chips is very high because of the sophisticated technology used and high cost of frying medium and electricity. Therefore, these processed products are not within the reach of common man. In order to make the processed potato products available to a large section of the population at an affordable price, simpler methods of potato processing are needed.

Dehydrated potato products

Preservation of foods by drying is perhaps the oldest method known to man. Drying results in the lowering of moisture content, leading to lesser chances of microbial growth. As a result, the product has a longer shelf life. The reduction in moisture content is accompanied by a reduced bulk which facilitates storage, transportation and packaging. Potato slices can be dehydrated in the sun, stored and, fried and consumed when required. Dehydrated potato chips are the most common processed potato product in the rural and semi-urban areas in the country (Sukumaran and Verma, 1993). Dehydration of potatoes offers advantages *viz.* it does not require much investment and is labour intensive, therefore, can be taken up at the level of individual families. Further, dehydrated chips have longer shelf life at room temperature and can be stored easily. Solar dehydration of potatoes is very popular in several states like Maharashtra, Gujarat, U.P. and M.P. and, remains a viable alternative to expensive and sophisticated potato processing technology. Dehydrated potato chips are quite popular because they are much cheaper than potato chips produced by the organized sector. Companies such as Potato King, Merino Industries Ltd. and Satnam Agri are leading in manufacturing of dehydrated potato products such as potato flakes and flour. More than 1% of the potato production is consumed by potato flakes and flour industry in India. In 2007-08, the total quantity of dehydrated products manufactured in India was 31,000 tonnes (Rana, 2011).

Potato starch

Potato is a good source of starch and contains 14-16% starch. Potato starch is utilized for special applications in industries such as pharmaceutical, textile, food and paper. Properties of potato starch are superior than maize and cassava, which leads to its higher price compared to maize and cassava starch. The method of starch extraction from potatoes is quite simple and easy. Moreover, the advantage of using potatoes for starch production is that the potatoes which are not suitable for table or processing and potatoes which are damaged and are too small or too large can be used for starch extraction. Presently, potato starch manufacturing industry is almost non-existent in India. The major reason behind this is the low cost of imported potato starch from Europe and China. This coarse starch is further refined and exported to other countries. Processing the potato into coarse starch with conventional method increases the value of potato by 30%, if processed with lactic acid using state-of-the-art technologies value increased by three folds, value of potato increased eight times if it is modified into water absorbent resin, and it increased to twenty times if processed into cyclodextrin and almost thirty times if processed into fine chemical products.

Scenario of potato processing

Organized sector

Potato chips production by the organized sector increased rapidly after the introduction of the liberalization policy of the government of India. The organized sector in the country produces about 30,000 tonnes of potato chips per year (Rana, 2011). Popularization of potato chips by companies such as Frito Lay India has led to fast incremental growth in potato chip manufacturing capacities.

Frito Lay India and ITC retained approximately 31.55% share of potato chips market during 2010-11. Besides these, some other organized players as of today at National and regional levels include, Haldiram, Kishlay, Balaji, Uncle chips, *etc*. The demand for potato chips is likely to increase further, because of its increasing popularity as a convenient fast food especially in urban areas. The second popular product is potato *lachha* and *lachha* market is also dominated by branded manufacturers. *Aloo bhujia*, which contains potato only as one of the ingredients, is also quite popular and the *Aloo bhujia* market is dominated by Lehar and Haldiram.

Unorganized sector

Potatoes are being processed at small scale in rural India. Potato processing in the unorganized sector is of considerable importance in a country like ours where majority of the population cannot afford to purchase potato chips produced by the organized sector which costs Rs. 300 per kg. Potato chips produced by the unorganized sector costs Rs. 150 to Rs. 200 per kg. In India approximately 377 thousand tonnes of potato chips are prepared by unorganized sector (Rana, 2011). In Kolkata alone there are about 200 small units producing processed products from potatoes. Small scale potato processors are found in almost all cities and towns in the country and they produce a variety of products like potato chips, potato *lachha*, potato *bhujia*, potato pops, potato *papads*, dehydrated potato chips *etc*. Dehydrated potato chips and potato flour are being produced by several small scale potato processors in western UP.

Although the price of the potato chips produced by the unorganized sector is low, the quality of the chips is poor compared to that produced by the organized sector. The chips produced by the unorganized sector can have up to 80% browning, up to 50% broken pieces and 10% green chips.

REFERENCES

Anonymous (2019) Area and production of horticultural crops: All India. National Horticulture Board, New Delhi, India.

Burton WG (1989) The potato, 2nd Edition, Longman, Essex: 742p

Burton WG, van Es A and Hartmans KJ (1992) The physics and physiology of storage. In, The Potato Crop, 2nd Edition, Harris PM (ed.), Chapman and Hall, London: 608-727.

Ezekiel R, Mehta A, Kumar D and Das M (2003) Potato Storage. In, The Potato-Production and Utilization in Sub-tropics. Khurana SMP, Minhas JS and Pandey SK (eds.), Mehta Publishers, New Delhi, India: 323-335.

Marwaha RS, Uppal DS, Kumar D and Sandhu SK (2003) Potato Processing. In, The Potato-Production and Utilization in Sub-tropics. Khurana SMP, Minhas JS and Pandey SK (eds.), Mehta Publishers, New Delhi, India: 336-346.

Rana RK (2011) The Indian potato processing industry: global comparison and business prospects. *Outlook on Agriculture* 40(3): 237-243.

Singh B and Ezekiel R (2003). Influence of relative humidity on weight loss in potato tubers stored at high temperatures. *Indian Journal of Plant Physiology* **8** (2): 141-44.

Singh BP, Rana Rajesh K and Govindakrishnan PM (2014) Vision 2050. ICAR-Central Potato Research Institute, Shimla-171 001, HP: x+26p.

Sukumaran NP and Verma SC (1993) Storage and Processing. In, Advances in Horticulture, Vol. 7-Potato, Chadha KL and Grewal JS (eds.), Malhotra Publishing House, New Delhi, India: 701-732.

Talburt WF and Smith O (1987) Potato Processing, 4th Edition, Van Nostrand Reinhold, New York: 796p.

20

Issues, Strategies and Options for Doubling the Income of Potato Producers in India

Pynbianglang Kharumnuid[1] and NK Pandey[1]

[1]ICAR-Central Potato Research Institute, Shimla, Himachal Pradesh, India

INTRODUCTION

Potato plays a vital role in food security worldwide. It is the third most important food crop, after rice and wheat. It is a nutrient-rich crop which provides more calories, vitamins and nutrients per unit area than any other staple crops. FAO declared potato as the crop to address future global food security and poverty alleviation during 2008. In Indian agriculture, potato plays a very important role as it alone contributes about 21 percent of the total vegetable area and 26 percent of total vegetable production of India (DAC&FW, 2017a). As per FAOSTAT data for the year 2017, India with 48.6 million t is ranked second in potato production in the world, only behind China with 99.2 million t (Figure 1).

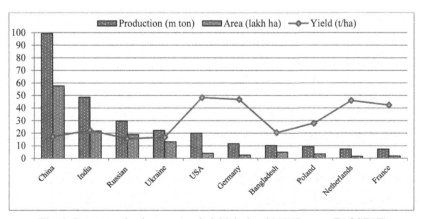

Fig. 1: Potato production, area and yield during 2017 (Source: FAOSTAT)

TRENDS IN AREA, PRODUCTION AND PRODUCTIVITY OF POTATO IN INDIA

At the time of inception of ICAR-Central Potato Research Institute (ICAR-CPRI), Shimla, India, in the year 1949, India produced only 1.54 million t potatoes from 0.234 million ha area at an average productivity level of 6.58 t/ha. As per the 1st advance estimates by National Horticulture Board, the potato production in India during 2018-19 was 52.6 million t from 2.18 million ha area with a productivity of 24.13 t/ha (Figure 2). Over the period of 1949-50 to 2018-19, the compound annual growth rate (CAGR) in potato area, production and yield were 3.3 per cent, 5.4 percent and 2.0 per cent per annum, respectively. The productivity in India is higher than in China and Russia, the third largest potato producer. However, the productivity is lower than most of the developed European countries. In India, yield of potato is hovering around 20-24 t/ha since last 9 years which is far below the productivity of European countries and other developed countries. Scientists working at ICAR-CPRI reported that at present level of farm management practices, India actually is able to harvest only 42-45% of the achievable yield, which could be improved to 80% by efficient and effective dissemination and implementation of improved technologies.

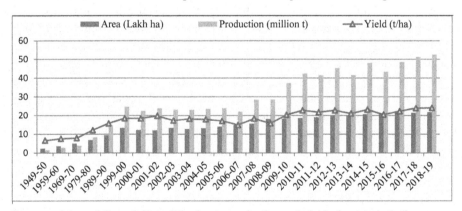

Compound Annual Growth Rate (%) of area, production and yield of potato in India

Year	Area	Production	Yield
2009-10 to 2018-19	2.1	3.2	1.1
1949-50 to 2018-19	3.3	5.4	2.0

Fig. 2: Trends in area, production and yield of potato in India (Source: DES, MOA, GOI and NHRDF, New Delhi)

CONDITION OF FARMERS IN INDIAN AGRICULTURE

India is predominantly an agriculture based country in which agriculture and allied activities contributes about 17 percent to the country's GVA and provides employment to about 54.6 percent of the population (DAC&FW, 2017b). Unfortunately, majority of the farming communities in the country get very less remuneration for their hard work. The recent National Sample Survey Office assessment on farmer situation reported that about 40 percent of farm households would like to exit agriculture but they remain in it because of limited opportunities outside the ambit. After independence, the policies for agricultural development mainly focussed on increasing agricultural production and ensuring food security. As a result of Green Revolution and Rainbow Revolution, India has become not only self sufficient in agriculture, but also a net exporter of many crops. However, the strategies for raising the income of well deserved farmers have not been given much emphasis by the policy makers and the government. It is with this backdrop, the Government of India has set a policy target of doubling farmers' income by 2022.

POTENTIAL OF POTATO FOR DOUBLING FARMER'S INCOME

Potato has a very high potential in doubling farmers' income as it can be used for table purpose (as vegetables), seed purpose and processing purpose. In India, potato is mostly grown (about 85%) in Indo-Gangetic plains for short duration (about 90 days). Thus, it is highly amenable to adjustment and fits well in various cropping systems. The demand of potato and its products is rising at a faster rate. It is estimated that by the year 2050, India would require about 125 million t of potato from an area of 3.62 million ha with a CAGR of 3.2% up to the year 2050 (CPRI Vision 2050). Potato is highly nutritious, thus will play a significant role in food and nutrition security for the ever increasing Indian population. Several options may be available for increasing incomes of potato growers. Some of them are (i) increasing the productivity, (ii) decreasing the cost of production, (iii) ensuring remunerative price for the produce and (iv) reducing harvest and post-harvest losses.

Increasing the productivity

In potato cultivation, productivity can be increased by adopting various technologies and strategies like:

Adoption of improved varieties

The adoption of improved potato varieties is critical for achieving sustainable and higher productivity and production. So far, ICAR-CPRI has developed 61 potato varieties for different agro-climatic regions of the country and majority of them are for north Indian plains. Potato varieties developed by CPRI are very popular among farmers and cover nearly 95% of total area under potato. Four varieties, viz. Kufri Jyoti, Kufri Bahar, Kufri Pukhraj and Kufri Chipsona-

1 altogether contributed around 75% of total area under potato. However, despite the fact that many recent improved varieties have been released in India which give more yield, longer storage duration and of higher qualities, many farmers still continue to grow old varieties and local varieties. These varieties give low yield and also are highly susceptible to diseases. The reasons may be high cost of seed, non-availability of seed of required varieties, poor seed quality and lack of information about improved varieties. Old varieties and local cultivars should be replaced by recent varieties by making farmers aware about the benefits of new varieties over older ones. This could be achieved through organization of training, laying out of FLDs and other awareness programmes as well as supply of quality seed of improved potato varieties.

Adoption of suitable potato based cropping system

Choosing of suitable crop sequence is very important for improving the productivity of each crop in the sequence. The cropping system varies with different agro-ecological zones. The ICAR-CPRI has identified as many as eight distinct potato growing regions. As more varieties are released every year and changing climatic conditions, continuous research should be conducted for finding the suitability of the recent varieties in different cropping systems and agro-ecological zones. Best cropping systems should be popularised among the farmers for achieving higher and sustainable productivity.

Supply of good quality seeds

Use of low quality seeds for planting is one of the main reasons of low productivity of potato. Majority of farmers use low quality farm saved seeds and seed purchased from local traders. Thus, the government should ensure regular supply of good quality seeds to the farmers. Progressive farmers may be motivated and seed villages could be established for seed production on a larger scale so that regular supply of seed is achieved.

Forecasting for efficient and effective management of Late Blight

Late blight of potato can cause huge loss to potato production. In India, the reduction in potato production due to late blight ranged between 5 and 90 percent depending upon climatic conditions, with an average of 15 percent across the country (Collins, 2000). Information regarding occurrence of late blight in advance will enable farmers to apply fungicides timely, effectively and efficiently, resulting in substantial reduction of losses due to late blight and also reduce cost of purchasing fungicides. The state government should make use of the late blight forecasting model like Indo- Blightcast model and the information should be made aware to all categories of farmers.

Enhancing farmers' knowledge and skill

Low adoption of improved potato technologies also account for low productivity in few states. This is because farmers are still unaware of the improved package of practices and are not competent enough to use the scientific package of practices. Thus, they are still relying on their own traditional practices. Thus, institutional and financial arrangement should be made for dissemination and diffusion of new technologies and enhance the knowledge and skills of farmers on scientific potato cultivation through trainings, demonstrations, technical advisories, etc for enabling them to use the technologies in their own fields.

Use of ICTs for agro-advisory services and decision making

Information and Communication Technologies (ICTs) should be promoted among the farmers as they play very important role in providing agro-advisory services to the farmers, thus helping them in deciding wide range of agricultural related activities, including marketing of the produce. ICAR-CPRI has developed many ICT tools like computer aided advisory system for crop scheduling, Indo-Blightcast, potato growing season descriptor, potato pest manager, nutrient recommendation for raising potato, potato variety, potato weed manager and potato master mobile app. These should be made aware among extension functionaries and farmers for enhancing their knowledge and decision making in agriculture.

Decreasing cost of production

Cost of production is inversely proportional to the income of farmers. Thus it needs to be lowered down in order to get higher returns from potato cultivation. Cost of cultivation can be decreased by adopting the following technologies/ strategies:

Judicious use of fertilizers and plant protection chemicals

Fertilizers and plant protection chemicals account for about 25-35 percent of the total cost of potato cultivation. Many studies reported that farmers use higher doses of fertilizers and plant protection chemicals than recommended level. The use of higher doses may be on account of lack of awareness and knowledge about ill effects of these chemicals on plant, soil and environment. Therefore, awareness programmes should be organized in study area in order to control rampant use of chemical pesticides.

Cultivation through True Potato Seed

Seed is the most expensive input which accounts about 30-40 per cent of the total cost of potato cultivation. Lack and high cost of quality seeds was one of the major constraints in potato cultivation in almost all the states of India. Potato production through TPS not only reduces the production cost, but also increases the net profit of the farmers. Some of the advantages of potato cultivation through TPS are disease free planting, low cost planting material and easy and cheap storage and transportation.

Organic Cultivation of Potato

Organic potato cultivation not only fetch premium prices but also reduces the cost of cultivation to a large extent as it relies mainly on crop rotations, FYMs, off farm organic wastes and biological insects and diseases management, which are sustainable and lower in cost. There is also a good scope of exporting organic products from India. Supports in the form of incentives and subsidies of important organic inputs should be given to farmers for encouraging them to take up organic potato farming.

Production of farmers' quality seeds

Since seed cost is the highest cost component in potato cultivation, the government should encourage farmers to grow their own good quality seed which could be used for planting in the next season or selling to the market at higher prices.

Collectivization of famers into FPOs

Farmer Producer Organisations help its member producers, especially marginal and small farmers to sell their produces in bulk at a price higher than when sold individually. The farmers will also get the benefits of purchasing production inputs at a price below the market prices and hiring of machineries and tools locally at low cost which is affordable by marginal and small farmers.

Ensure remunerative price for the produce

Better income for farmers can be ensured by adopting the following approaches/ ways:

Through production of quality seeds

Lack and high cost of quality seeds is one of the main problems in potato cultivation. The market rate of potato seeds is very much higher when compared with table potatoes. In most of the states in India, majority of farmers procured seeds from Punjab and other states at higher price and higher transportation cost. Thus, there is a big scope for farmers to start up seed production venture. Therefore, efforts should be made by the state government and SAUs to train the progressive farmers, NGO members, youths, etc for production of quality seeds.

Avoiding middlemen through direct marketing and contract farming

Direct marketing and contract farming enables farmers to sell their produce directly to the consumers/contracting firms without the involvement of the middlemen. Cases of direct marketing like Apni Mandis in Punjab and Haryana, Rythu Bazaars in Andhra Pradesh and Uzhavar Santhaigal in Tamil Nadu have enabled farmers to get remunerative prices. Many studies reported that contract farming provides many advantages to the farmers such as assured price, assured market, credit, technologies, inputs, extension services, risk sharing, employment generation and reducing the cost of production and transaction. Thus, direct marketing and contract farming may be adopted by the state for increasing farmers' income.

Production of processing potatoes

Many processing varieties like Kufri Chipsona 1, Kufri Chipsona 4, Kufri Frysona, Lady Rosetta, FC1, etc. are procured by many MNCs like McCain, PepsiCo and many other medium and small scale industries for processing of chips, French fry and other processed products. The export of fresh potatoes and other processed products is increasing at a significant rate. In India, the demand for processing quality potatoes is expected to rise to 25 million t during the year 2050 (CPRI Vision 2050). Thus, farmers need to be encouraged and trained to adopt processing potatoes for selling at higher price.

Value addition/processing of potato

With ever increasing urbanization, the demand of processing potato in India as well as from other countries will be increasing at a rapid rate. As per CPRI Vision 2050, the pattern of Indian potato industry suggests that the demand for processing potatoes is expected to rise rapidly over next 40 years for French fries (11.6% CAGR) followed by potato flakes/powder (7.6%) and potato chips (4.5%). Potatoes can be processed into various forms such as chips, fries, dehydrated products (dehydrated chips, dice or cubes, papads, flakes, granules and flour) potato starch, cookies, etc. Thus, farmers and organized sectors like SHGs, NGO members, etc may be encouraged to take up processing business for improving their income.

Reduce harvest and post-harvest losses

Post harvest losses in potato are to the tune of 15-18 % of the total production. Harvesting, sorting/grading, transportation, storage at wholesaler and retailer levels are the main operations and channels where losses were found to be high. Farmers need to be made aware and trained about good harvesting practices and post-harvest operations. Very less number of farmers in India uses improved machineries and tools for scientific harvesting and post-harvest operations of potato. This results in heavy losses (cut, bruises, etc.) during harvesting and post-harvest. Many types of machineries like diggers, graders, seed treatment machines and handling machineries/tools have been developed by agricultural research institutes and companies, which significantly reduce harvesting and post harvest losses of potatoes. Farmers have to be made aware about these machines and their skills have to be improved for operation of these machineries. Potato processing centres need to be established in every major potato producing areas/blocks, so that farmers can sell their produce to the processing centres at higher prices.

Similarly, cold storages have to be established in every major potato producing block in the district in order to reduce post-harvest losses as well as transportation cost. Majority of the cold storages are not upgraded and non-functioning. Government should see the possibility of upgrading and repairing the non-functioning storages besides constructing new ones so that farmers could store potatoes in cold storages at times when there is a glut in the market. CPRI Shimla has modified the traditional heap storage of farmers. This modified heap method is low cost and work efficiently for 2-3 months. Weight loss and rottage are significantly reduced (<10%) in potatoes stored in modified heaps as compared to those stored at room temperature (Mehta et al., 2007). Potato farmers need to be made aware and trained about these improved heap storage.

CONCLUSION

There are many means to enhance the farm income like adoption of improved varieties, following suitable potato cropping system, enhancing farmers knowledge and skill about the latest technologies, which will certainly help in doubling farmers income. Government intervention and policy makers have to ensure the remunerative prices for their produce and marketing strategy for reducing middlemen and to promote direct marketing and contract farming.

REFERENCES

Collins WW (2000) The global initiative on late blight- alliance for the future. In, Potato Global Research and Development. Khurana S.M.P., Shekhawat G.S., Singh B.P., Pandey S.K., (eds), Vol I, Indian Potato Association, CPRI, Shimla, India: 513-524

CPRI (2015) Vision 2050. ICAR-Central Potato Research Institute, Shimla, Himachal Pradesh, India.

DAC&FW (2017a) Horticultural Statistics at a Glance 2017. Department of Agriculture, Cooperation and Farmers Welfare, Ministry of Agriculture and Farmers Welfare, Government of India, New Delhi.

DAC&FW (2017b). Annual Report 2016-17. Department of Agriculture, Cooperation & Farmers Welfare, Ministry of Agriculture & Farmers Welfare, Government of India, New Delhi.

DES (2019). District wise crop production statistics. Directorate of Economics and Statistics, Ministry of Agriculture and Farmers Welfare, Govt. of India, New Delhi. Retrieved on March 12th, 2019 from https://aps.dac.gov.in/APY/Public_Report1.aspx\

Mehta A, Ezekiel R, Singh B, Kumar D and Pandey SK (2007) Modified heap and pit storage for table and processing potatoes. ICAR-Central Potato Research Institute, Shimla, India Technical bulletin-82

NHRDF (2019) State wise area and production data. National Horticultural Research and Development Foundation, New Delhi. Retrieved on April 25th, 2019. http://nhrdf.org/en-us/Area And Productiion Report

Printed in the United States
by Baker & Taylor Publisher Services